Surface Analysis of
PAPER

Surface Analysis of PAPER

Edited by

Terrance E. Conners
Associate Professor of Forest Products
Mississippi State University

Sujit Banerjee
Professor of Chemistry
Institute of Paper Science and Technology

CRC Press
Boca Raton New York London Tokyo

Library of Congress Cataloging-in-Publication Data

Surface analysis of paper / edited by Terrance E. Conners, Sujit Banerjee.
p. cm.
Includes bibliographical references and index.
ISBN 0-8493-8992-5 (alk. paper)
1. Paper--Surfaces. 2. Surfaces (Technology)--Analysis. I. Conners, Terrance E. II. Banerjee, Sujit, 1959– .
TS1121.S87 1995
676'.27--dc20 95-23091
CIP

International Standard Book Number 0-8493-8992-5
Library of Congress Card Number 95-23091
Printed in the United States of America 2 3 4 5 6 7 8 9 0
Printed on acid-free paper

PREFACE

Paper owes much of its versatility to its surface characteristics, as these can be tailored to a variety of applications. Characterizing and controlling these properties is central to product development and control, and the outgrowth of new analytical techniques in chemistry and microscopy provides us with a rich array of sophisticated tools. Our intent in compiling this book is to provide the collective experience of people who use many of these tools in the paper industry. We hope that the book will allow the reader to quickly match the tool to a given problem and also to appreciate the range of problems and situations that can be probed with a given technique.

This book has been organized into three sections: I. Physical Characterization of Surfaces; II. Spectroscopic Methods; and III. Emerging Technologies. Each chapter in this book represents an area or a technique. The focus is on practice; the emphasis lies on examples of what the method can do, and where and why it might be practical to use it. As a method can often best be appreciated from an understanding of its fundamentals, however, just enough theoretical background is presented to anchor the applications. Most of the authors participating in this volume have extensive hands-on experience in the paper industry, but in some cases a technique may not have been applied extensively to paper, and here the author's affiliations lie outside the industry. Several chapters were solicited by the editors specifically because there was little or no information about the application of these analytical techniques to pulp or paper, and we hope that you find these to be useful.

Finally: our thanks and acknowledgements. As editors, we have had the privilege of working with people who actually delivered manuscripts when they said they would and who revised them cheerfully according to our dictates. Organizing this book has been a pleasure; to the authors go our thanks. We would also like to thank the staff at CRC Press and our editors (all of them!) for their help and support during the preparation of this volume.

Terrance E. Conners
Forest Products Laboratory
Mississippi State University
P.O. Box 9820
Mississippi State, MS 39762
USA
conners@fpl.msstate.edu

Sujit Banerjee
Institute of Paper Science and Technology
500 10th Street, NW
Atlanta, GA 30318
USA
s.banerjee@ipst.edu

List of Contributors

Umesh P. Agarwal
USDA Forest Products Laboratory
One Gifford Pinchot Drive
Madison, WI 53705-2398

Timothy B. Arnold
Dyer Energy Systems, Inc.
141 Middlesex Road
Tyngsboro, MA 01879

Rajai Atalla
USDA Forest Products Laboratory
One Gifford Pinchot Drive
Madison, WI 53705-2398

Sujit Banerjee
Institute of Paper Science and Technology
500 10th Street, NW
Atlanta, GA 30318

Marie-Claude Béland
PAPRICAN
570 St. John's Boulevard
Pointe Claire, Qc
H9R 3J9
Canada

Until 1998, on leave at:
Institute of Optical Research
KTH
S-100 44 Stockholm
Sweden

Gianluigi Botton
University of Cambridge
Department of Materials Science and Metallurgy
Pembroke Street
Cambridge, U.K. CB2 3QZ

Kenneth L. Busch
School of Chemistry and Biochemistry
Georgia Institute of Technology
Atlanta, GA 30332-0400

James J. Conners
Boehringer Ingelheim Pharmaceuticals
900 Ridgebury Road
P.O. Box 368
Ridgefield, CT 06877

Terrance E. Conners
Forest Products Laboratory
Mississippi State University
P.O. Box 9820
Mississippi State, MS 39762

Glynis de Silveira
University of Cambridge
Department of Materials Science and Metallurgy
Pembroke Street
Cambridge, U.K. CB2 3QZ

Lisa D. Detter-Hoskin
Electro-Optics, Environmental and Materials Laboratory
Georgia Tech Research Institute
925 Dalney Street
273 Baker Building
Atlanta, GA 30332-0827

Glenn L. Dyer
Dyer Energy Systems, Inc.
141 Middlesex Road
Tyngsboro, MA 01879

Frank M. Etzler
Boehringer Ingelheim Pharmaceuticals
900 Ridgebury Road
P.O. Box 368
Ridgefield, CT 06877

Paivi Forsberg
Kymi Paper Mills, Ltd.
Research Center
FIN-45700 Kuusankoski
Finland

Michael A. Friese
Appleton Papers, Inc.
P.O. Box 359
Appleton, WI 54912

Derek G. Gray
PAPRICAN and Department of Chemistry
Pulp and Paper Research Centre
3420 University Street
McGill University
Montreal, Qc
H3A 2A7
Canada

Shaune J. Hanley
PAPRICAN and Department of Chemistry
Pulp and Paper Research Centre
3420 University Street
McGill University
Montreal, Qc
H3A 2A7
Canada

William K. Istone
Champion International Corporation
Technical Center
West Nyack Road
West Nyack, NY 10994

Patrice J. Mangin
Université du Québec à Trois-Rivières
C.P. 500
Trois-Rivières, Qc
G9A 5H7
Canada

From August 1995 to August 1997:
Visiting Scientist, Stora Chair
Department of Paper Technology
KTH
S-100 44 Stockholm
Sweden

Jon S. Martens
Conductus
969 West Maude Avenue
Sunnyvale, CA 94086

John M. Pope
Byk-Gardner
2435 Linden Lane
Silver Springs, MD 20910

Arthur J. Ragauskas
Institute of Paper Science and Technology
500 10th Street, NW
Atlanta, GA 30318

John F. Waterhouse
Institute of Paper Science and Technology
500 10th Street, NW
Atlanta, GA 30318

Table of Contents

III. Emerging Technologies

1

Three-Dimensional Evaluation of Paper Surfaces Using Confocal Microscopy

Marie-Claude Béland and Patrice J. Mangin

PAPRICAN, Pointe Claire, Quebec, Canada

INTRODUCTION

The evaluation of paper surfaces is important to better understand the nature and the behavior of paper during conversion and end-use processes such as coating and printing. Paper surfaces can be characterized in terms of roughness, porosity and compressibility. These properties can be measured many ways, but the measurements obtained depend to a large extent on the measuring method used.[1] Traditional means of measuring roughness for instance, are indirect and often only provide a single figure for the roughness of a sheet. Renewed interest in direct measurement of paper surfaces by profilometry is linked to the availability of more powerful software packages capable of generating two-dimensional (2D) maps. Three-dimensional (3D) surface maps can be generated using confocal laser scanning microscopy (CLSM) data and surface reconstruction software. The images thus obtained can be used for both qualitative and quantitative characterization.

The first patent describing confocal microscopy dates back to 1957 and was filed by Marvin Minsky. Interested in trying to work out the wiring diagram of solidly packed cells in the tissue of the central nervous system, he faced one major obstacle: he had no method to visualize a three dimensional region of the sample. In particular, he had to find a way to control light scattering. "One day it occurred to me (Minsky) that the way to avoid all that scattered light was to never allow any unnecessary light to enter in the first place."[2]

Confocal laser scanning microscopy still follows the basic principles established by Minsky. The source pinhole, the illuminated spot on the object, and the detector pinhole all have the same foci at different locations on the optical path and hence are "confocal" with respect to one another (see Figure 1). Essentially, confocal microscopy allows the acquisition of single focal planes: light (whether reflected, fluorescent or transmitted) originating from the object can be acquired in slices of limited thickness. Consecutive

0-8493-8992-5/95/$0.00+$.50

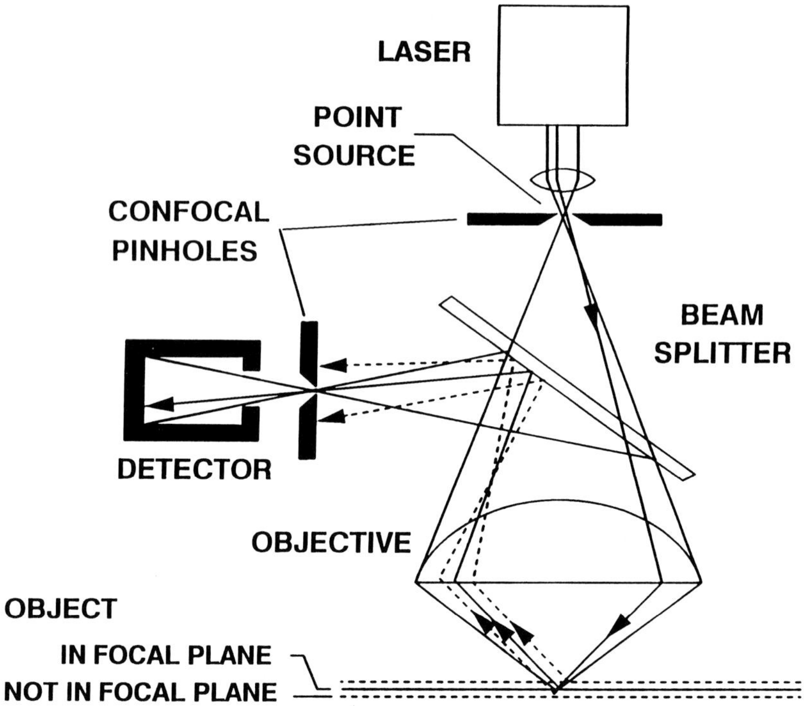

Figure 1. Confocal principle

images are obtained by moving the acquisition plane deeper into the object. A 3D surface map of that object can be reconstructed from the acquired reflected light image stack.

The advent of powerful computers has undoubtedly fostered the expansion of confocal technology. Several types of confocal systems with different configurations are now available.[3,4,5,6,7] Applications of confocal microscopy are already numerous in biology and medicine[8] and are increasing in materials sciences.[9,10] Applications in the pulp and paper industry are also becoming more numerous, ranging from investigations of fiber properties[11,12] and paper drying processes[13] to paper surface characteristics.[14,15]

This chapter describes confocal microscopy and how this technique can be applied to measure and characterize paper surfaces, both qualitatively and quantitatively. Applications include surface physics, pulping, papermaking, and printing science. In addition to the structure of the paper surface, its behavior under compression is important for relating paper surface characteristics to end-use performance and printability. Since confocal microscopy is relatively new to pulp and paper science, applications discussed here are by no means exhaustive and many more will undoubtedly follow as confocal technology improves and as image analysis capabilities are further developed.

CONFOCAL PRINCIPLE

In conventional microscopy, light reflected from an object point which is not in the focal plane produces out of focus-blur in the image. Thus the image of the specimen that is in the focal plane suffers from the superimposition of out-of-focus images from planes above and below the focal plane. In contrast, confocal microscopy uses pinholes to generate a point source and to limit the light entering the detector to only that reflected from the focal plane. Thus, light originating from outside the focal plane is not detected. The ability to obtain distinct optical sections through the object is the main advantage of confocal microscopy over conventional microscopy.

Figure 1 illustrates the confocal principle for the Leica CLSM configuration. An argon ion laser provides an intense and stable light source, with a wavelength range of 488 nm to 514.5 nm. At the light source, a pinhole of fixed size is used to obtain point source illumination. The light passing through the objective and reflected from the object is directed at the detector by a beam splitter. A variable-size pinhole in front of the detector allows only reflected light from the focal spot on the object to reach the detector. A point-by-point image is obtained by scanning the beam over an XY plane. The XY image thus obtained has a limited thickness in the Z direction which depends on the opening of the detector pinhole: the more open the pinhole, the thicker the focal plane. When the pinhole is fully opened, the image resembles that obtained with a conventional light microscope.

This sectioning ability is illustrated in Figure 2 where the normalized intensity of a point object placed on the optical axis of the lens is plotted as a function of the distance from the focal plane of the lens. In a conventional microscope, the intensity of such a point object is constant with increasing distance and the object is not discernible in the Z direction. In a confocal microscope, a point object in the focal plane produces a sharp intensity peak. Since the intensity of the object decreases as the distance from the focal plane increases, it is possible to localize the object in the Z-direction. The confocal curve shown corresponds to a small detector pinhole size. As the detector is opened, the peak gradually broadens and eventually becomes more like the conventional system. Images of single focal planes can therefore be acquired and sectioning of the object is possible by moving the location of the focal plane through the Z direction of the object. This results in a stack of images, each image originating from a different height in the object. A detailed 3D topographical map of the surface of the object can then be reconstructed from this stack. This way of mapping the surface is particularly interesting for paper surfaces since, unlike most other methods, it is non-contacting. Two modes of image acquisition are available on the Leica system: reflected light and epi-fluorescence. Topographical maps are reconstructed from reflected light images.

Since the effective point spread function of the image is squared for the confocal microscope, it generates images that are sharper in the XY plane than those obtained with a conventional light microscope.[16] The point spread function in essence describes the

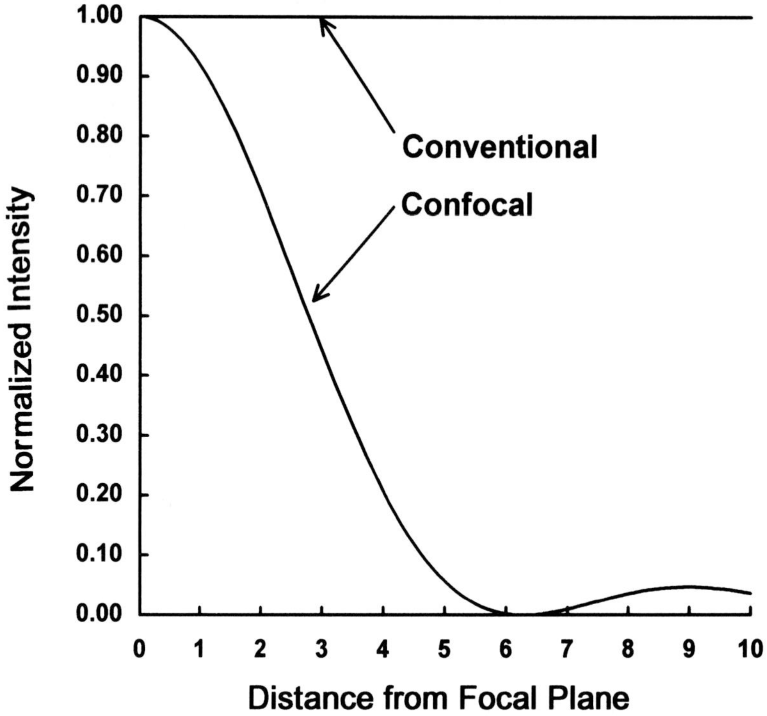

Figure 2. Normalized intensity of a point object placed on the optical axis of the lens as a function of the distance from the focal plane.

shape and intensity of the beam spot used for imaging. The effective point spread function is given by the product of the point spread functions of the two imaging lenses, the objective lens and the condenser. In the confocal microscope, one lens serves both purposes, since an epi-illumination configuration is used. Therefore, each confocal image is also clearer in the XY plane than if acquired with conventional microscopy. This has been discussed in more detail by Wilson.[17]

Confocal microscopy does have its limitations. For example, very thick paper samples require a large Z-directional range which may cause image acquisition problems. Likewise, the positioning of a focal plane at the bottom of a thick object leads to a weaker signal. Thus, acquiring images through an entire paper sample to reconstruct the bulk structure of the sheet has proven difficult, particularly when a coating layer is present. However, the reconstruction of the surface topography is not affected. Density of the object is also a factor impeding proper image acquisition. Limits are mainly related to the size of the object or volume under observation, the positioning of any one focal plane within that object or volume and the quantity of information reaching the detector.[18] This last factor is particularly important when trying to detect a fluorescent signal. It should also be noted that for materials science purposes, the field of view is rather small and that coupling the z micro-moving stage with an automatic xy scanner would be useful for examining larger areas. New systems with larger fields of view and with larger memory and computing power could alleviate these limitations. Using longer wavelengths could reduce stray light when imaging in depth but at the cost of some resolution.

3D TOPOGRAPHICAL MAPS VERSUS 3D PROFILOMETRIC MAPS

Confocal microscopy provides advantages not afforded by either mechanical or optical profilometers. In these instruments, a diamond tip stylus or the spot of a laser beam records irregularities of the paper surface and produces profile lines. By stacking individual profile lines acquired by incremental displacements perpendicular to the scanning lines, a computer generates oblique views of the paper surface.[19,20] Mechanical and laser profilometers can produce only such oblique views or 3D profilometric maps – they do not produce images of the complete surface. Such a profilometric map is emulated by the CLSM in Figure 3a, where an uncalendered lightweight coated (LWC) paper has been imaged. Interpreting a profilometric map can be difficult and identifying the features of the paper surface, like surface fibers or coating defects, requires some experience.[12,13,21] Figure 3b shows a 3D topographical map of the same area. The image is color-coded for height, where the lighter colors (whiter areas) represent the higher regions and the darker colors (dark gray to black areas) the lower regions of the surface. This 3D topographical map is reconstructed from 16 focal planes. Fibers are now easily identifiable. It should be emphasized that 3D topographical maps are not true pictures of the surface fibers but computer-generated digital images, color-coded for height.

Profilometric maps only partially sample the surface, the distance between the scan lines being relatively large. Confocal microscopy allows for the entire surface to be sampled, from the top of the highest fiber, down to the bottom of the deepest open pore. The sampling is complete because image planes are acquired that include all of the surface, as well as information originating from slightly below the surface, something that other profilometry techniques cannot provide. Subsequently, only the information corresponding to the surface is extracted from the data set to reconstruct the topographical map. This completeness of sampling is a substantial improvement over more conventional ways of obtaining quantifiable maps of paper surfaces. For quantification purposes, the parameters derived from complete data sets are more accurate descriptors of a paper surface than those calculated from profilometric maps. The pixel mapping in the XY-plane, and the pixel intensity extracted from 3D topographical maps readily provide XYZ information. Furthermore, these images are digital and therefore lend themselves very well to image analysis.

Paper surface applications presented here utilize both qualitative and quantitative evaluation of 3D topographical maps. The quantitative evaluation of 3D topographical maps includes the analysis of the pixel frequency distribution as a function of the gray level intensity. For example, such histograms may be readily transformed to represent the distribution of the paper surface area as a function of depth and may provide a number of statistical parameters that describe the surface. In addition, topographical maps obtained using confocal microscopy are digital images of the entire surface, making them useful for the quantification of small-scale spatial distribution of surface features as well as for other image analysis applications.

(a)

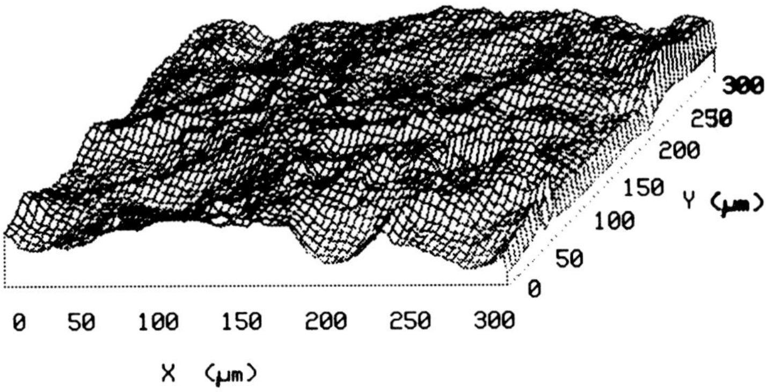

(b)

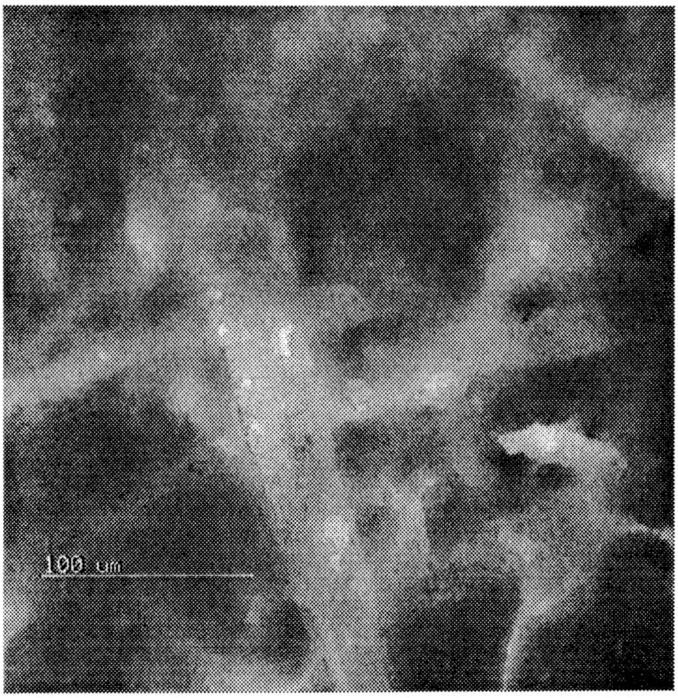

Figure 3. (a) Oblique view, or 3D profilometric map, of an uncalendered lightweight coated paper generated from a CLSM 3D topographical map. (b) 3D topographical map representing the same area as in (a). The 3D map is reconstructed from imaging 16 planes parallel to the paper surface. The total scanning range in depth is 30 μm. The measured area is 0.313 mm x 0.313 mm, with a 16X (air) objective. The intensity range includes 256 gray levels. Each gray level corresponds to a depth of 0.12 μm. The light tones represent the highest points and the dark tones the lowest points of the paper surface.

IMAGE QUALITY

Many variables, such as pinhole size, image format and size of the 3D data set (pixel size and number of images acquired in depth) and signal-to-noise ratio, affect the quality of reflectance mode confocal images. The way in which these interact depends on the particular application. For instance, optimal conditions for observing two paper surfaces, one uncoated and the other coated, may differ greatly. It should be noted that the optimization of image quality does involve subjective judgement and the variables which must be optimized may vary from system to system. The most important factors affecting the resulting image will be discussed separately, though they are all interdependent.

PINHOLE SIZE

Detector pinhole size is critical since it determines the thickness of the XY slices that are acquired. To illustrate this point, Figure 4 shows images of the same area on an uncoated paper surface acquired with two different pinhole sizes. Notice that the information present in Figure 4b also includes that present in Figure 4a. However, the larger pinhole size makes the image "thicker" in appearance. That is, the image in Figure 4b suffers from the superimposition of out-of-focus information coming from planes above and below the focal plane. This renders the plane brighter and less uniform in focus, making it less suitable for topographical reconstruction.

Optimum pinhole size may be determined for each objective using a plane mirror object. However, the value thus obtained merely approximates the best pinhole size for objects such as paper surfaces which are far from ideal plane mirror surfaces. Furthermore, other variables must be held constant to make this measurement and this usually is not possible when observing surfaces other than mirror-like planar objects. A general rule is that the smallest possible pinhole should be used in order to maximize Z resolution. In practice, however, the thickness of the object must be considered as well as the signal level reaching the detector. A smaller pinhole reduces the amount of light reflected from out-of-focus planes but also reduces the signal originating from the focal plane itself.

IMAGE FORMAT

The most common format for scanning a single plane image is a 512 x 512 pixel array. This format results in square pixels, ensuring that the lateral resolution is equal in both X and Y directions. In the Z direction, the inter-plane spacing ideally should be chosen to match the length of the pixel side, making the resulting voxel (3D equivalent of a pixel) cubical in shape. However, limitations in computing power and storage capabilities make such an approach impractical. The Nyquist theorem[22] states that the inter-plane spacing should be approximately 2.3 times less than the axial resolution of the objective being used. For most paper surfaces, aiming for cubical voxels would likely result in too many planes being acquired. For surface reconstruction, doubling the number of planes may have no influence on the quality of the topographical map, since the Z value calculated

(a)

(b)

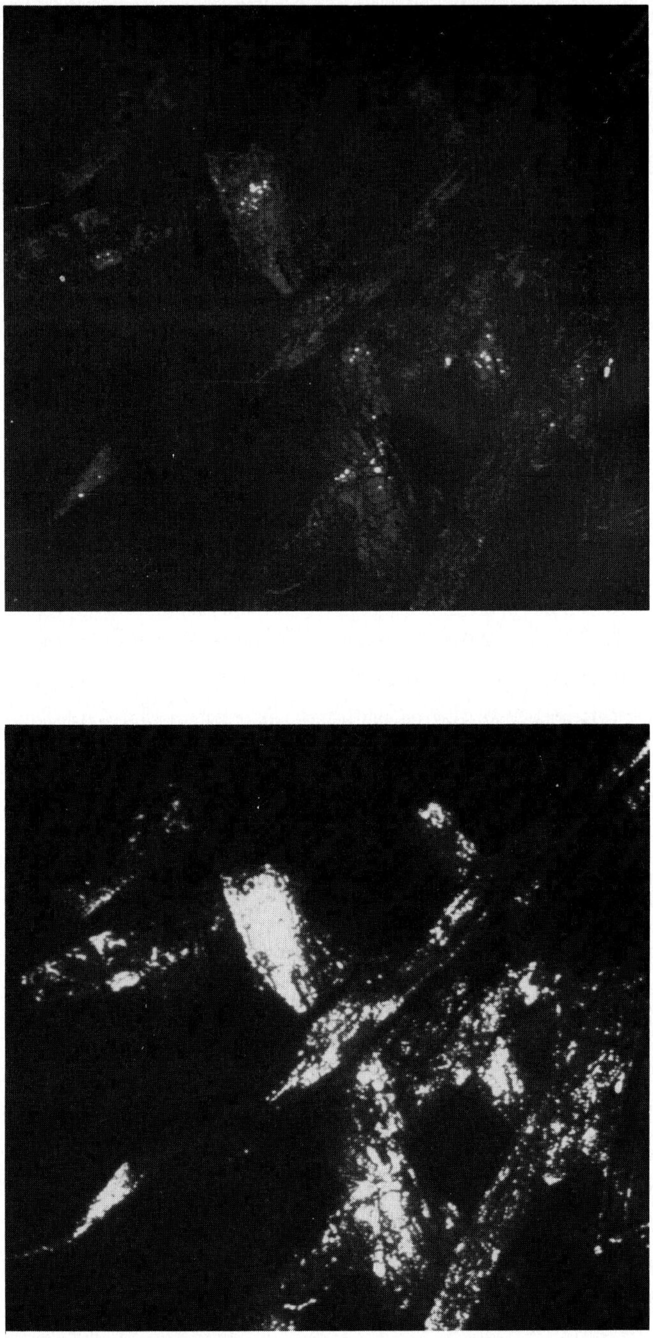

Figure 4. View of an uncoated paper surface (313 μm x 313 μm) to show effect of pinhole size on acquired image. (a) Pinhole value 40; (b) Pinhole value 80.

for the reconstruction already takes into account neighboring planes, as will be discussed in the next section. For most of the applications discussed here we used thirty-two planes, and the step size between the planes was adjusted to encompass the entire surface.

The number of planes acquired in the Z-direction has an effect on the quantitative nature of the data set. When insufficient information is acquired, either for each image or for the 3D surface reconstruction to be complete, under-sampling may result in artifacts in the frequency distribution of the image. For instance, peaks may appear in the intensity distribution of the topographical map at intervals corresponding to the step size. This can usually be avoided by using smaller step sizes and a greater number of focal planes.

RESOLUTION

Resolution determines the size of the finest structural details that can be distinguished. Both lateral resolution (in the XY plane) and axial resolution (in the direction of the optical axis, *i.e.*, Z-direction) are used to characterize a single focal plane. Resolution can be calculated theoretically from wavelength of the light and the numerical aperture of the objective used. In practice, many other factors are involved, making the actual resolution difficult to determine. The term Z-discrimination will be reserved for 3D topographical maps which are reconstructed from several focal planes. Z-discrimination, therefore, is the resolution of a topographical map.

The actual resolution of a focal plane is highly dependent on particular acquisition conditions (*e.g.*, image format, wavelength of light source, objective used, pinhole size, depth of the focal plane within the object, and signal-to-noise ratio). Consequently, it is not easily determined. The theoretical resolution is often given in terms of the full-width-half-maximum (FWHM) value of the intensity curves obtained for a point form object. The FWHM represents the distance from the point form object where the intensity has decreased to half its peak value. The performance of Leica objectives in the reflection mode is shown in Figure 5 for a wavelength of 488 nm. Resolution improves with increasing numerical aperture (larger lens opening). At a constant numerical aperture, the axial resolution is better for an air objective than for an oil immersion objective. Using a 16X air objective with a 0.45 numerical aperture gives a lateral resolution of about 0.5 μm and an axial resolution of approximately 2.5 μm. Changing the magnification changes these calculated resolution values.

Several equations exist to calculate both axial and lateral resolution,[22] but most fail to show the dependence of resolution on factors such as pinhole size used, depth of the focal plane in the object, and changes in the effective wavelength as the focal plane is moved through the object. The effect of pinhole size and focal plane position are illustrated by Wilson.[17] Such factors make resolution an object-specific variable difficult to calculate. Resolution is also dependent on the accurate alignment of the pinhole on the optical axis, both in the axial and the transverse directions,[23] and on the signal-to-noise ratio of an image under given acquisition settings.[24] Furthermore, once the individual focal planes

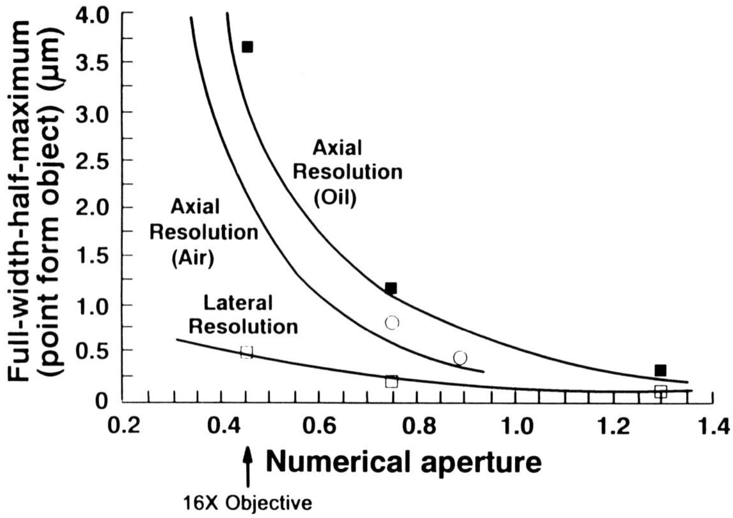

Figure 5. Full Width–Half Maximum values as a function of numerical aperture. Lateral resolution, axial resolution in air and axial resolution in oil are shown for a wavelength of 488 nm. Points on the graph show the measured values for Leica objectives. (Graph courtesy of Leica).

have been acquired in the Z-direction of an object, it is the 3D reconstruction of the surface of that object that matters and must be quantified. Depending on acquisition settings and the mode of reconstruction, this 3D topographical map will have a Z-discrimination which is likely to be better than that of individual planes because it is calculated from information over several planes.

Z-DISCRIMINATION

As the intensity of a given XY pixel position in the image stack changes from the top to the bottom of the data set, an intensity *vs* Z curve is obtained for each pixel. Figure 6a shows an example of how the intensity of one pixel varies as a function of the position in the image stack (from plane 0 to plane 15) and consequently also as a function of depth (0 μm to 48 μm). For every XY pixel, the exact surface location needs to be determined in order to reconstruct the topographical map.

On the Leica system, one of four algorithms can be used to calculate this location. These are: *i*) maximum intensity; *ii*) center of mass; *iii*) modified center of mass; and *iv*) the first occurrence of maximum greater than specified threshold. The depth of the corresponding pixel in the 3D reconstruction is given in Figure 6b for these four types of reconstructions. The maximum intensity mode could be acceptable if enough planes are acquired, ensuring that all maxima on the object sampled. However, this may not always be the case. The mode of first occurrence greater than a specified threshold can be used to isolate information corresponding to a limited depth range. The center of mass mode takes into account the entire profile, which could be useful if observing an object whose

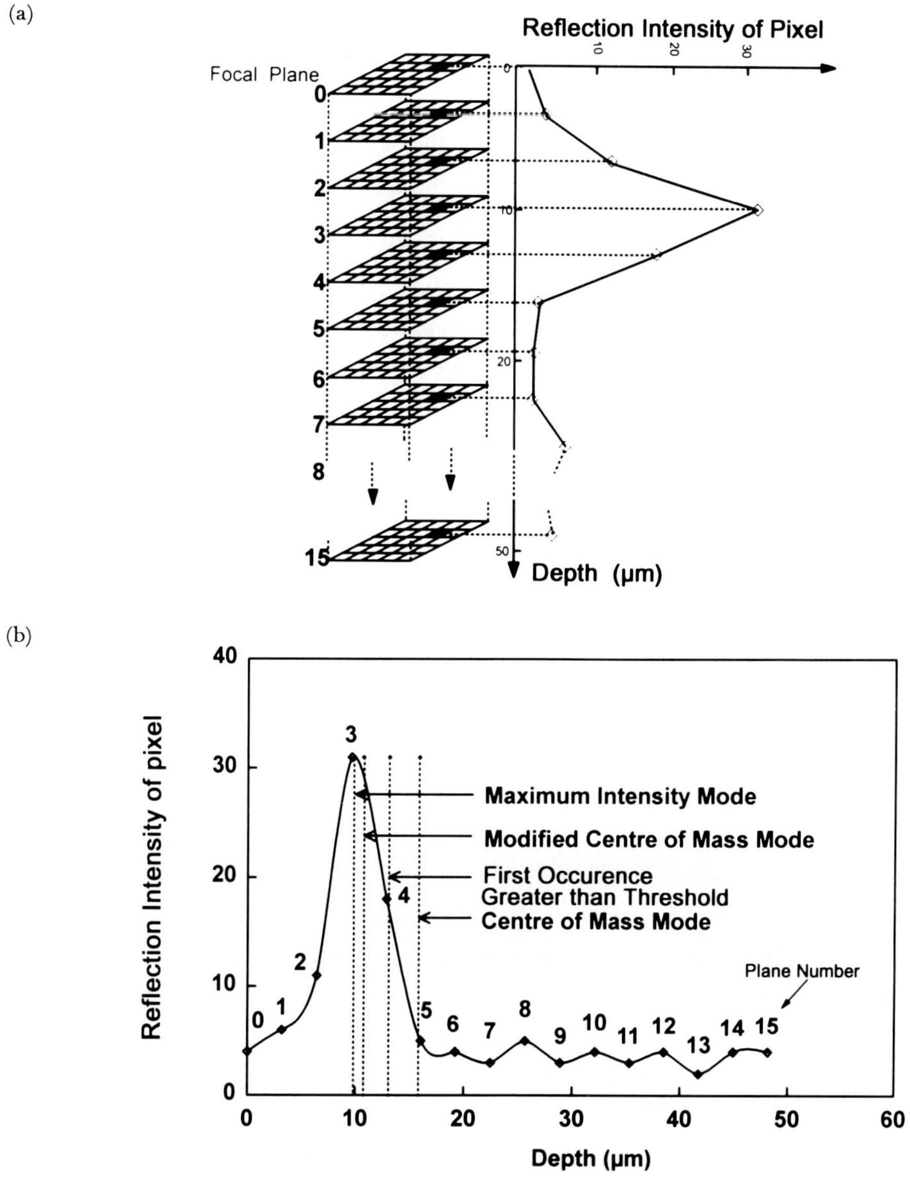

Figure 6. (a) Variation of the intensity of one pixel in XY as depth is increased, planes are numbered 0 to 15; (b) Corresponding Z intensity of this pixel that would appear on the reconstructed topographical map. Four types of 3D reconstruction algorithms can be used, all of which are available on the Leica system. Arrows point to the topographical depth (*i.e.*, the depth on the topographical reconstruction for this particular pixel) that would be calculated from this intensity curve using each of the four modes.

intensity was relatively constant throughout its thickness. However, in order to reconstruct surfaces from images acquired in reflection, the preferred mode of acquisition is the modified center of mass mode, which gives greater importance to the intensity peak than to the background. It also accounts for the fact that the point in depth corresponding to the true highest intensity may not have been sampled. Figure 6b illustrates this point, where the depth calculated with the modified center of mass algorithm falls to the right of the acquired maximum (plane three) because plane four has a higher intensity than plane two.

In this example, there would be 0.19 μm per gray level (48 μm/256) on the topographical map. The actual Z-discrimination of this topographical map would lie between 0.19 μm (ideal) and half the step size used (*i.e.*, 1.5 μm) (worst case).

SELF-SHADOWING

When acquiring images for surface reconstruction, it is the light reflected from the surface that is important. If very thick objects are imaged, self-shadowing occurs as the depth of the focal plane increases within the object. This phenomenon is illustrated in Figure 7, where 7a shows a focal plane close to the surface. The intensity of the reflected light from the surface of these fibers is unobstructed. Figure 7b shows a plane taken 12 μm deeper within the paper sample. For the areas directly below the fibers seen on the upper plane, light must pass down through these fibers and be reflected back up again in order to be detected, and the reflected light intensity is thus greatly diminished. The upper fibers therefore appear to cast a shadow on the fibers beneath them until at some point an image is no longer obtainable under constant operating conditions. Nevertheless and most importantly for surface analysis, the shadowing phenomenon does *not* affect surface reconstruction. In the reconstruction of 3D topographical maps, reflection intensity from the surface, regardless of depth, predominates over that originating from underneath the fiber layers. This also implies that the reconstruction of the surface is not affected by sheet density, as self-shadowing occurs within the sheet, not at its surface. Thus, while self-shadowing does not hamper the reconstruction of a 3D map of the paper surface, it does prevent direct 3D reconstruction of the interior paper structure. However, technological progress has been rapid and this type of difficulty could soon be alleviated.

IMAGE QUANTIFICATION

A 3D topographical map of a paper surface is the reconstruction of the surface, as viewed normal to the surface. The surface is defined from the top of the highest fiber to the bottom of the deepest open pore. This is illustrated schematically in Figure 8.

Once the 3D topographical map is obtained, several parameters are readily available. A topographical map is a matrix of pixels, *i.e.* a series of consecutive pixel lines, each line representing a height profile on the surface. ISO roughness parameters may be calculated from profile sets corresponding to the images.[25,26,27] These surface roughness parameters are associated with profile irregularity forms. The arithmetic mean deviation, R_a, is given by Equation (1),

(a)

(b)

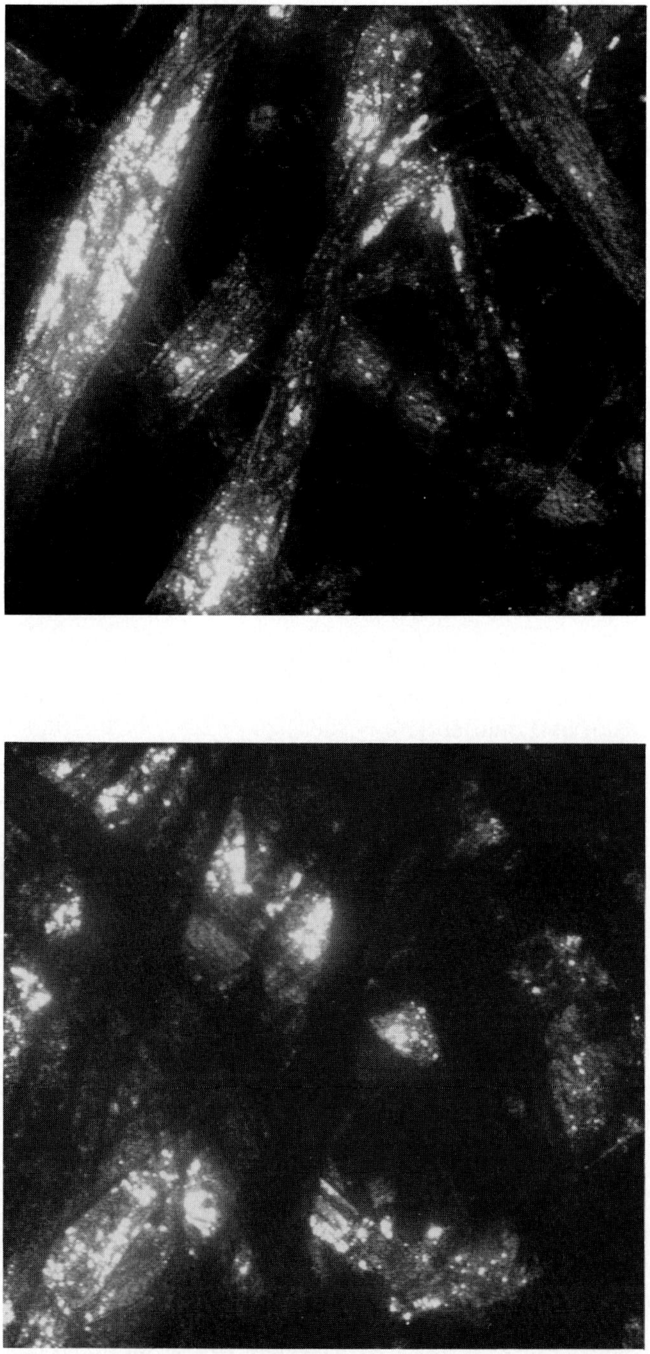

Figure 7. Self-shadowing phenomenon as depth of focal plane is increased: (a) focal plane close to the top of the surface; (b) focal plane approximately 12.3 μm deeper into the paper sample (313 μm x 313 μm).

Figure 8. Cross section of a paper surface and the corresponding topographical profile, as reconstructed from a 90° angle to the plane of the surface.

$$R_a = \frac{1}{L} \int_0^L |y(x)| \, dx \tag{1}$$

where L is the sampling length, y is the distance between a profile point and a reference line, and n is the discrete number of profile deviations measured. The root mean square deviation, R_q, is given by Equation (2):

$$R_q = \left[\frac{1}{L} \int_0^L y^2(x) \, dx \right]^{0.5} \tag{2}$$

The skewness associated with a profile, or as in a topographical map, a set of profiles, is given by Equation (3):

$$S_k = \frac{1}{R_q^3} \cdot \frac{1}{n} \sum_1^n y_i^3 \tag{3}$$

where S_k represents a measure of asymmetry of the distribution density of profile deviations. In addition, the values corresponding to the distance between the lowest and highest points of a profile and the reference line, as well as the difference between these two values are also available. The algorithms used in the CLSM software are based on Equations (1–3) and are adapted to the 3D topographical maps they characterize.

In addition, the average intensity and standard deviation may be obtained for whole images or portions thereof. Iso-contour lines can be defined with a varying distance between levels, as shown in Figure 9, where the lines determine regions on the surface that are separated by 40 gray levels, corresponding to approximately 8 μm in height. The corresponding areas, perimeters and roughness of a particular contour may be calculated.

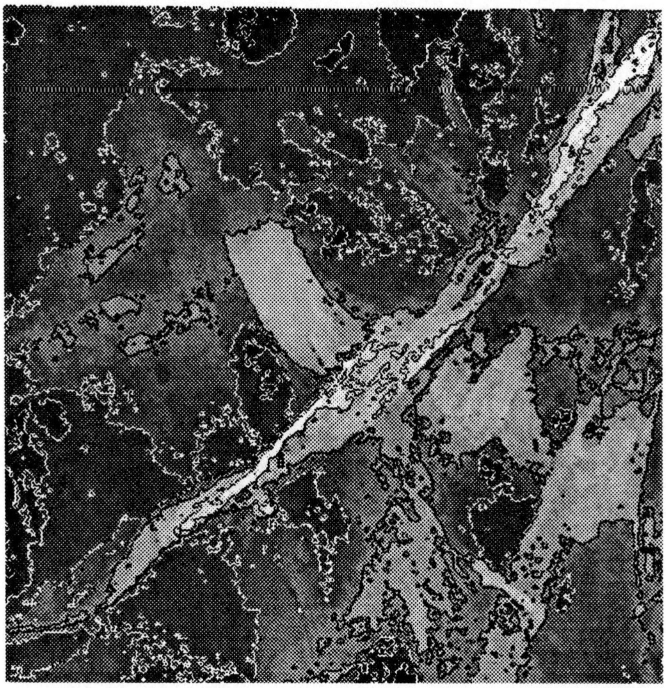

Figure 9. Iso-contour lines at every 40 gray levels, or about every 8 μm apart, superimposed onto the corresponding topographical map (313 μm x 313 μm).

These may be valuable in providing texture information about the surface under observation. In addition, such iso-contour lines could lend themselves very well to fractal analysis,[28,29,30] which could potentially distinguish different types of paper structures.

On a topographical map, the intensities of adjacent pixels taken three by three to form a triangle, the intensity being representative of height. Depending on the surface roughness, such a triangle will generate an area that is relatively small if the surface is smooth or relatively large if the surface is rough. The ratio of the sum of all the triangular areas formed from a given topographical map to that of the projected flat surface provides yet another measure of roughness, the bearing ratio. The higher the ratio, the rougher the surface. Direct volume calculations on the 3D data set are also possible over various ranges.

In addition to topographical maps, extended focus images, which represent what would be seen in a conventional microscope without the out-of-focus blur, can be useful for evaluating the texture of surfaces. Figure 10 shows an example of such an image as well as the topographical map of the same area. This example also illustrates the capability of the confocal microscope to reconstruct the surface of a dark sample which would otherwise be difficult to view in a conventional microscope. The qualitative differences between these two images of the same surface can be useful to distinguish between structure and texture of surfaces.

(a)

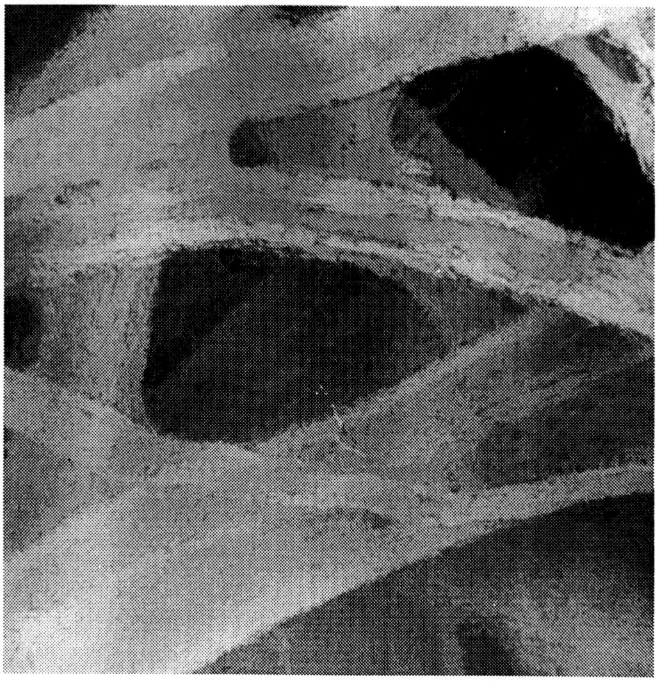

(b)

Figure 10. Black cardboard surface (125 μm x 125 μm). Topographical map (a) and extended focus image (b) of the same area as in (a). The extended focus image represents what would be seen in a conventional microscope without the out-of-focus blur.

The frequency distribution as a function of intensity level may be obtained from the topographical data. Since the raw data may not always begin at an intensity level of 256 (depth = 0), the effective reference point (point on the distribution that is set at depth = 0) of these curves has been arbitrarily set at that intensity level which represents 0.01% or more of the total surface. In other words, intensity levels associated with 25 pixels or less in a 512 x 512 format are ignored. The frequency histogram is easily converted to an area *vs* depth curve, since frequency is related to surface area and intensity level is related to depth information. The area *vs* depth curve can in turn be converted to a surface pore size distribution curve or an Equivalent Surface Pore function.[20,31] Equations (4) and (5) are used to do this.

$$D_i = \Delta R - (\Delta R/256) \cdot I \tag{4}$$

$$A_i = \frac{S}{f} \cdot F_i \tag{5}$$

where D_i is the depth at a gray level I, ΔR is the total Z range of the acquisition, A_i is the area that the pixels at gray level I occupy, S is the image area as given by the objective used, f is the number of pixels on the image as given by the image format and F_i is the frequency value corresponding to the gray level I. This data is further transformed to normalized cumulative area *vs* depth by dividing each area value by the total area of the sample and summing the results.

The nth order topographical equation is given in its general form in Equation (6).[31] For most paper roughness measurements, like with the Parker Print-Surf air-leak instrument or the model pore approach,[20,30] the calculated roughness corresponds to the third order roughness. Likewise, it is possible to calculate a roughness value (G_3) from the normalized cumulative area curve according to Equation (7),

$$G_n = \left[\int_0^1 z^n \cdot dF(z)\right]^{\frac{1}{n}} \tag{6}$$

$$G_3 = \left[\sum_0^{256} D_i^3 \cdot \left(Ac_{i+1} - Ac_i\right)\right]^{1/3} \tag{7}$$

where Ac_i is the normalized cumulative area at intensity I.

AREA AVERAGING

Many quantitative parameters can be obtained using confocal microscopy. Due to the existence of local structural variations in paper of both high and low frequency, averaging is necessary when quantifying paper surfaces with confocal images as these represent relatively small areas (313 μm by 313 μm with a 16X objective). In addition to random area averaging, tiling of adjacent images can be used in order to better evaluate the lower frequency variations.

REPEATABILITY

The repeatability of the Leica CLSM has been assessed previously.[32] Results are easily reproduced since pinhole size, voltage, offset (gain) and step size are recorded for all saved images. Instrumental variation was measured on a paper sample by imaging the same area ten times. For the first data set acquired, the pinhole, voltage, offset and step size values were noted. For all subsequent measurements, the sample was brought out of focus by changing the macro-position of the stage. The voltage, pinhole, and offset values were also randomly increased and/or decreased before being brought back to the initial values. For each data set, a topographical map was reconstructed in order to verify the variation of the dispersion of the data around the mean intensity. The variation caused by instrument settings was found to be 2.8 %.

APPLICATIONS

SURFACE PHYSICS

Macro-Structure and Micro-Structure: The intensity distributions that are obtained for each image can be quantified for roughness and for shape. Two curves that have the same G_3 roughness do not necessarily have the same shape, mode, or width of distribution. In addition, the objective used determines a certain field of view. Observing a surface at different magnifications allows different surface characteristics to be isolated. With a low magnification, the field of view is larger and roughness of the surface can be described on a "macro" scale. At higher magnification, the field of view is reduced and the surface can be described on a "micro" scale. This combination of dimensions for evaluating paper surfaces may allow the separation of the effects of surface treatment in terms of fiber network effects *vs* fiber surface effects. One recent advancement of confocal technology is the development of a scanning laser microscope/macroscope capable of acquiring images on two different scales.[33] The system possesses a sliding holder that can put either a confocal microscope lens or a macroscope telecentric lens into the optical path, thereby allowing images of different dimensions to be acquired on the same specimen. In its present form, the macroscope images are limited by the axial resolution of the lens used ($>150\ \mu$m), which means that an entire sheet of paper can be contained in one focal plane and that a high resolution axial reconstruction of the surface is difficult. Developments to improve this are ongoing. The *lateral* resolution is high, as illustrated in Figure 11, which shows a 1 cm^2 area of a handsheet. The image was acquired using 2048 x 2048 pixels, resulting in a pixel size of about 5 μm x 5 μm. Images as large as 7.5 cm x 7.5 cm can be acquired. The microscope mode allows one to zoom in on an area of interest and in this mode both the lateral and the axial resolution are high and comparable to other confocal systems.

Surface Compressibility: PAPRICAN has recently developed an apparatus designed to apply static compressive stress to the paper surface while still allowing CLSM imaging to be possible.[34] In addition to characterizing the paper surface, its behavior can also be characterized since the same area of the paper surface can be imaged under increasing compressive stress.

To illustrate, an uncalendered commercial sheet made from 100% bleached kraft was imaged at different pressure levels. Figure 12 shows the same paper surface in the uncompressed state (Figure 12a) and compressed at a pressure of 3.60 MPa (Figure 12b). Surface height data at five pressure levels were obtained and the G_3 roughness values calculated. An exponential decay of surface roughness with pressure was observed. The static compressibility is defined as the slope of the roughness as a function of the logarithm of the applied pressure. This compressibility is related to both the fiber

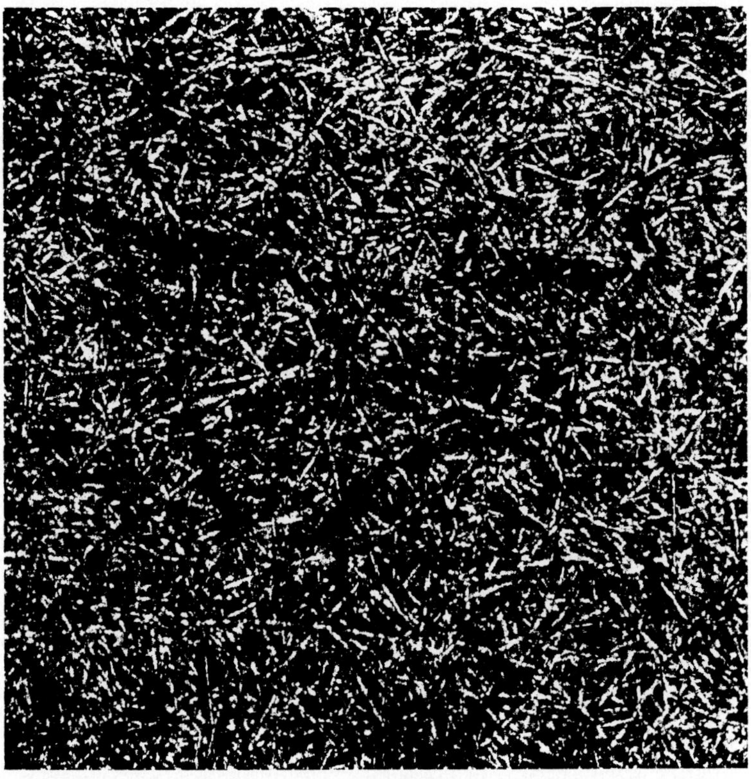

Figure 11. Handsheet surface of 1 cm x 1 cm acquired as a 2048 x 2048 pixel image using the microscope/macroscope in macroscope mode. Fields as large as 7.5 cm x 7.5 cm can be acquired.

(a)

(b)

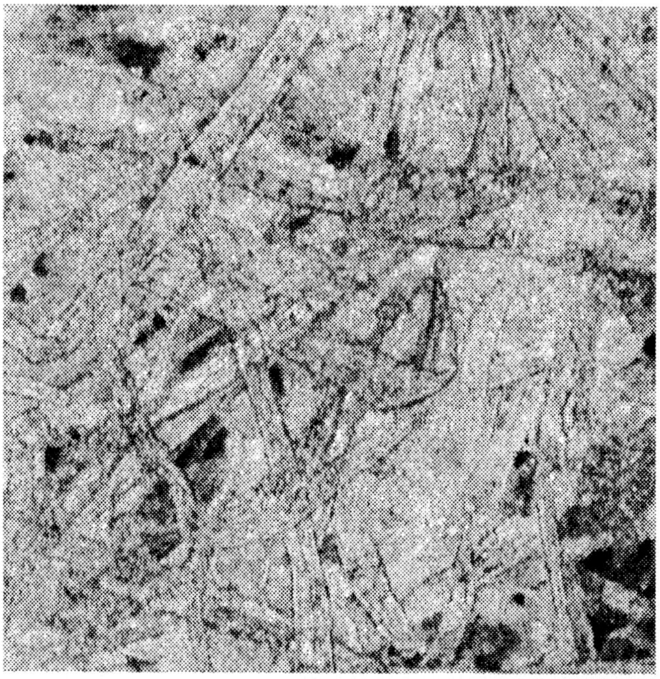

Figure 12. Same area of a commercial 100% bleached kraft (hardwood) paper surface that has been conventionally calendered to a Parker-Print Surf (S10) value of 3.36 μm. (a) Uncompressed, G_3 value of 7.78 μm; (b) Compressed at a pressure of 3.60 MPa, the G_3 value is now 3.97 μm (313 μm x 313 μm).

characteristics and the structural characteristics of the paper surface. Upon calendering, the local compressibility of the paper surface decreases. The surface compressibility after calendering depends both on the calendering process and on the furnish. Both the initial surface roughness and the surface compressibility are important for gravure print quality.[15]

Two samples having the same uncompressed roughness can lead to different print qualities. Figure 13 shows the area versus depth distributions for two calendered paper samples, one a conventionally calendered TMP sample, the other a soft-nip calendered kraft sample. Both have an uncompressed G_3 roughness of 13.1 μm; their surface area distributions, however, are quite different. The kraft sample has a narrower distribution than the TMP sample indicating that, even at the same roughness level, the samples differ in their surface structure. When the samples are compressed to approximately the same pressure, the two samples show different roughnesses (3.30 μm for TMP-CC and 5.15 μm for K-SN) and different distributions. In the compressed state, it is the TMP sample that has the narrower distribution, and because of its higher compressibility its roughness is lower under compression than that of the kraft sample. The TMP sample has more large pores that collapse upon compression while the kraft sample still has many small pores that do not collapse. The size distribution of the surface pores under compression could also help explain the differences in the number of missing dots (at 2 MPa, TMP-CC: 45 and K-SN: 873). A compressed sheet that has a higher percentage of flat area will allow more dots to be transferred to it. Similarly, the quality of the dots transferred are likely to depend on the shape of the pores that do not collapse under load.

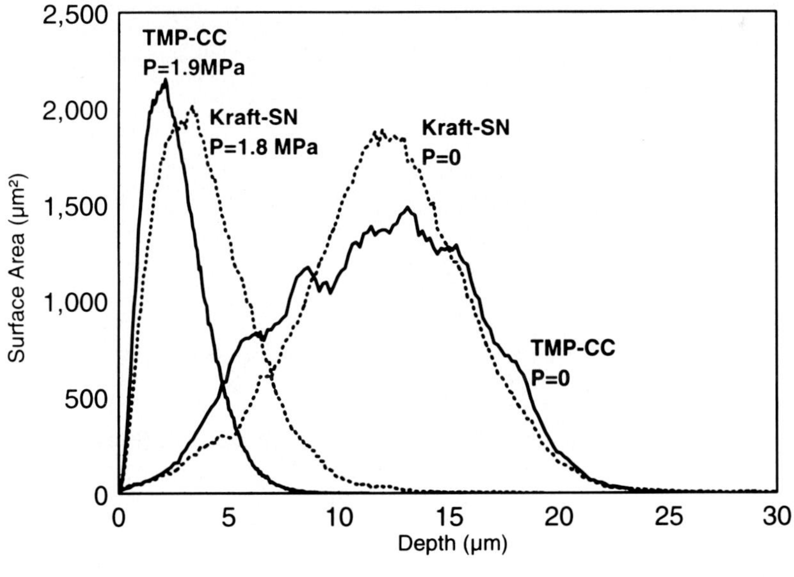

Figure 13. Surface area distributions for two samples having the same uncompressed roughness (G_3 = 13.1 μm), but different roughnesses under compression: TMP-CC G_3 = 3.30 μm; and K-SN G_3 = 5.15 μm.

PAPERMAKING APPLICATIONS

Calendering: Calendering improves the surface finish of the sheet, modifying the distribution of the surface pore areas in the process. The same type of calendering affects the surface differently if the basestock furnishes are different. Figure 14 shows the changes in area distributions for two samples (one a TMP sheet, the other a kraft sheet) conventionally calendered to the same PPS roughness. The initial non-calendered samples are very different in nature and thus will react differently to calendering. Changes in both the shape of the distributions and the mean depths can be observed and quantified for each sample. Surface changes are better shown by changes in the distributions than by changes in PPS values.

Fillers: Mineral fillers, which have a higher light scattering coefficient than fibers, improve the optical properties of paper and can also affect its printing properties. The effects of fillers on printing properties, however, are not straightforward as the printing process must also be considered. For instance, offset printing requires a strong surface while gravure printing requires a smooth closed surface. In his work, Roehr[36] underlines the importance of the surface structure of filled papers. Improved printing properties with high filler content is sometimes claimed[37] but seldom substantiated with data.

Figure 15 shows the G_3 roughness measured with confocal microscopy as a function of the filler content in the sheet. The filler used is synthetic clay (sodium-alumino silicate) with an average pigment size of 1.65 μm. The modification of surface smoothness by fillers

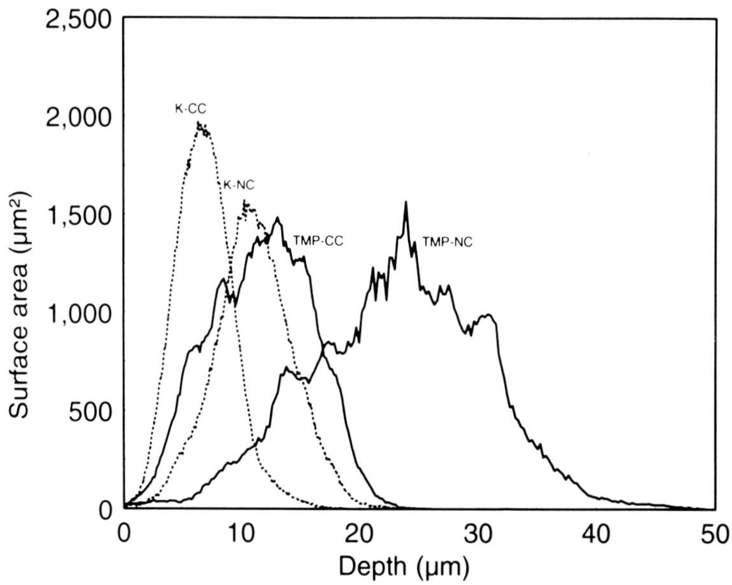

Figure 14. Surface area distributions for a TMP sheet (solid lines) and a kraft sheet (dotted lines), conventionally calendered to the same target PPS-S10 value (measured values were TMP-CC: 3.65 μm; K-CC: 3.35 μm).

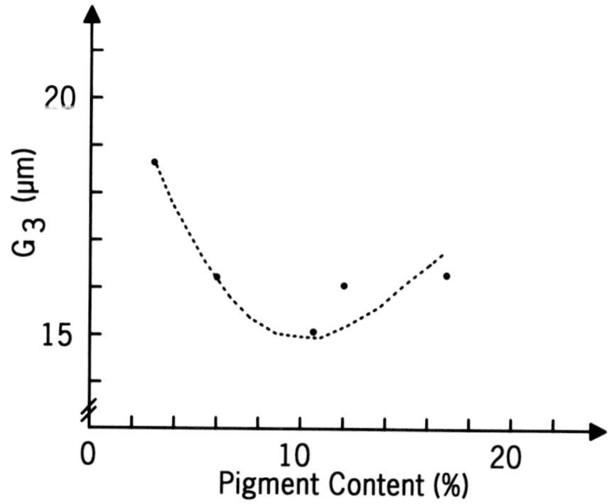

Figure 15. Third-order roughness G_3 roughness as a function of filler content. The pigment used is sodium alumino-silicate of average size 1.65 μm.

is more complex than first expected. The surface roughness varies as a function of the filler content and attains a minimum around 10% filler content. Other ISO roughness parameters such as R_a and R_t[25,26,27] as well as the bearing ratio were measured and show the same trend.

Qualitative evaluation of the three-dimensional maps of the paper surface reveals that the paper surface pores are first filled with pigments. Then, as filler content increases, pigments agglomerate and deposit not only within the surface pores but also on top of fibers. As a result, the surface roughness increases again. This behavior has a direct effect on the printing properties of paper. Although ink transfer to paper is not affected, a maximum in print density as a function of filler content also occurs at around 10% filler content. Indirect roughness evaluation methods, such as air-leak roughness, did not reveal the phenomenon due to the compression of the sample during measurement. This illustrates the importance of direct evaluation as performed with confocal microscopy.

Characterization of Coated Surfaces: Coating improves the finish of paper by filling in the surface pores and also by changing the surface structural behavior. To understand how coating and calendering affect the distribution of the surface pores, one paper basesheet (PBS), the same basesheet coated before supercalendering (BSC), and after supercalendering (LWC2) are shown in Figure 16. Figures 16a, b and c present the 3D topographical maps of these surfaces. The corresponding surface pore distributions are given in Figure 17. The quantitative data obtained from the distributions are given in Table 1. The mode, skewness, kurtosis, and calculated roughness values, G_3, are averages over eight different areas for each paper sample. The standard error of these measurements appears in square brackets.

The distribution mode is shifted towards the surface and the width of the distribution also decreases as the basesheet is coated then supercalendered. The coating has covered

(a)

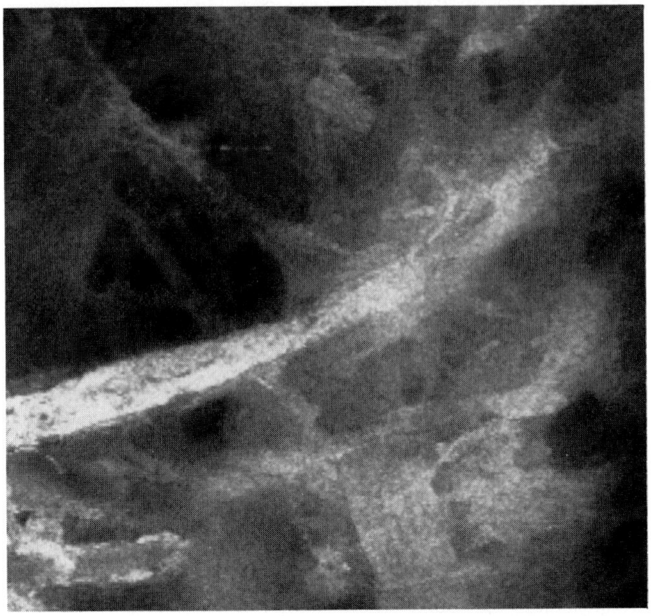

(b)

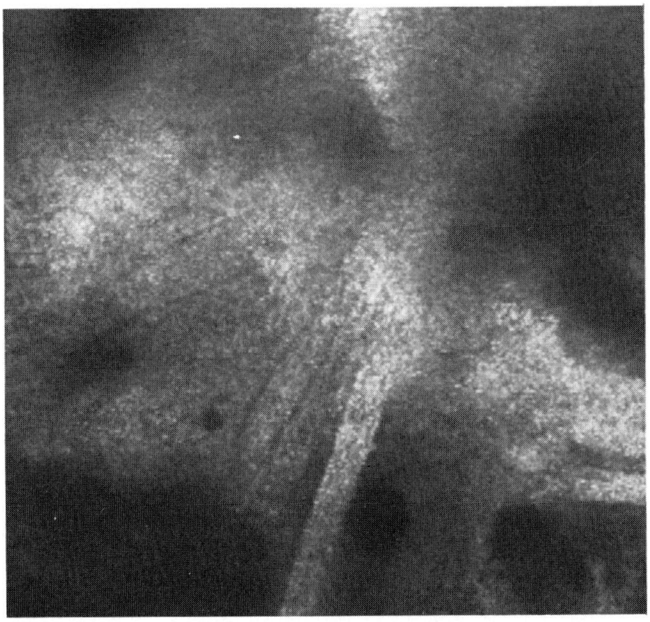

Figure 16. 3D topographical maps of (a) a paper coating basestock; (b) the same paper coated; then (c) (next page) supercalendered. The Z-scanning depth is 51 μm with a 16X (air) objective (0.313 mm x 0.313 mm).

most of the surface fibers and partially filled in the surface pores. This is clearly seen in Figure 16 and also in the left shift of the curves in Figure 17. Supercalendering has reduced the pore depth by pressing the coated surface fibers into the paper and has smoothed the surface. This is reflected both in the skewness increase from coated to supercalendered and

(c)

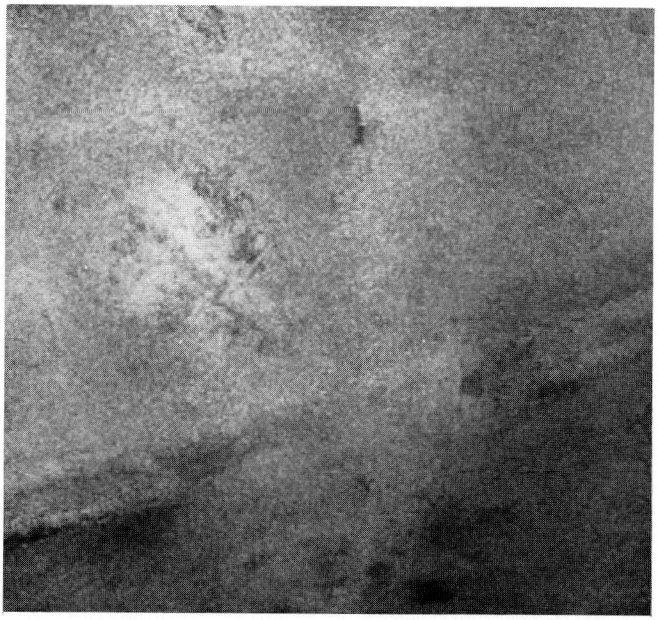

Figure 16, continued. (c) Supercalendered coated paper.

in the smoother appearance of the left hand side of the histogram for the supercalendered surface. The skewness can be interpreted in terms of the distribution of the peaks and valleys that compose the surface. Thus, a relatively smooth surface that shows valleys will have a positive skewness. Likewise, a relatively smooth surface that shows the presence of peaks will have a negative skewness. Despite the surface treatment imparted to the paper surface and the improved roughness, the skewness increase indicates that the deeper pores have remained relatively unaffected by supercalendering. The high kurtosis confirms the presence of structural elements in depth. These parameters are useful for quantifying the effects of a specific coating applied to a basestock and subsequent calendering. Structural modifications of the paper surface caused by coating and supercalendering could therefore be differentiated for basesheets with different surface structures and for different coating formulations. The structural information obtained may also be directly related to the end-use performance of the paper sample.

The normalized cumulative area as a function of depth curves are shown in Figure 17b. These give another representation of the changes occurring at the surface and are used for the G_3 calculations. A clockwise 90° rotation of this graph around the origin gives the profile of the Equivalent Surface Pore for each surface *i.e.* the profile of the average pore representing the surface.

Coating Layer Characterization: The CLSM has the unique ability to make optical cross-sections without damaging the sample under observation. The technique has been applied successfully by Jang *et al.*[38] to image cross-sections of single fibers dried on a glass plate.

(a)

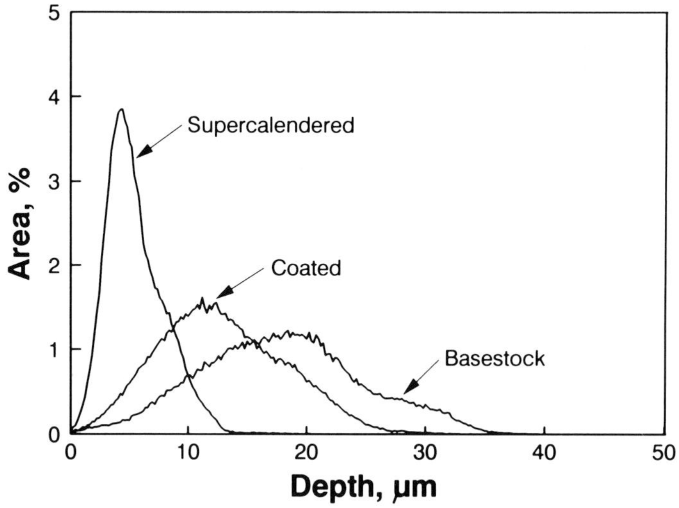

(b)

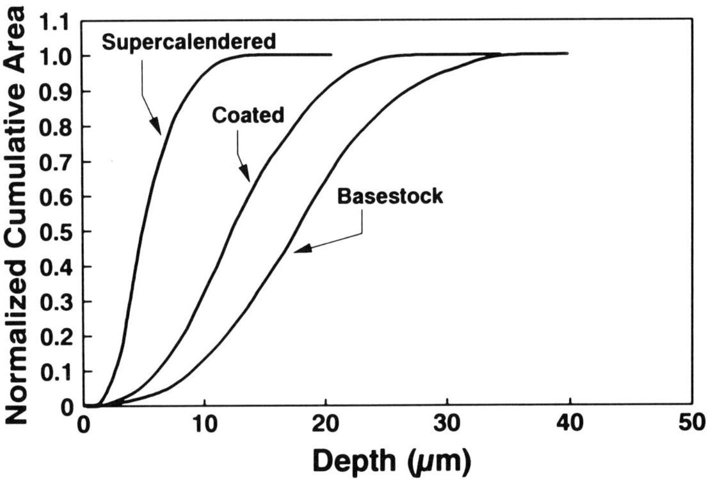

Figure 17. (a) Distribution of surface pore areas of the three paper samples imaged in Figure 16. The distribution modes are 18.4 μm, 11.2 μm and 4.4 μm, while the G_3 roughness values are 20.1 μm, 14.7 μm and 6.5 μm respectively; (b) Distribution of normalized cumulative area *vs* depth for these same samples.

However, as can be seen in Figures 18 and 19, images of CLSM optical cross-sections of paper do not yet match the quality of standard light microscopy images of mechanical cross-sections.[39] Figure 18 illustrates a two-layer coated paper with an adhesive polymer. The left hand side of the image is the result of a horizontal averaging of the pixel intensities. This could be used to measure the average thickness of coating layers ***in situ***, without any mechanical cutting. However, the image acquisition settings for this figure were optimized for the coating layers resulting in image saturation for the paper layer

Table 1
Average Data For the Three Samples of Figure 16

Sample	Mode (μm) [std error]	Skewness [std error]	Kurtosis [std error]	G_3 (μm) [std error]
Basestock	16.80 [1.40]	21.46 [11.05]	-0.36 [0.44]	18.97 [0.68]
Coated	11.40 [0.70]	36.96 [11.55]	0.50 [0.39]	13.03 [0.75]
Supercalendered	4.80 [0.41]	81.89 [25.22]	3.83 [2.08]	5.53 [0.22]

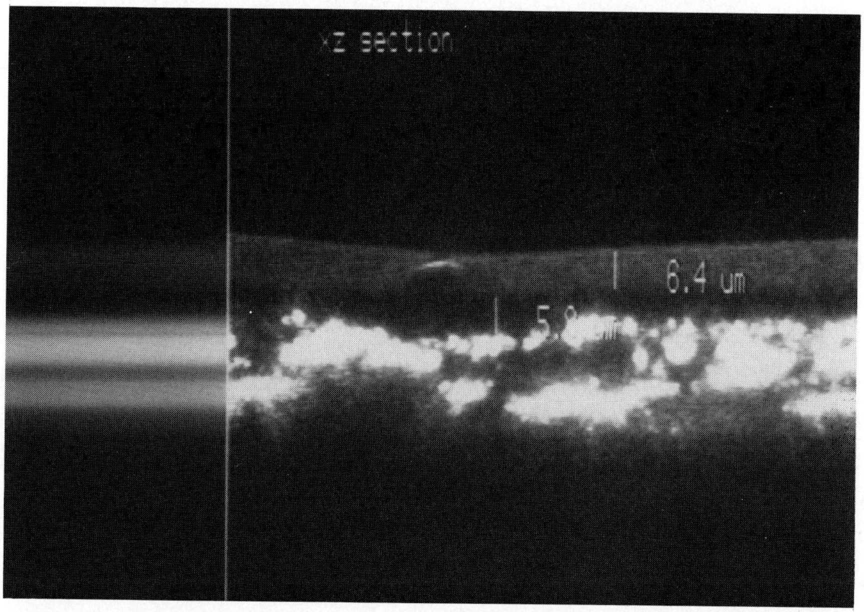

Figure 18. Optical cross-section of a two-layer coated paper obtained with a 40X (oil) objective. Averaging of the layers, according to density, is shown on the left part of the picture. Due to contrast adjustment, the top layer (polyethylene, 30 μm) has not been imaged. The middle layer (polyvinylidene chloride, 6.4 μm) and the adhesive polymer layer (5.9 μm) are well defined.

underneath the coatings. The paper structure is not resolved and interpretation of thelower part of the image is not possible. Figure 19 is an optical cross-section of a coated board. The density line shows optical density variations within the coating layer. We surmise that these are related to variations of the pigment concentration since coating pigments and pigment agglomerates reflect light differently.

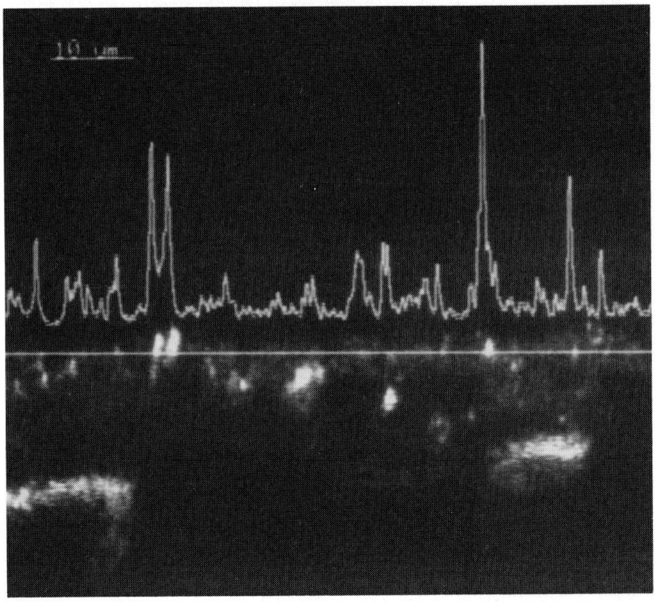

Figure 19. Optical cross-section of a coated board obtained with a 63X (oil) objective. The line through the surface shows intensities that might be directly related to the coating layer density variations.

Coated Surfaces Under Compression: As mentioned earlier, paper surfaces under compression have been quantified in order to relate the surface characteristics with end-use properties. Using the aforementioned compression apparatus, the compressibility of commercial uncoated papers made from different furnishes was found to be an important parameter for predicting gravure printability.[15]

As expected, the compression behavior of lightweight coated (LWC) paper is much different than the compression behavior of uncoated paper. The decrease in roughness for an LWC was found to be much less than that of its corresponding basestock. In addition, the levelling off of the roughness occurred at lower pressures than for uncoated samples. Quantitative analysis of the compressibility of these papers could be related to printing characteristics and might lead to the development of predictive printability models.

PRINTING SCIENCE

The paper surface is critical in the transfer of ink to paper during printing.[40] The uncompressed paper surface structure, the paper surface structure in the printing nip, and the compression recovery are all important to the final print quality. Moreover, the compression recovery also affects how the print is perceived. Confocal microscopy was used to elucidate ink location and transfer phenomena, and a study of the behavior of the paper surface under compression was also used to derive a gravure printability equation.

Ink Mapping: During printing, ink is transferred to the paper surface, which is a three-dimensional structure having irregularities of different sizes. In ink-paper interaction studies, ink mapping has been hampered by the lack of 3D data sets because standard image analysis provides XY coordinates for the ink pigment but no information on its Z-location in depth. Confocal microscopy, however, can provide three-dimensional data, even on printed surfaces. Figure 20 shows two registered images (both images are of the same area) of a slightly calendered newsprint printed at a low ink level. The topographical map in Figure 20a was reconstructed from reflectance mode images whereas Figure 20b is an epi-fluorescence image of the same area.

The reflectance mode image in Figure 20a is able to show complete topographical information because the carbon black particles in the ink film reflect enough light to trigger the photo-multiplier tube. On the topographical map, the ink film is indistinguishable from the surroundings because it is so thin and its topography follows that of the paper surface (illustrated by the profile line). With epi-fluorescence imaging, as in Figure 20b, the fibers, which are slightly fluorescent, are clearly shown while the carbon black particles (which are not fluorescent) are not detected. Therefore, carbon black appears black in the epi-fluorescence image and the ink film can be located by overlaying this image onto the topographical map. Three-dimensional ink mapping is performed by a side-by-side comparison of the two matching images.

Ink Transfer Mechanisms: To explain ink transfer, the paper surface is usually modelled as a series of hills and valleys.[41,42] Ink transfer models are founded on the hypotheses that at low ink levels, ink deposits on the fibers which contact the printing plate and that little ink flows towards the pores. At high ink levels, it is supposed that ink does flow from the hills into the valleys. Three-dimensional ink mapping now provides the needed direct evidence to support these hypotheses.

Figures 20 and 21 show examples of the same newsprint printed at low and high ink levels. At low ink level, there is no evidence of ink flow toward the pores. Ink is located on the high areas. However, the uppermost portions of these areas are not inked. Therefore, we infer that a localized flow must occur. At high ink level, the presence of carbon black pigment in the pores provides evidence that ink flows from the high areas towards the low areas. Again, the uppermost portions of the fibers are not completely covered, showing that ink has been squeezed out of those areas in intimate contact with the printing plate.

Gravure Print Quality: Figure 22 shows the surface of a slightly calendered newsprint printed in gravure (Heliotest NC). The image combines a 3D topographical map obtained in the reflectance mode with a registered image taken in the epi-fluorescence mode. The epi-fluorescence imaging conditions were adjusted to show only the gravure ink, which contrary to black offset ink, is more fluorescent than the mechanical pulp fibers. The height and fluorescence intensity profiles based on the same reference line illustrate that the image contains both topographic and ink location information. As expected for a rough surface, no ink was transferred to the area corresponding to the large pore seen in the

(a)

(b)

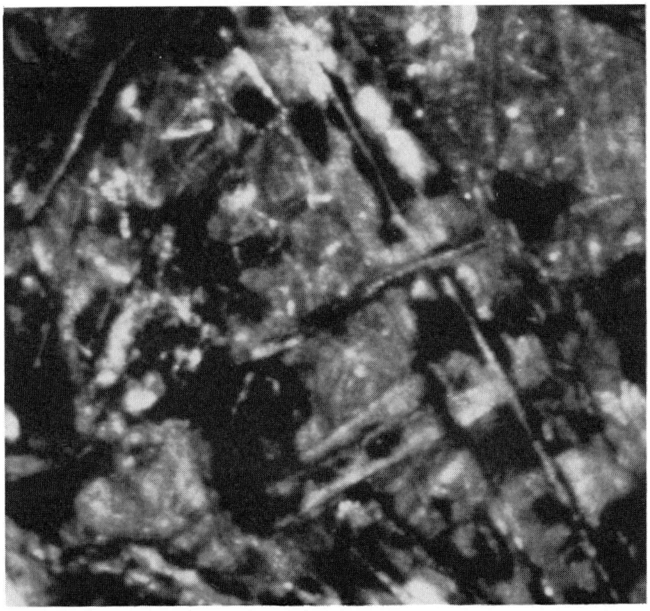

Figure 20. (a) 3D topographical map of a slightly calendered newsprint (S10 = 4.15 μm) printed at a low ink level. The average ink weight on the paper surface is 0.15 g/m^2. The measured area is 250 μm x 250 μm, with a 20X objective. The image is generated from reflected light, and the intensity profile along the reference line shows that the ink layer is imaged as well as the paper surface and does not hamper the topographical map from being generated; (b) Epi-fluorescence image of the same printed area as in (a). The fibers (slightly fluorescent) are imaged while the carbon black pigment (non-fluorescent) is not imaged. Carbon black appears black on this image.

(a)

(b)

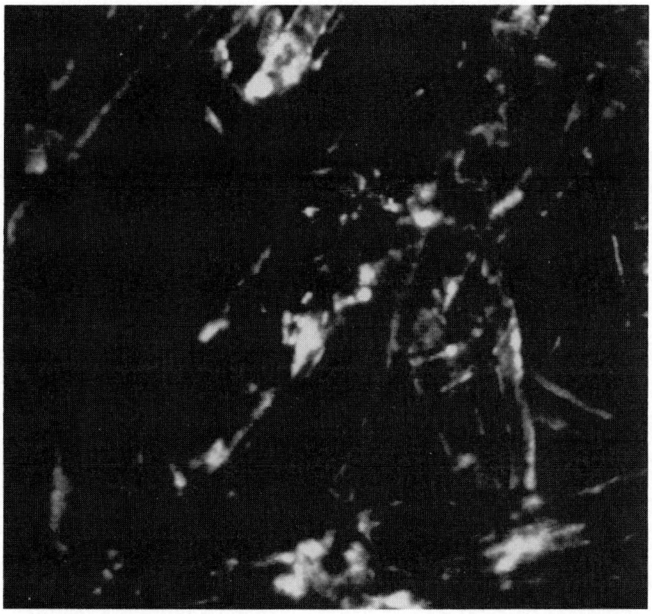

Figure 21. (a) 3D topographical map of the same newsprint as in Figure 20, this time printed at a high ink level. The average ink weight on the paper surface is 3.6 g/m^2. The measured area is 250 μm x 250 μm, with a 20X objective. The image is generated from reflected light; (b) Epi-fluorescence image of the same printed area as in (a).

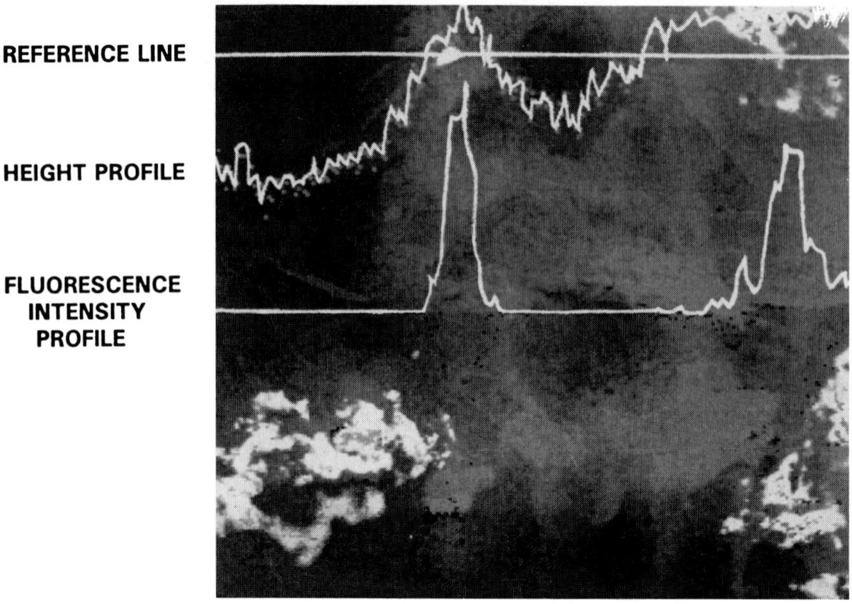

Figure 22. Addition of reflected light and fluorescence images of the surface of a slightly calendered newsprint in gravure (Heliotest NC). The height and the fluorescence intensity profiles correspond to the reference line. At the top left corner of the image, a depression in the height profile represents a paper surface pore. The fluorescence intensity profile, representing gravure ink, is flat at this location. This indicates that no gravure ink has been transferred, hence the missing dot on the print.

upper left corner, thus leading to a missing dot. For the three dots that are printed (the bright spots in the corners), ink transfer was non-uniform leading to dots that are not well-formed. When the newsprint was calendered to a high surface finish, the same technique showed that both dot density and circularity were more uniform.

Electrophotographic Printing: The adhesion of toner printed onto paper depends on surface energy, on acid–base relationships between the toner and the paper surface, and on the local roughness of the paper surface. To assess the influence of roughness on the adhesion of half-tone dots in electrophotographic printing, registered images were taken of the same area, before and after the sample underwent a peel test. The topographical map (Figure 23a) clearly shows that the depression in the paper surface has a size similar to the diameter of the half-tone dot. After the peel test, only a small portion of the dot has been removed, as was measured on the fluorescent image. Similar sets of images on other areas reveal that, when the half-tone dots rest on high areas in the paper surface, they are almost completely peeled away during the test. This demonstrates the importance of the pore size distribution of the sheet surface. Confocal microscopy also provides the non-destructive imaging technique that is necessary for this kind of analysis.

(a)

(b)

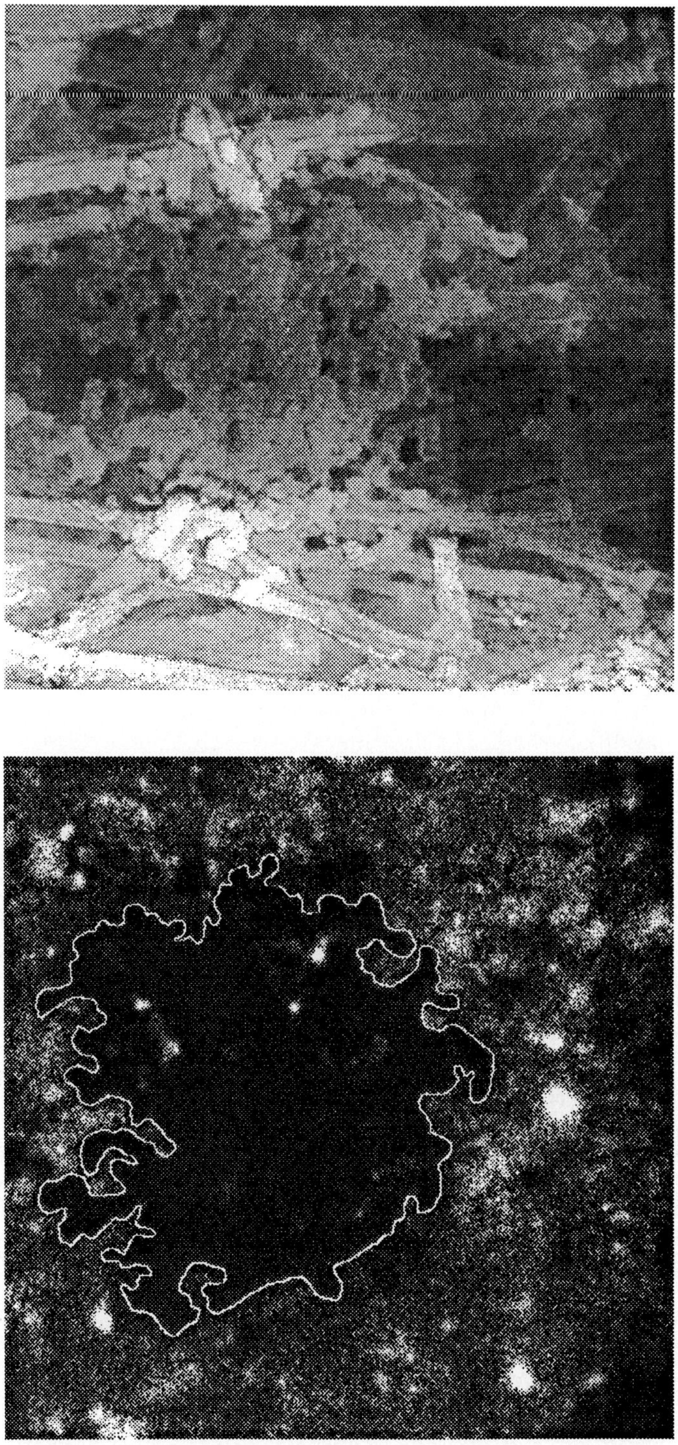

Figure 23. (a) Reflected light image of an electrophotographic half-tone dot printed on bond paper. The diameter of the dot is 150 μm; (b) Fluorescence image of the same area as in (a), showing toner localization on the surface.

(c)

(d)

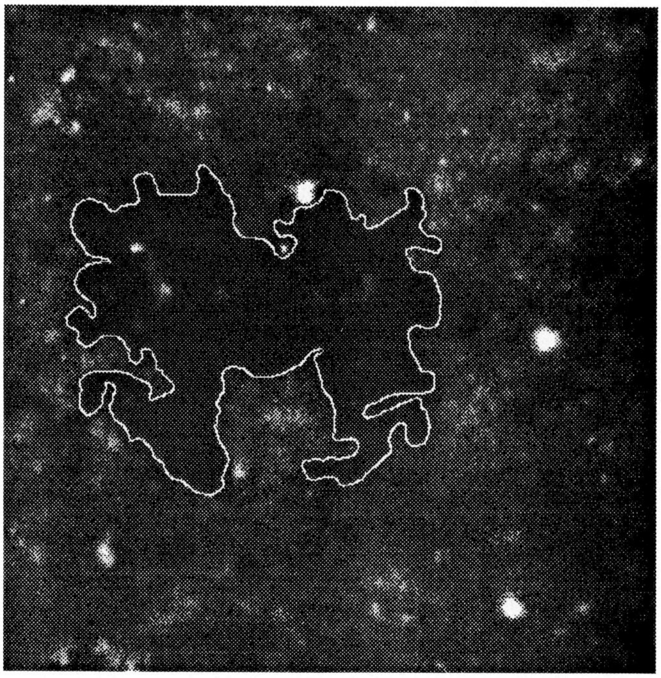

Figure 23, continued. (c) Reflected light image of the same dot, after the print has undergone a peel test; (d) Fluorescence image of the same area as in (c), showing where the toner has been detached from the paper surface.

Relationship Between Topography and Gloss Variations: Print gloss uniformity is an important print quality factor for coated paper. Recently, a method for imaging and measuring small-scale print gloss variations was applied to a series of commercial printed papers.[43] To understand how much of the variation is due to topography, a print was imaged using confocal microscopy. The gloss image was also obtained *on the same area*. Figure 24a shows the gloss image obtained on a small region of a commercially printed sample. In this image, glossy areas are bright and non-glossy areas are dark. The gloss image tells nothing about the relative heights of the glossy areas or about the relative roughness of the non-glossy ones. It does however, provide information about the size of the glossy facets as well as their spatial distribution on the paper surface. Figure 24b shows the corresponding topography of that same surface. The direct superposition of the gloss image onto the topographical map reveals that much of the gloss variation is caused by facets facing too far away from the mean specular reflection plane. Some of the variation was also due to local micro-roughness.[44]

(a)

(b)

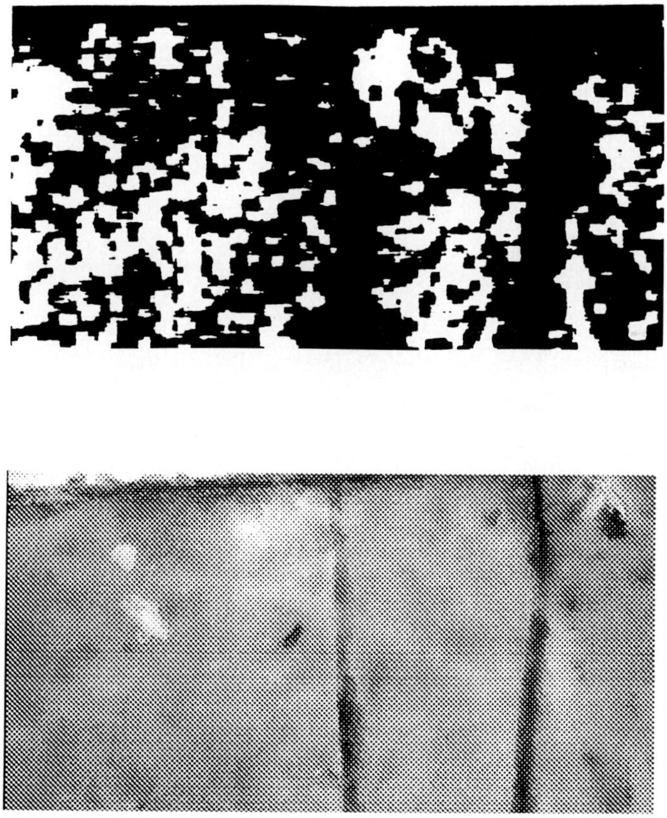

Figure 24. (a) Gloss variation image of an area on a printed LWC showing large amounts of gloss variation; (b) Registered 3D topographical map of the surface shown in (a). The image is composed of several tiles, for a total area of 0.626 mm x 1.252 mm.

CONCLUSIONS

Confocal microscopy has the potential to provide answers to many fundamental questions related to paper surface structure and behavior. This technique offers several advantages for generating three-dimensional topographical maps of surfaces. The method is non-contacting and non-destructive. Surfaces can therefore be observed and quantified before and after various tests and surface treatments. Imaging a surface that has been observed using other techniques is also possible and so-called registered image stacks[45] can be obtained and can elucidate causal relationships between topography and other surface properties. Furthermore, since the laser scans the entire surface, sampling is more complete than with traditional profilometers. Consequently, the parameters obtained are more representative of the paper surface. Images can be obtained from either reflected light or fluorescence and observing the same surface in different ways can lead to a better understanding of that surface. By varying the magnification, the surface structure can be characterized on different scales and this may reveal differences in both the structure of the surface and its variation. Paper surfaces under compression have been imaged and quantified. This recent innovative use of confocal microscopy resulted in the calculation of a local static compressibility parameter. Application of this technique in turn led to the development of a gravure printability equation. In conjunction with the evaluation of surface compressibility, it is proposed that surface conformability also be studied as a characteristic of the paper surface deformation under pressure.

Applications in papermaking and printing have led to some interesting conclusions about the surface characteristics of the samples. Calendering and coating could be studied in terms of process variables that influence the changes in surface pore distributions. The evaluation of filled papers showed an effect related to filler content that was not evidenced by indirect roughness evaluation methods.

The use of confocal microscopy to study entire paper cross-sections is still in its infancy. Nevertheless, preliminary studies show that evaluating the coating layer nondestructively on the paper surface is a possibility that deserves attention. Future developments in confocal technology will undoubtedly encourage advancement in this area. Though paper cannot be imaged easily in its entire thickness, improvements in confocal systems in terms of the lasers used and the different light path configurations may soon allow the entire three-dimensional structure to be reconstructed.

Confocal microscopy has been applied to printing science, where it allowed the verification of hypotheses made in ink transfer models in both the gravure and the offset printing process. Electrophotographic prints have been evaluated and registered images of a single area before and after a peel test provided insight into the influence of roughness on the adhesion of the half-tone dots. The possibility of obtaining registered images of the surface prior to and after a given treatment is a substantial advantage that will allow the elucidation of underlying mechanisms that affect end use performance. Finally, confocal microscopy has also been used to obtain the underlying topography corresponding to a gloss variation image. By understanding the role that topography plays in the printing process, some of the factors that control print quality could be identified.

The description of the surface structure can be complemented with that of the surface behavior under compression in order to view and quantitatively assess the characteristics of that surface under conditions similar to those used in printing. Both the change in roughness as a function of pressure as well as the change in shape of the distribution curves as a function of pressure are important for the calculation of the surface compressibility and the surface conformability. The elaboration of these techniques will lead to more descriptive and predictive models of printability and will also help elucidate the important factors for preventing unwanted reactions of paper during printing.

Confocal microscopy has opened many new horizons in paper and printing research and it is certain that many applications of this technology still remain to be investigated. As progress is made in the technology, it is expected that confocal microscopy will play an increasing role in pulp and paper research.

ACKNOWLEDGEMENTS

Throughout the course of this work, many useful discussions have taken place and the authors gratefully acknowledge the many friends and colleagues at PAPRICAN that contributed to improving this manuscript. In particular, the authors would like to acknowledge Derek Page and Tetsu Uesaka for stimulating discussions and helpful comments. The authors also thank Josiane Chasle for the experimental work on the effect of fillers; Karl Duarte for his cooperation and helpful advice in photography; Dr. A.E. Dixon (Univ. Waterloo) and Dr. S. Damaskinos (Univ. Waterloo) for providing the macroscope images. The authors would also like to thank Per-Åke Johansson (STFI) and Michael MacGregor (Voith Sulzer) for the gloss variation image and helpful discussions. Dr. M. Hoppe, Dr. C. Thoni and Dr. W. Knebel (Leica LaserTechnik GmbH) are gratefully acknowledged for technical assistance and resolution information. Michael MacGregor (Voith Sulzer) and Dr. Johann Engelhardt (Leica Lasertechnik) are also thanked for kindly agreeing to review this manuscript and making many enlightening suggestions. This work was written while the authors were employed at PAPRICAN.

REFERENCES

1. Mangin, P. J. 1993. A structural approach to paper surface characterization. Proceedings of the TAPPI Process and Product Quality Conference, Atlanta, October 31 to November 3, pp. 17–23.

2. Minsky, M. 1988. Memoir on inventing the confocal scanning microscope. *Scanning* 10:128–138.
3. Harris, R. 1993. Fluorescence without fading? Two-photon fluorescence in confocal microscopy. *MSC–SMC Bulletin* 21(3):20–22.
4. Dixon, A. E., Damaskinos, S. and Atkinson, M. R. 1991. Transmission and double-reflection scanning stage confocal microscope. *Scanning* 134:299–306.
5. Juskaitis, R. and Wilson, T. 1992. Surface profiling with scanning optical microscopes using two-mode optical fibers. *Applied Optics* 31(22):4569–4574.
6. Gmitro, A. F. and Aziz, D. 1993. Confocal microscopy through a fiber-optic imaging bundle. *Optics Letters* 18(8):565–567.
7. Dabbs, T. and Glass, M. 1992. Fiber-optic confocal microscope: FOCON. *Applied Optics* 31(16):3030–3035.
8. Shotton, D. M. 1989. Confocal scanning optical microscopy and its applications for biological specimens. *J. of Cell Science* 94:175–206.
9. 1992 3D Imaging Sciences in Microscopy Conference Abstracts, Amsterdam, The Netherlands.
10. Verhoogt, H. Van Dam, J., Posthuma de Boer, A., Draaijer, A. and Houpt, P. M. 1993. Confocal laser scanning microscopy: A new method for determination of the morphology of polymer blends. *Polymer* 34(6):1325–1329.
11. Jang, H. F., Robertson, A. G. and Seth, R. S. 1991. Optical sectioning of pulp fibers using confocal scanning laser microscopy. *Tappi J.* 74(10):217–219.
12. Hamad, W. Y. and Provan, J. W. 1993. On the fracture and fatigue characterization of single wood pulp fibres. Proceedings of the 5th International Conference on Fatigue and Fatigue Threshold, Montreal, May 3–7, Vol.3:1429–1434.
13. Nanko, H. and Ohsawa, J. 1990. Scanning laser microscopy of the drying process of wet webs. *J. of Pulp and Paper Science* 16(1):J6–J12.
14. Mangin, P.J. and Béland, M.-C. 1994. Three-dimensional evaluation of paper surfaces using confocal microscopy: Applications to research and development. *In*: Advances in Printing Science and Technology. Banks, W. H., ed. Pentech Press, London, Volume 22:159–186.
15. Mangin, P. J., Béland, M.-C. and Cormier, L. M. 1993. A structural approach to paper surface compressibility – Relationship to printing characteristics. Transactions of the 10th Fundamental Research Symposium, Oxford, England, Vol. 3.
16. Engelhardt, J. and Knebel, W. 1993. Leica TCS – The confocal laser scanning microscope of the latest generation; technique and applications. *Scientific and Technical Information* X(5):159–168.
17. Wilson T. 1990. Optical aspects of confocal microscopy. *In:* Confocal Microscopy. Wilson, T., ed. Academic Press, London. Chap.3.

18. Pawley, J. B. 1990. Fundamental limits in confocal microscopy. *In:* Handbook of Biological Confocal Microscopy. Pawley, J. B., ed. Revised Edition, Plenum Press, New York. Chap 2.
19. Kent, J.J., Climpson, N.A., Coggon, L., Hooper, J.J., and Gane, P.A.C. 1986. Novel techniques for quantitative characterization of coating structure. *Tappi J.* 69(5): 78–83.
20. Mangin, P.J. 1990. The measurement of paper roughness: A 3D profilometry approach. *In:* Advances in Printing Science and Technology. Banks, W.H., ed. Pentech Press, London, Volume 20:218–235.
21. Gervason, G., Wehbi, D., Roques-Carmes, C. and Cheradame, H. 1986. Calandrage du papier. Etude par caractérisation de la topographie de sa surface. *ATIP* 40(8): 433–443 (in French); and *Cellulose Chemistry and Technology* 20:465–482.
22. Webb, R.H. and Dorey, C.K. 1990. The pixelated image. *In*: Handbook of Biological Confocal Microscopy. Pawley, J.H., ed. Plenum Press, Revised Edition, New York. Chapter 4.
23. Cogswell, C.J. and Sheppard, C.J.R. 1990. Confocal brightfield imaging techniques using an on-axis scanning optical microscope. *In*: Confocal Microscopy. Wilson, T., ed. Academic Press, London. Chapter 8.
24. Sheppard, C.J.R., Gu, M., and Roy, M. 1992. Signal-to-noise ratio in confocal microscope systems. *J. Microscopy* 168(3):209–218.
25. ISO 4287/1. 1984. Surface roughness – Terminology – Part 1: Surface and its parameters.
26. ISO 4287/2. 1984. Surface roughness – Terminology – Part 2: Measurement of surface roughness parameters, First Edition.
27. ISO 4288. 1985. Rules and procedures for the measurement of surface roughness using stylus instruments. First Edition.
28. Mandelbrot, B.B. 1983. The Fractal Geometry of Nature. W.H. Freeman, New York, pp. 25–28.
29. 1992 Fractals in Engineering Conference Abstracts, June 3 to 5, Montréal.
30. Kent, H.G. 1990. The fractal dimension of paper surface topography. Preprints of the 1990 Printing and Graphic Arts Conference, pp. 73–78.
31. Mangin, P.J. and Geoffroy, P. 1989. Printing roughness and compressibility. Transactions of the 9th Fundamental Research Symposium, Cambridge, England, Volume 2:951–978.
32. Svoboda, K.K.H. 1991. Comparative studies of Biorad and Leica confocal laser scanning microscopes. Proceedings of Scanning 91, pp. 1–91.
33. Dixon, A. E., Damaskinos, S., Ribes, A., Seto, E., Béland, M.-C., Uesaka, T., Dalrymple, B., Duttagupta, S. P., Fauchet, P. M. 1995. Confocal scanning beam laser microscope/macroscope: Applications requiring large data sets. Proceedings of the 3-

Dimensional Microscopy: Image Acquisition and Processing II Conference, San Jose, February, SPIE Vol. 2. (in press).

34. Hamel, J., Kreklewetz, W., Régev, S., Béland, M.-C., and Mangin, P.J. 1993. Novel apparatus for the direct evaluation of paper surfaces under compression. Unpublished work.
35. Mangin, P. J., Béland, M.-C., Cormier, L. M. 1994. Paper surface compressibility and printing. Proceedings of the 1994 International Printing and Graphic Arts Conference. TAPPI Press, Atlanta, pp. 19–31.
36. Roehr, W.W. 1966. Measurement and reduction of ink strike-through in newsprint. *Tappi J.* 49(6):255–259.
37. Davidson, R.R. 1965. Experiments on loading paper with low refractive index fillers. *Paper Technology* 6(2):107–120.
38. Jang, H.F., Robertson, A.G., and Seth, R.S. 1991. Optical sectioning of pulp fibers using confocal laser scanning microscopy. *Tappi J.* 74(10):217–219.
39. Lepoutre, P. and de Silveira, G. 1991. Examination of cross-sections of blade and roll-coated LWC paper. *J. Pulp and Paper Sci.* 17(5):J184–J186.
40. De Grâce, J.H. and Mangin, P.J. 1984. A mechanistic approach to ink transfer. Part I: Effect of substrate properties and press conditions. *In*: Advances in Printing Science and Technology, Banks, W.H., ed., Pentech Press, London. Volume 17:312–332.
41. Bery, Y. 1982. Three-equation ink transfer model. *In*: Advances in Printing Science and Technology. Banks, W.H., ed., Pentech Press, London. Volume 2:206–234.
42. Mangin, P.J. 1988. Etude de la déstructuration de la surface du papier en zone d'impression offset. Ph.D. thesis, Institut National Polytechnique of Grenoble, France (in French), pp. 227–247.
43. MacGregor, M.A. and Johansson, P.-Å. 1991. Gloss uniformity in coated paper – Measurements of commercial paper. Proceedings of the 1991 TAPPI Coating Conference, TAPPI Press, Atlanta, pp. 495–504.
44. MacGregor, M.A., Johansson, P.-Å., and Béland, M.-C. 1994. Small-scale gloss variation in prints – Paper topography explains much of the variation. Proceedings of the 1994 International Printing and Graphic Arts Conference. TAPPI Press, Atlanta. pp. 33–43.
45. MacGregor, M.A., 1990. Image analysis techniques for studying 'orange peel' gloss effects in a LWC paper. Proceedings of the 1990 Coating Conference, TAPPI Press, Atlanta, pp.125–138.

2

Scanning Electron Microscopy: A Tool For the Analysis of Wood Pulp Fibers and Paper

Glynis de Silveira, Paivi Forsberg,[*] and Terrance E. Conners[**]

University of Cambridge, Cambridge, U.K.
[*]*Kymi Paper Mills, Ltd., Research Center, Kuusankoski, Finland*
[**]*Mississippi State University, Mississippi State, Mississippi, U.S.A.*

INTRODUCTION

The scanning electron microscope (SEM) was originally developed in the early 1950s at the University of Cambridge, U.K. by a group of graduate students working with Oatley and Everhart;[1] it was subsequently improved by one of Oatley's students, K.C.A. Smith.[2] An SEM is basically composed of three elements: *i*) a vacuum chamber with an electron gun at the top and a sample holder at the bottom; *ii*) one or more signal detectors pointing at the specimen; and *iii*) a console of electronics to raster a focussed electron beam over the sample and to acquire signals from the detectors. There are a number of reasons why this instrument is so popular, in spite of the fact that images are produced only in black and white: *i*) its depth of field is greater than that of ordinary light microscopes and gives a three-dimensional appearance to the specimen image; *ii*) a minimum amount of sample preparation is required; *iii*) the microscope can provide higher magnification and greater resolution than ordinary light microscopes; and *iv*) one can rapidly view relatively large areas of a sample.

The first SEM dedicated to the science of pulp and paper was installed at the Pulp and Paper Research Institute of Canada in 1958. Since then, SEMs have been frequently used to examine the topographical features of a paper sheet and the changes sustained by pulp fibers as a consequence of different pulping processes. However, the distribution of inorganic filler particles and coating pigments can also be studied using various specialized detectors and x-ray detectors can provide semi-quantitative information about the

0-8493-8992-5/95/$0.00+$.50

constituent elements in a sample. X-ray applications are discussed in another chapter in this book.

This chapter will describe the use and advantages of the SEM in the study of pulp fibers and related materials; particular emphasis will be placed on the special requirements of non-conductive samples such as paper. Separate sections will also deal with the development and use of variations of the conventional SEM, namely the Low Temperature SEM (LTSEM) and the Environmental SEM (ESEM).

BEAM/SPECIMEN INTERACTIONS

In-depth descriptions of the various components, functions and capabilities of a SEM are available from a number of authors[3,4,5] and would therefore be redundant in this book. However, as most writers are primarily concerned with highly conductive metallurgical specimens, the special requirements of nonconducting samples such as paper and fibers merit specific consideration here. Nonconducting specimens are more prone to damage in an electron beam than are conductive materials.

Backscattered and Secondary Electrons: In any electron microscope there is a primary source of electrons, the gun. When these *primary electrons* impinge upon a specimen in a vacuum a number of interactions occur which scatter the electrons (Figure 1). These interactions can be either *elastic* or *inelastic* in nature.

Elastic scattering results from the collision of primary electrons with the nuclei of the atoms of the specimen. The trajectory of the beam electrons is affected, but negligible amounts of kinetic energy are transferred. From the Rutherford model[6] which describes the cross-section (*i.e.*, probability) for elastic scattering, it is possible to conclude that elastic scattering is more probable in high atomic number (Z) materials and at low beam energy. After interacting with the atoms of the specimen about 30% of the primary electrons reemerge from the sample, giving rise to the backscattered electron signal (BE).

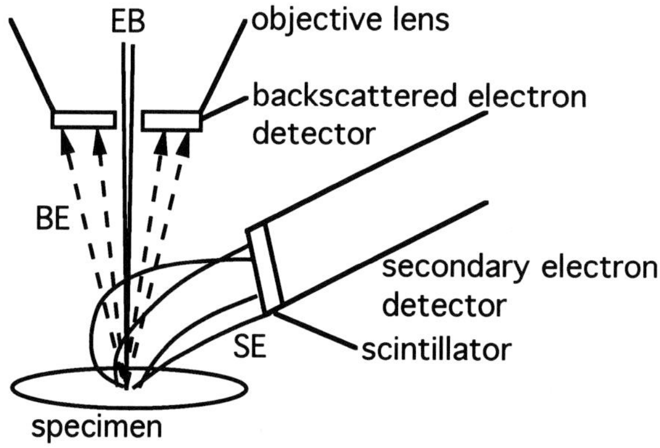

Figure 1. Schematic diagram of the interior of an SEM chamber showing the relative positions of the specimen and the detectors, the deflected trajectories of the SE due to the positive bias of the Everhart-Thornley detector and the straight-line trajectories of the BE.

Heinrich[7] and Bishop[8] used Monte Carlo simulations to show that backscattering events increase with increasing atomic number. They also found that the backscattered fraction is quantified by the backscattered coefficient which increases monotonically with Z. This forms the basis for atomic number contrast, also referred to as *Z contrast*.

Inelastic scattering transfers energy from the beam electrons, decreasing their kinetic energy, to the atoms of the specimen. Interaction of the primary electrons with the atoms that constitute the specimen leads to the transfer of a few electron volts of energy and the ejection of loosely bound electrons. These low energy (0–50 eV) ejected electrons, referred to as secondary electrons (SE), are produced at a shallow sampling depth as a consequence of the low kinetic energy with which they are generated. In addition to the generation of secondary electrons, inelastic scattering also leads to the generation of characteristic x-rays, Bremsstrahlung or continuum x-rays, Auger electrons, etc.

Conventional SEMs are often fitted with detectors for the secondary and backscattered electrons and perhaps with one or more detectors for x-rays; the other signals are usually detected and processed using separate pieces of dedicated equipment.

Interaction Volume: Monte Carlo calculations simulate the trajectories of the primary electrons within the samples and calculates them in a stepwise manner (Figure 2). The length and angle of each scattering event are determined through the use of random numbers, chosen to produce a distribution of scattering events similar in behavior to that of real electrons. The sum of the all the inelastic (negligible angular deviation) and elastic (between 0–180 degree angle) scattering events results in an interaction volume.[9]

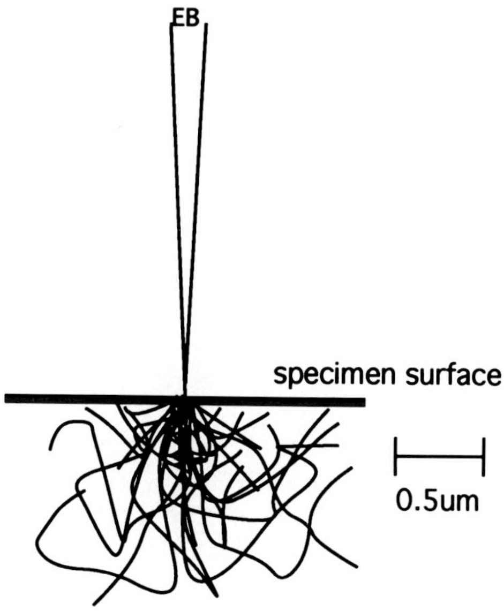

Figure 2. Monte Carlo simulation of the trajectories of elastically scattered electrons interacting with the atoms of the specimen. A projection, on a plane perpendicular to the specimen's surface, of several hundred electron trajectories. A three dimensional plot is used to predict the interaction volume.

The depth and shape of the interaction volume created by incident electrons striking a sample depends on the accelerating voltage, the atomic number of the elements constituting the specimens and the angle of tilt between the incident beam and the specimen. More to the point, the interaction volume (depth and breadth) affects image resolution, as signals originating from points below or adjacent to the spot on the specimen surface where the electron beam is focused will contribute to the signal being collected (whether secondary electrons, backscattered electrons, or x-rays). In high atomic number specimens (such as metals and their alloys) the interaction volume is relatively small even at accelerating voltages of 20–40 keV, and the elastic scattering events produced within the sample are restricted to a confined semi-hemispheric volume (Figure 3a) surrounding the area around the incident beam, close to the surface. On the other hand, the interaction volume for electron beams and samples with a low average atomic weight (such as organic materials and polymers) is quite different. High beam voltage electrons can penetrate quite deeply[10] because of their initial higher energy and the lower rate at which this energy is lost. As a result of elastic scattering, electrons deviate from their initial trajectory and the angular lateral scattering contributes to an increased interaction volume whose shape resembles a pear (Figure 3b). To decrease the size of the pear-shaped interaction volume (and to increase the image spatial resolution), low Z number specimens should be analyzed at low accelerating voltages (*e.g.*, less than 10 keV, and preferably less than 5 keV).

Monte Carlo simulations of electron beam trajectories in which the sample is tilted in relation to the incident beam indicate that the interaction depth becomes smaller as the specimen is tilted due to the forward scattering of electrons. While normal incident electrons (0° tilt) propagate down into the specimen, in tilted specimens the electrons propagate close to the surface. Many electrons exit after a single-angle scattering event, generating an interaction volume with a reduced depth dimension (Figure 4).

Due to the difference in depth at which elastic and inelastic events occur the information obtained about the specimen is specific to the energy levels at which the scattering events take place. This implies that the higher energy backscattered electrons which are produced deep within the sample have a chance to interact with other elements within the specimen before they leave the surface and are detected (Figure 5a).[11] These multiple collisions lead to the production of secondary electrons and backscattered electrons which exit the sample some distance away from the area scanned by the incident beam (Figure 5b). As these SE_{II} are the result of backscattering, they provide lateral and depth information similar to BE (which is inherently a low-resolution signal); the majority of the BE_{II} are not detected by the backscattered electron detector (BED).

Topographic and Chemical Information: Secondary electrons (SE_I) are produced at low kinetic energies by the top few layers (5λ, where λ is the mean free path for SE) of the sample near the incident beam, thereby providing information about the topography of

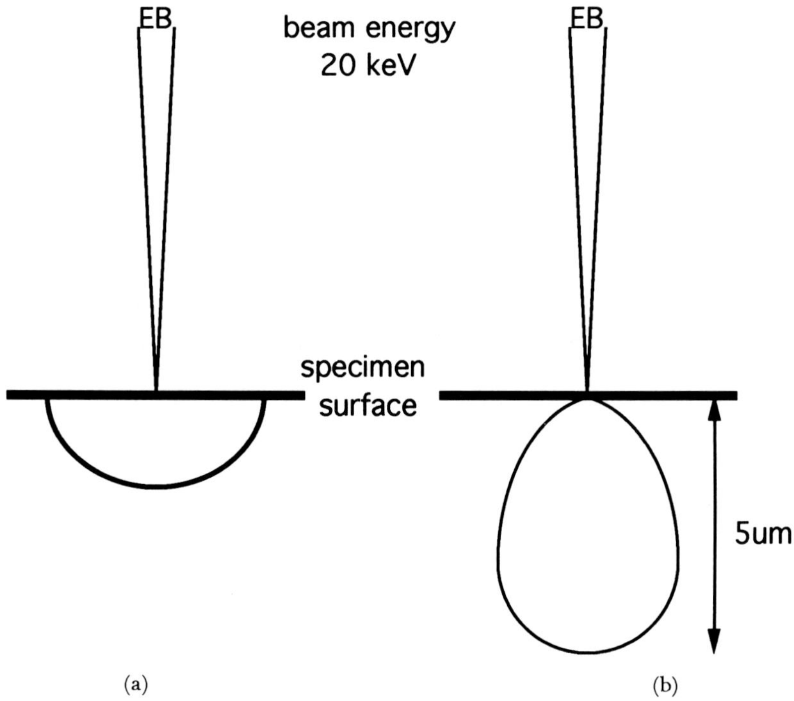

Figure 3. Influence of the average atomic weight of the sample on the interaction volume, at the same beam energy. Note the difference in shape and depth of penetration of the interaction volume: (a) high average density (*e.g.,* metal); (b) low average density (*e.g.*, polymer).

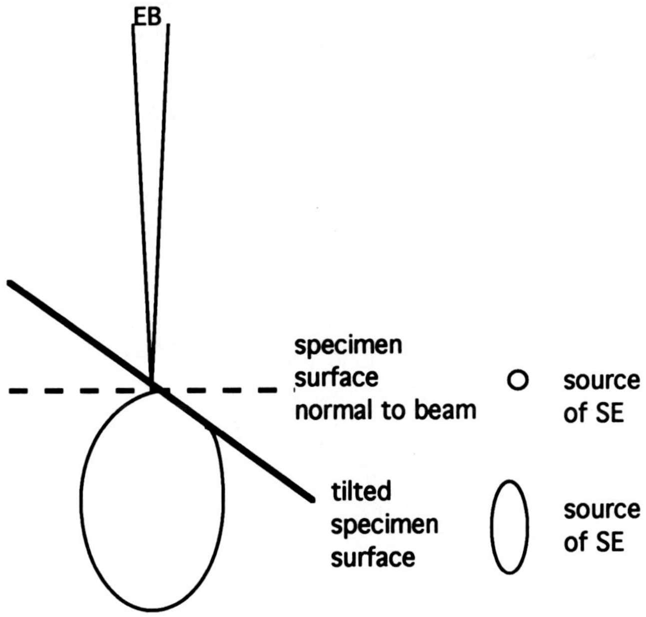

Figure 4. Tilting a flat sample in relation to the beam causes the interaction volume to intersect the surface so that the BE are generated closer to the surface and the SE_{II} originate from a much greater area. The area of impingement of the incident beam becomes oval and the SE_I are therefore produced by a larger area.

(a)

(b)

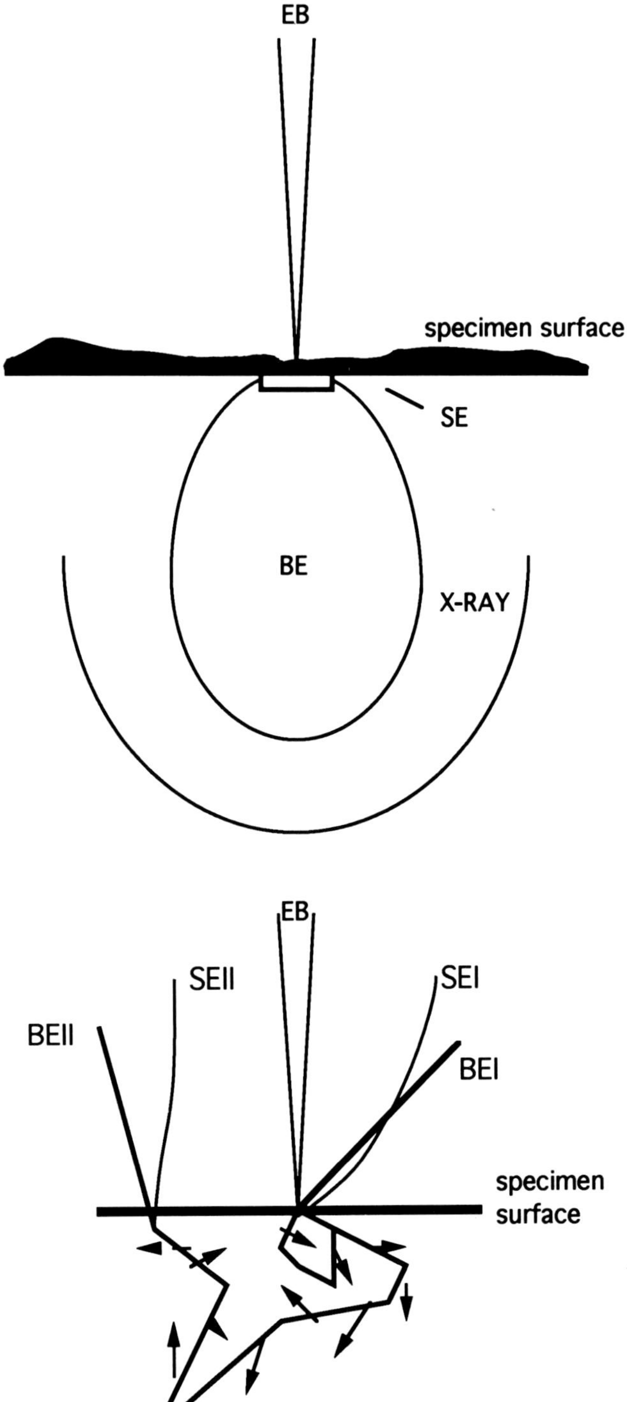

Figure 5. Schematic illustration of the interaction volume showing: (a) the origin and the depth at which SEI, BE and x-rays originate; (b) the generation of secondary electrons (SEI) by the incident beam around the impingement area and the generation of secondary electrons (SE_{II}) by backscattered electrons (BE_{II}) as they leave the sample some distance from the area of incidence.

the sample. Even if SE_I are generated deeper within the sample (due to high accelerating voltages) they cannot escape the solid because they must overcome the surface potential barrier which requires several electron volts. The presence of the SE_{II} adds a certain amount of compositional contrast to the topographic (or shape contrast) information provided by the SE_I signal.

At a given accelerating voltage only the topmost surface layer of a polished metal sample is imaged by the SE. However, a rough surface such as uncoated paper exposes several layers of partially superimposed fibers to the primary electron beam. Between these fibers exist deep inter-fiber spaces. Additionally, the surface of dried wood pulp fibers is not smooth. Features such as fibrils, crimps and surface wrinkles resulting either from the pulping process or from drying stresses also produce secondary electrons (Figure 6); the relatively sharp edges of these features facilitate the production and escape of SE, causing these edge-like features to appear highlighted. In the case of high-energy incident beam electrons, fine or thin surface details such as fibrils or pieces of fiber wall will be transparent to the beam because the image-forming secondary electrons originate from the underlying structures (Figures 7a and 7b). Quite often, even the surface rugosity of dried fibers may not be seen because of the relatively deep penetration of the incident electrons. Occasionally this effect proves to be an advantage; for example, a piece of lightweight coated paper can be viewed at successively higher accelerating voltages (up to about 60 keV) to uncover some information about coating uniformity and basesheet appearance.

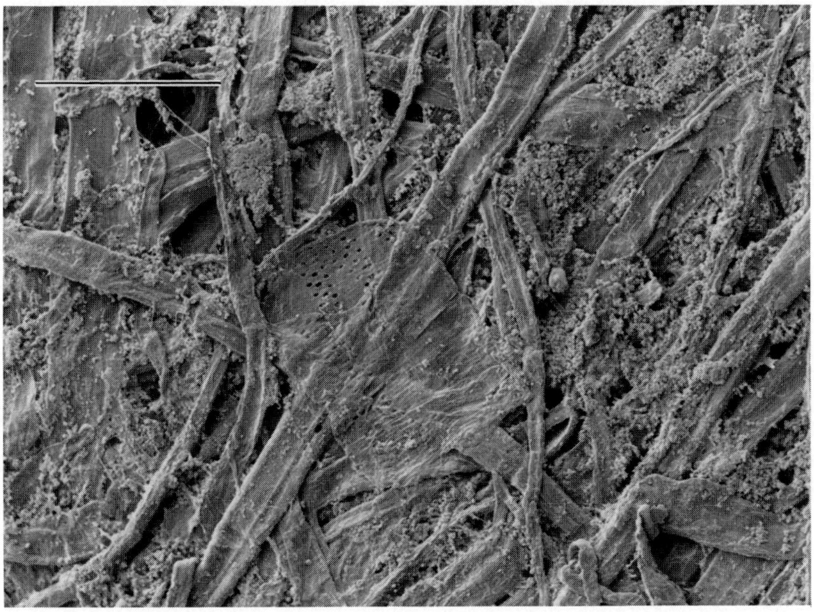

Figure 6. SE image of the surface of a filled paper ($CaCO_3$ pigment) made from a bleached hardwood pulp. The high pigment charge (20% by weight) results in considerable deposition of filler pigment particles on the paper surface causing filling of the void spaces (bar equals 100 μm).

(a)

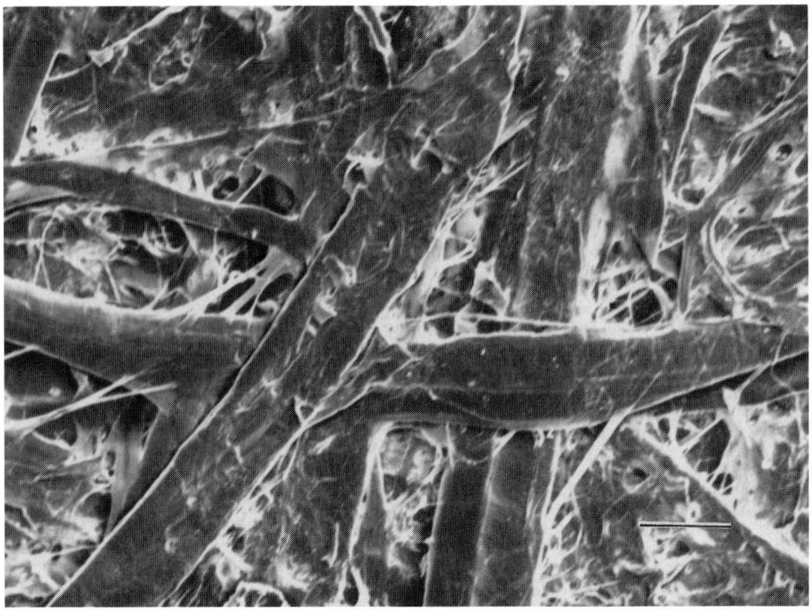

(b)

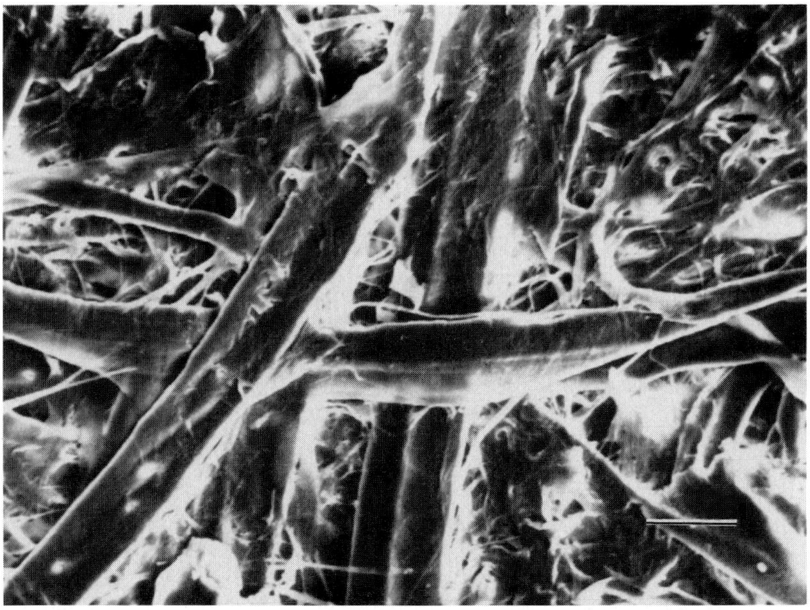

Figure 7. SEM micrographs of a newsprint (NP) surface showing the effects of beam voltage on the amount of topographic information generated (sample perpendicular to the incident beam); (a) at a low voltage (5 keV) showing details of fibrils and fines; (b) at high voltage (20 keV) showing less information about the surface of the fibers as well as the fibrils and fines (bars equal 30 μm).

Backscattered electrons are produced deeper within the sample, from a volume that depends on the energy of the primary electrons and the average atomic weight of the elements involved. The amount, the energy and the direction of travel of the backscattered electrons emitted by the sample is indicative of the topography and the composition (Z contrast). Because BSE travel in straight lines as they are emitted from the specimen, they provide shape contrast due to local variations in the specimen surface inclination (*i.e.*, facets facing the detector will be imaged, while those inclined away from it will not be seen). Quantitative evaluation of the chemical composition can be accomplished using specialized image analysis-based software, but this ideally requires a smooth polished specimen placed perpendicular to the beam so that the gray level contrast is the result of elastic scattering events and not the consequence of specimen tilt or local area inclination.

With a polished specimen normal to the beam, higher atomic weight elements will appear brighter on the CRT screen because they will emit a greater number of backscattered electrons. Two elements that are adjacent on the periodic table (*e.g.*, aluminum [Z=13] and silicon [Z=14]) will appear as two regions with distinguishable gray levels. The brighter (whiter) region will correspond to the silicon. In the same way, if two compounds with different atomic weights are present in the sample (Figure 8a), the compound with the higher average atomic weight will appear brighter. This information is very useful in determining the presence of metal particles in pulp fibers or the contamination of unfilled and uncoated papers by pigment particles (refer to Figure 8b).

Sample Imaging and Sample Charging: Low average atomic weight specimens such as paper and pulp fibers can not be clearly imaged when bombarded by a beam of primary electrons[12] due to an insufficient yield of secondary electrons from the elements that constitute cellulose, lignin and hemicellulose. Additionally, insulating samples (such as most polymers) do not permit the negatively charged incident electrons to escape to ground (positive sample stage) and electrons accumulate on the sample causing it to charge. This causes the image intensity to oscillate as the sample charges (and discharges) repeatedly. Both the imaging and charging problems can be overcome by coating the sample with a very thin layer (5 to 20 nm) of a conducting heavy metal (such as gold, palladium or a mixture of both), thereby creating a conductive pathway for the electrons between this metal layer and the sample stage. This metallic layer is commonly applied in a vacuum with a commercially available instrument known as a "sputter coater", and the metal to be used is largely a matter of personal preference. If an x-ray analysis is to be conducted on a sample, an evaporated carbon coating may be substituted for the metal to avoid potential spectroscopic difficulties, but the image quality will be inferior.

Specimen cleanliness is important. Contaminant particles of dust which are present on the surface of conducting specimens will charge if a conducting pathway between the surface of the particle and the conducting substrate (usually a coated specimen glued onto

(a)

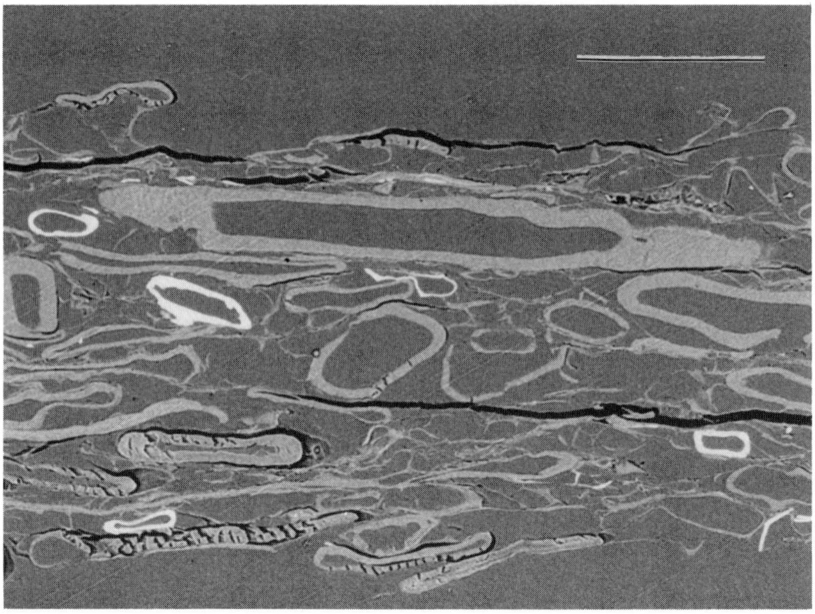

(b)

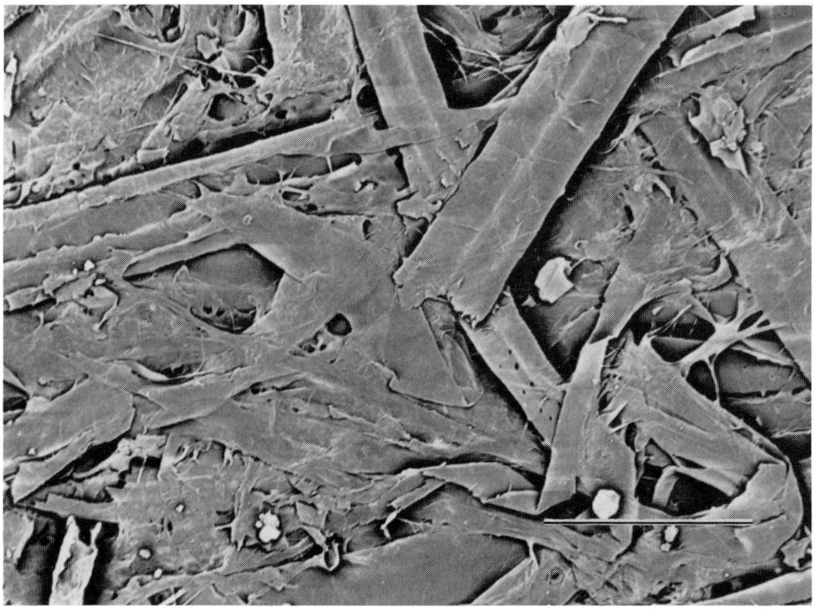

Figure 8. Chemical information obtained in BE mode: (a) Cross-section of a mechanical pulp handsheet prepared by halogenating a small fraction of the pulp before recombination, in order to change its average atomic weight. A sample of the paper sheet was embedded in polymer, microtomed, coated with carbon and analyzed in the SEM. The different gray levels indicate decreasing average atomic weights, such that the white areas correspond to halogenated pulp, the lighter gray to the untreated pulp, the darker gray to the polymer and the black areas to the void spaces (probably the lumen of fibers into which the polymer has not penetrated) (bar equals 50 μm). (b) Surface of a recycled paperboard showing the presence of mineral particles (bar equals 100 μm).

a stub) does not exist. As the charge builds up the incoming electrons (which also have a negative charge) are repulsed by the accumulated charge and a black circle is visible surrounding the charging particle (Figure 9).

SPECIMEN PREPARATION AND IMAGING FOR A CONVENTIONAL SEM

Wood pulp fibers and paper samples are hygroscopic, radiation sensitive, thermolabile samples of low contrast and weak emissivity and, most of all, poor conductors. Much attention should, therefore, be paid to stabilizing the specimens by immobilizing any fluid and coating the specimens with a thin layer of conducting metal before examining them in a SEM. Strict guidelines should be followed to insure that no finger grease or contaminant particles are deposited on the surface of the specimen. These can not only be imaged but also have a deleterious effect on the microscope vacuum, liquid nitrogen (LN_2)-chilled x-ray detectors and the small beam-collimating apertures between the electron gun and the specimen.

EXAMINATION OF WOOD PULP FIBERS

Conventional SEMs are not the ideal analytical instrument for the study of hydrated pulp fibers, as samples will dehydrate in the column vacuum (10^{-3} to 10^{-6} torr). Partially hydrated pulp fibers can, however, be imaged at low magnification if the sample preparation step is bypassed and a fresh uncoated specimen is examined in the SEM at less than optimal conditions (low voltage, low beam current, short record time, decreased signal to noise ratio and poor microscope vacuum).[13,14,15] The resolution is degraded, however, and may be unsatisfactory at higher magnifications when fiber fractions, fiber walls and microfibrils are to be observed. Depending on the specimen to be examined and the magnification which is to be employed it may be best to dehydrate and stabilize the specimen for viewing, but there are alternatives available to the microscopist which might be preferred in some circumstances.

Air drying is the simplest method of preparing pulp samples for the SEM but it introduces a number of artifacts, particularly in wet samples due to surface tension effects. Attempts to overcome these effects have led to the development of such techniques as: *i*) critical point drying (considered to be consistently reproducible), *ii*) freeze substitution and *iii*) freeze drying, which were routinely used by Parham[16] and Sachs[17,18] for pulp fibers. These techniques are very useful, but they are not without their attendant drawbacks. One alternative is the use of Low Temperature SEM (LTSEM) (discussed later) to observe pulp fiber structures.[19] Another alternative is the Environmental Scanning Electron Microscope (ESEM) in which the atmosphere in the sample chamber can be controlled either by the introduction of water vapor (thereby increasing or decreasing the water content of the sample) or other gases to simulate specific conditions.

(a)

(b)

Figure 9. SE image of an uncoated newsprint sample showing: a) a charged dust particle on the surface observed at a relatively high accelerating voltage (10 keV). Note the dark ring surrounding the particle where the incident electrons are repelled by the accumulated charge (bar equals 200 μm); b) at much lower accelerating voltages (1.2 keV) the fiber surface does not charge and individual fibrils and microfibrils can be clearly observed (bar equals 100 nm).

Air Drying of Wood Pulps: Air drying is the most straight forward dehydration method, and it consists of leaving small amounts of the sample to dry at room temperature under dust free conditions. In the case of aqueous pulp suspensions the usual practice is to spread a few drops of the suspension on a light microscope cover slip and then dry mount the cover slip on a SEM specimen holder. However, air drying causes severe shrinkage, collapse and the distortion of delicate surface structures due to surface tension forces as the water evaporates. These forces can be very large in narrow interstices and apertures. Gradual replacement of the water by a nonpolar liquid such as acetone or ethanol (*i.e.*, solvent exchange drying through a graded series) causes less damage due to lower surface tension, but soluble extractives may be removed from the wood fibers, thereby introducing a different set of artifacts. This cautionary statement probably does not apply to most chemical pulps, as the extractives will have been removed during the pulping process.

Critical Point Drying: Critical point drying (CPD) is generally considered the best technique for most biological specimens because the phase change from liquid to gas occurs at a critical point, thus eliminating the deleterious effects of surface tension forces. Optimally, CPD requires that the sample be fixed (*e.g.*, with glutaraldehyde or osmium tetroxide) prior to dehydration to toughen biological structures and to increase the permeability of cells to the organic solvents and transitional liquids; good results may also be obtained without the use of fixatives for some specimens. The dehydration process is carried out through a graded series of hygroscopic solvents such as ethanol, followed by the substitution of acetone or amyl acetate for the dehydration agent. This intermediate solvent is exchanged for carbon dioxide under pressure and, after raising the temperature above the critical point, the CO_2 is bled to the atmosphere. Thus the pulp slurry is immersed and dehydrated in a number of solvents, including water; any component which is soluble in any one of these solvents is unavoidably removed from the specimen, and delicate surface structures are liable to be displaced or removed by the inevitable agitation during handling in these liquids.

The loss of some of the resinous material and fine structures, the initial swelling due to the presence of alcohol and the progressive shrinkage which occurs with the transitional fluids are some of the disadvantages of this technique.

Freeze Substitution: Freeze substitution is a chemical dehydration process in which ice in frozen hydrated specimens is removed and replaced by an organic solvent. It is always carried out at temperatures below 273 K. Organic solvents such as acetone and ethanol are used at or below recrystallization temperatures to dissolve the ice in frozen specimens. Once the substitution is complete the sample may be freeze dried or critical point dried for SEM, or it may be infiltrated with an embedding medium in order that thin sections may be cut for transmission electron microscopy (TEM). The resin infiltration can take place either at low temperatures or at room temperature or above.

Substitution fluids are chosen based on their ability to dissolve sample ice below the recrystallization temperature of ice (*i.e.,* 140 K). For pulp slurries the usual non-polar fluids are acetone and diethyl ether mixed with small amounts of glutaraldehyde (0.5%). In practice, frozen specimens are immersed in the substitution fluid at 173–193 K and remain at this temperature for several days. The temperature is then slowly increased either to the temperature at which the low temperature embedding is carried out (190–240 K) or to room temperature for conventional embedding and/or drying for SEM.

The lowest temperature at which freeze substitution can be performed is influenced by the melting point of the solvent and, to some extent, by the water content and size of the sample; the rate of substitution is a function of the temperature, water dissolving capacity, viscosity and the relative diffusion rate of the substitution liquid. Some disadvantages are inherent in this process, the most important being the length of time it takes for the substitution to proceed at such low temperatures. This may take from a few hours (*e.g.*, six to eight hours for TMP) to several days (*e.g.*, seven days for a highly bleached chemical pulp).

Freeze Drying: The freeze drying technique itself is similar to ordinary vacuum distillation, with one essential difference: the material to be dried must be solidly frozen below its eutectic point, before being subjected to a very low absolute pressure (high vacuum) and a controlled heat input. Under these conditions the water content (in the form of an ice matrix) is selectively removed by sublimation. The freeze drying system is a kinetic system in which a constantly changing state of imbalance exists between the ice and the pressure/temperature conditions within the system. The limit of imbalance is determined by the maximum amount of heat which can be applied to the specimen without causing a change from the solid to the liquid state ("melt back"). There are four conditions that are essential for proper freeze drying to occur (Figure 10):

- The specimen must be solidly frozen (usually a small amount of pulp stock is frozen in LN_2 slush at 63 K).
- A condensing surface at or below 233 K must be provided.
- The system must evacuate to an absolute pressure of 5 to 26 μm of Hg.
- A heat source (controlled between 233 K and 338 K) must provide the heat of sublimation (employed to drive the water from the solid to the gaseous state).

Some artifacts are caused by this method, especially fiber shrinkage, the collapse of non-rigid cell walls onto the underlying structure and cell wall rupture due to ice damage if the rate of freezing is not sufficiently rapid. For hydrated pulp fibers, a room

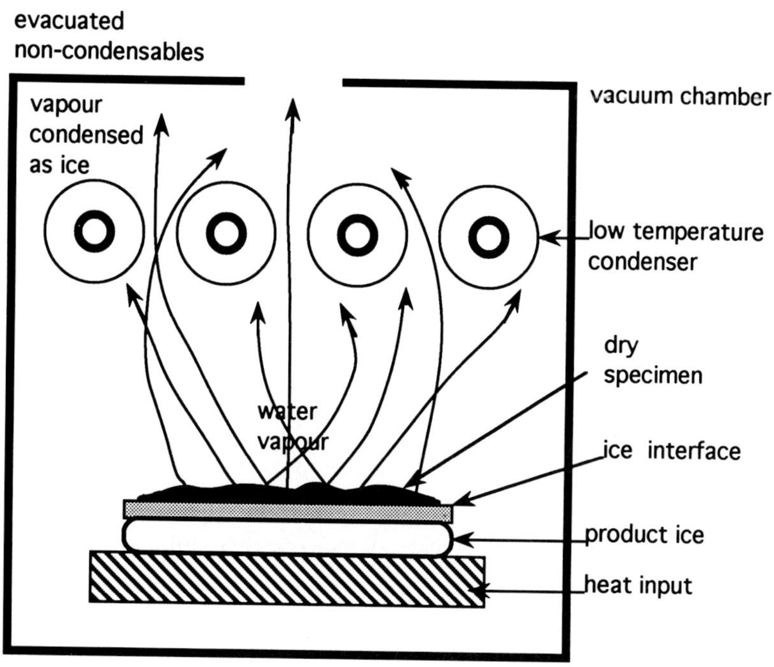

Figure 10. Diagram depicting the various processes comprising the freeze drying technique.

temperature chemical substitution of water with tert butyl alcohol (or methyl 1-2 propanol) in an especially designed apparatus followed by freeze drying yields relatively good results with a reduced number of artifacts generated.

EXAMINATION OF PAPER SAMPLES

Preparation of paper samples for the SEM is much simpler than the examination of hydrated wood pulp fibers. The relatively low moisture content of most papers (6 to 8% M.C.), even those conditioned under the stringent standard paper testing conditions imposed by TAPPI and CPPA (50% R.H. and 20 °C),[20] means that samples can be analyzed without further drying. The paper and the stub should always be handled with cotton or, preferably, latex gloves to avoid contamination of the sample and of the microscope. The procedure generally followed is detailed below because the non-conductive nature of paper requires that it should be handled with special care:

i. A sample of paper the size of the microscope stub used should be cut with a razor blade, taking care not to touch the surface with either fingers or an instrument that may compress the paper (Figure 11a). Avoid contaminating the paper surface or distorting its structure.

ii. The paper sample should be stuck onto the stub with double-sided tape, preferably a conductive carbon tape. For thick or rough specimens a conductive pathway of silver paste (commercially referred to as silver dag) or colloidal silver paint in an ethanol base should be made between the top surface of the paper and the metal stub (Figure 11b). Care should be taken to insure that the silver paint is not very dilute, as it can penetrate the paper by capillary action, thereby contaminating it.

iii. At this point it is preferable to leave the sample overnight in a desiccator to allow the solvent in the silver paint to evaporate and to remove any residual moisture that may be present.

iv. The sample is next coated in a sputter coater or an evaporation coater with a thin film (which should not exceed 30 nm in thickness) of a heavy metal such as gold, palladium or a mixture of the two. The thickness of the film should be such that surface details are not obscured; relatively smooth samples such as thermal gradient calendered (TGC) papers or coated papers should be coated with about 10 nm of metal. Samples with a rough surface may require thicker layers of metal in order to bridge small gaps and create a continuous conducting layer. (If the coater is not calibrated for layers of specified thickness, place a clean light microscope cover slip in the chamber with the specimen. The coating thickness can then be evaluated by comparing the coverslip to a new one.)

v. The sample is now ready to be viewed in the SEM.

Figures 12a and b show the same area of newsprint surface in which one was sputter coated with a 5 nm layer of gold and the other with a 30 nm layer. Note that Figure 12a more readily shows fine fiber surface details, although the image is noisy. The signal to noise ratio improves with a thicker layer of gold (*i.e.*, a greater yield of secondary electrons) but at the same time certain details are obscured. This is especially noticeable at higher magnifications, at which point the gold particles will be visible.

Z-direction information (*i.e.*, in the thickness direction) about paper structure can also be obtained in the SEM by preparing cross-sections of the paper sheets. The procedure consists of embedding the paper under vacuum in a polymer and subsequently cutting and polishing the resulting stub. Very often, however, this results in an image with poor SE contrast. If the aim is to acquire Z contrast information of the different compounds constituting the sample (as in Figure 8a), it should be coated with a thin film of carbon to prevent charging. Sputter coating with heavy metals should be avoided, as these can prevent the BSE from emerging from the surface of the specimen. As an alternative to the embedding procedure, "quick-and-dirty" cross-sections can be accomplished by freezing the specimen in liquid nitrogen, then fracturing it in LN_2 perpendicular to the surface between a chilled, cleaned, new, single-edged razor blade and a brass plate. This often gives useable results, although the cross-section is rough and compression artifacts may render this technique of little value for some types of observations.

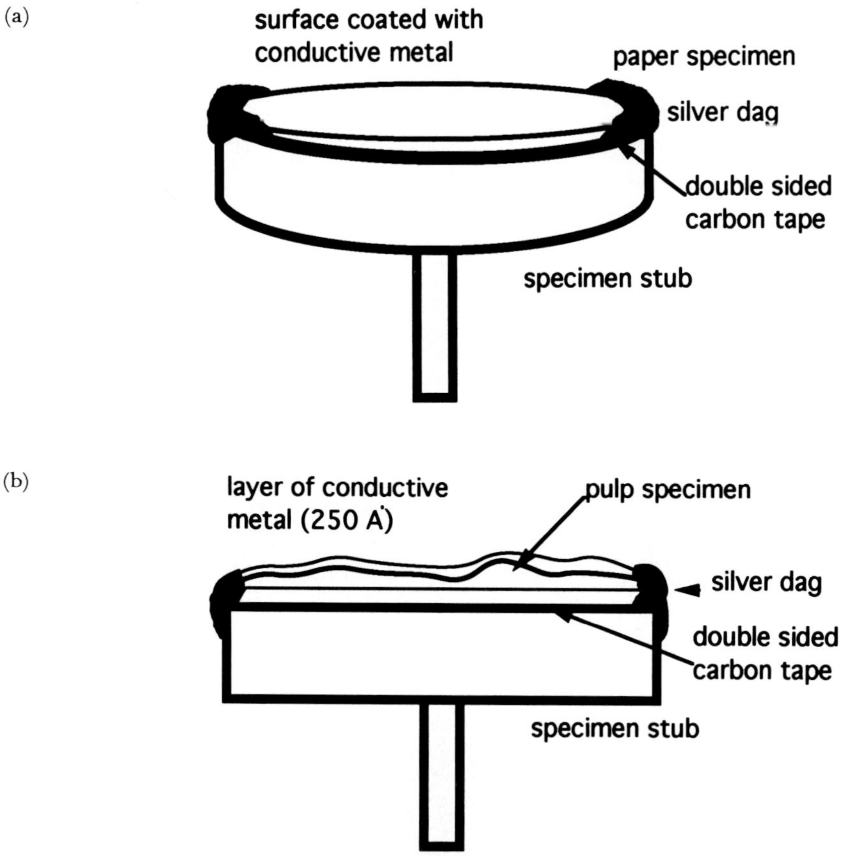

Figure 11. Diagrammatic representation of a pulp or a paper sample prepared for the SEM: (a) A paper sample stuck on a microscope stub with conductive carbon tape; (b) A pulp sample coated with a thin layer of gold and made conductive with the use of silver dag, with the same tape being used in this instance.

SEM Detectors: Most modern analytical instruments come equipped with both a secondary electron detector (SED) and a backscattered electron detector (BED). Often they also have two CRT monitors or a screen that can be divided in two (in addition to the recording CRT), so that both images can be viewed simultaneously. If both detectors are used at the same time, then a compromise has to be made regarding the choice of voltage (keV) and working distance (WD). With the SED it is possible to obtain images with a good depth of field at low voltages (3 to 5 keV), especially with a higher brightness LaB_6 or field emission gun and long working distances (20 to 50 mm WD). However, BEDs require higher voltages (15 to 20 keV) for a greater yield of backscattered electrons as well as shorter working distances (5 to 10 mm WD) so that a greater number of electrons can be detected.[5] The compromise involves finding conditions suitable for both detectors.

There are a number of advantages in being able to examine the same area with BED and SED simultaneously. Figure 13 shows a newsprint surface on which a ray cell and other fines are visible. The SE image (a) does not indicate how well the fines particles are

(a)

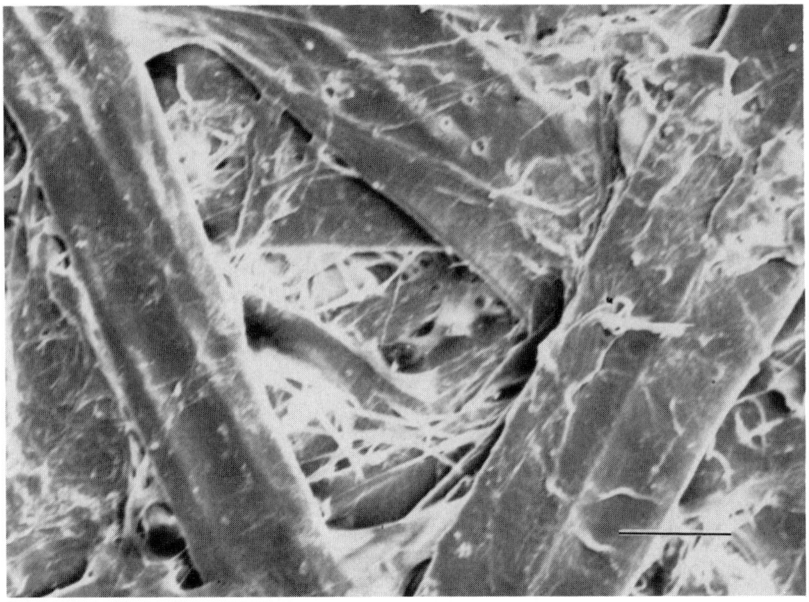

(b)

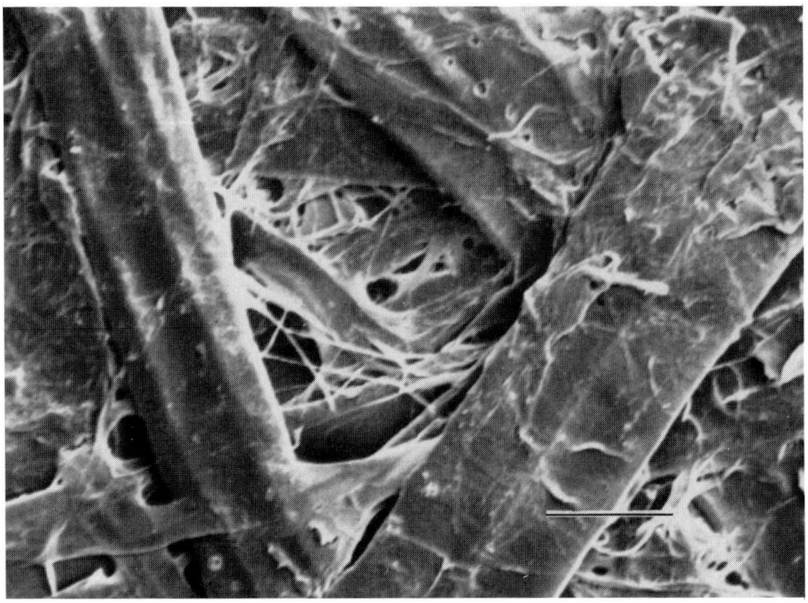

Figure 12. The effects of the thickness of the sputter coated metal layer are depicted at identical microscope conditions (15 keV): (a) The Au/Pd layer is 5 nm thick. Note the charging of the fines and fibrils that bridge the fibers with a greater amount of detail on the fiber surface; (b) The Au/Pd layer is 30 nm thick. Here, fibrillar details are clear but information about the fiber surface is lost. (Bars equal 30 μm).

bonded to the rest of the surface material, while the BE image (b) indicates that the majority of fines are well bonded to the surface but that the ray cell is not. The potential for linting of the ray cell can be determined (Figure 13b) because of the dark halo surrounding a particle that has not totally bonded to the substrate material. In the same way the SE image shown in Figure 14a provides information about the presence of fines on the surface of this newsprint. However, it is only by examining the same area with a BED (Figure 14b) that it is possible to say that some of the fines are actually mineral particles; these can be subsequently analyzed with x-ray spectroscopy to determine their chemical composition.

LOW TEMPERATURE SCANNING ELECTRON MICROSCOPY OR CRYO-SEM

Low temperature stages were initially developed five years after the first SEM became commercially available.[12] The singular advantage of the low temperature SEM (LTSEM) is that it is the only way to visualize the natural structure of biological materials, as the conventional preparative techniques remove the liquid phases and (potentially) distort the solid phases through extraction or dehydration. It is also a relatively fast technique for obtaining medium resolution (50–100 nm) information about a wide range of samples. Many small specimens composing larger samples are readily evaluated by LTSEM to determine the topological and topographical inter-relationships of the component parts. The somewhat larger specimens (*e.g.*, whole egg or embryo) that can be observed invariably suffer from problems of ice crystal damage unless chemical protectants are used.

Many researchers have discussed the fundamental problems of instrumentation and sample preparation when applying LTSEM to biological and non-biological samples[21,22,23] and a couple have discussed specific applications to wet pulp fibers.[19,24] Moss[25] and more recently de Silveira[26] have carried out in-depth studies of beaten hydrated fibers using LTSEM and attempted to determine the mechanisms implicated in artifact formation.

SAMPLE PREPARATION

The simplest method to prepare samples for the Cryo-SEM is to quench cool a small aqueous suspension of fibers (200 fibers in two to three ml of distilled water) in LN_2 (cooling rate 0.5 x 10^3 K/s). To effectively double the rate of cooling by avoiding the Leidenfrost phenomena and to ensure that the liquid within the fibers is also frozen, a LN_2 slush with a cooling rate of 1.2 x 10^3 K/s should be used. By using a cooled shroud which is also under vacuum the sample can then be transferred to the precooled stage in the microscope. Some cryogenic devices have a pre-chamber attached to the SEM sample chamber in which a limited amount of sample manipulation can be carried out. Fracturing or cutting with a precooled blade and even sputter coating with gold or carbon can be performed in the pre-chamber. In other cryogenic devices these preparatory steps are carried out away from the SEM prior to transferring the sample onto the microscope's

(a)

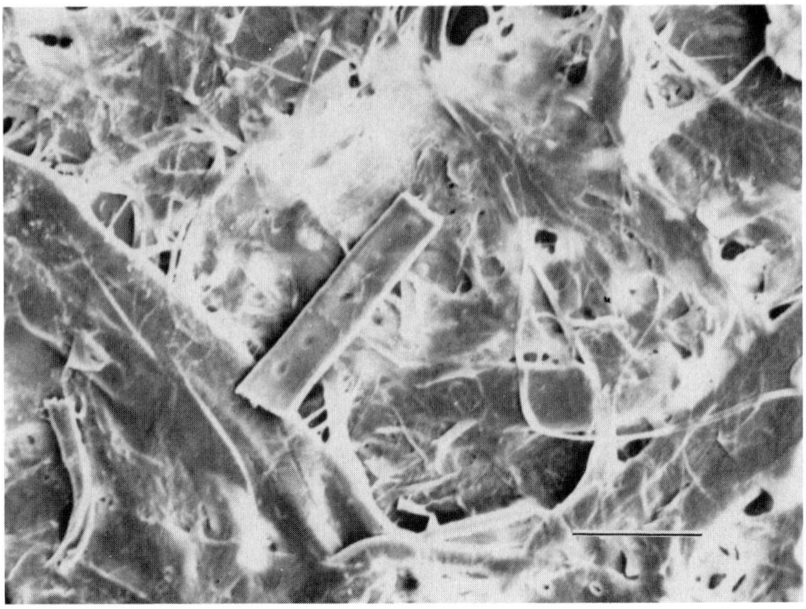

(b)

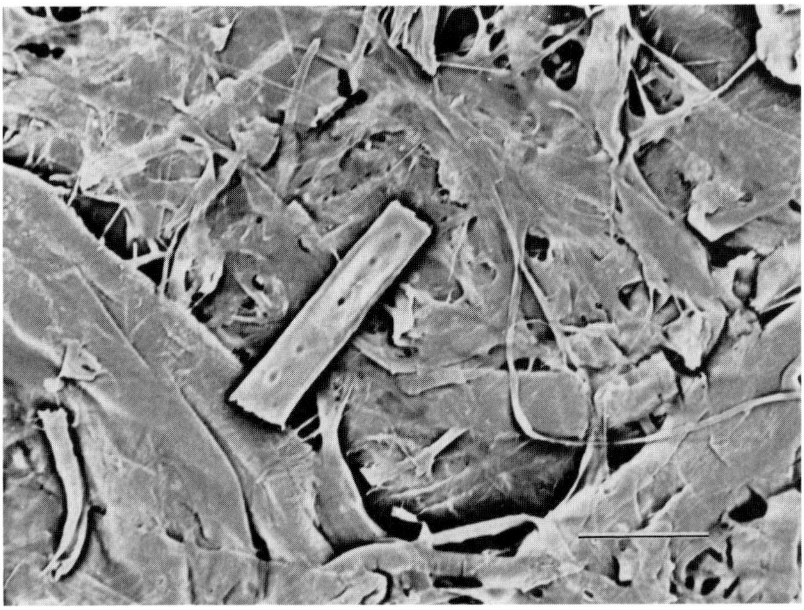

Figure 13. Micrographs of a newsprint surface with a ray cell present: (a) In SE mode no information can be obtained about the degree of bonding of the ray cell to the rest of the surface; (b) BE image showing a black halo surrounding the ray cell, indicating that most of it has not bonded with the rest of the fibers and fines. (Bars equal 1 μm).

(a)

(b)

Figure 14. SEM micrographs of the same area on the surface of a recycled linerboard showing: (a) In SE mode, surface details about the fibers and overall sheet roughness; (b) In BE mode, details about the presence of mineral particles, the degree of bonding of the fibers and, to some degree, the extent of the void spaces. (Bars equal 100 μm).

cryo-stage. Controlled sublimation of the surface ice can be achieved either before transfer to the SEM or during observation on the cryo-stage.

ARTIFACTS

All sample preparation methods lead to the formation of artifacts, although the causes of some types of artifacts are better understood because the procedures have been extensively used. New techniques such as the LTSEM and the ESEM have only been available for a few years and the artifacts they generate with hydrated fibers specimens have not been fully analyzed. Preliminary results indicate that the LTSEM provides a picture that is probably closer to reality than critical point drying, the most commonly used method.

The artifacts generated in LTSEM stem from the improper cooling and manipulation of the sample. Effective cooling of the aqueous media, the fibers and the water contained in the small interstices of the fiber wall requires that the cooling rate be high. To do this the sample should be very small, clean and the aqueous media should be free of soluble matter. Dissolved solutes act as nucleating agents during the cooling process leading to the formation of large ice crystals that may disrupt the fiber structure. At higher concentrations they crystallize at the boundaries of ice crystals creating a "lace" or "membranous-like" artifact which persists after the ice has been sublimated (Figure 15).

Figure 15. LTSEM micrograph showing the artifacts resulting from the presence of dissolved solutes in the aqueous matrix of a chemical pulp slurry, which was mixed with chemical inert synthetic fibers. Chemical analysis of the matrix indicated a considerable amount of dissolved lignin and hemicellulose. The artifacts develop as the ice crystals grow and the solutes crystallize at the boundary of the ice. When the ice is sublimated the remaining crystallized solutes resemble a lace-like or a membranous structure (bar equals 40 μm).

Transferring the frozen sample from the cooling apparatus to the microscope stage requires both speed (to prevent warming of the sample) and vacuum (to avoid condensation of atmospheric matter onto the cold surface to be examined). The vacuum and the temperature within the microscope pre-chamber and the SEM have to be carefully controlled to avoid contaminating the sample with ice crystals at the sputter coating and observation stages. If sublimation of the frozen aqueous media is to be carried out, the heat source should be placed above the sample to obviate heating the whole sample from below (carried out by heating the precooled sample stage within the microscope) which induces artifacts similar to the ones created by freeze drying .

RESULTS

Results obtained to date indicate that LTSEM is an ideal technique for the observation of pulp fibers. Fully hydrated specimens can be frozen with minimum sample manipulation or preparation and observed within ten minutes if the sample preparation techniques are attentively followed. Cryo-SEM micrographs (Figures 16a and b) of fully hydrated pulp fibers show details about the microfibrillar and cell wall structure that can not be examined in conventional SEM due to the collapse of these structures. Noticeable are the disruptions brought about by the pulping and bleaching processes to the external fiber walls in their swollen state.

ENVIRONMENTAL SCANNING ELECTRON MICROSCOPE (ESEM)

The low vacuum SEM was developed from the ideas of G.D. Danilatos.[27] Publication of his ideas led to the creation of a number of competing instruments (sometimes known as "wet SEMs" because images can be recorded in a poor vacuum, even one which contains water vapor). Most of the commercialized instruments presently use only a backscattered detector, but one we have experience with (the environmental scanning electron microscope, or ESEM, manufactured by ElectroScan in Wilmington, Massachusetts, USA) also uses a secondary electron detector of unique design which is based on gas ionization principles.[28] This microscope can be used to observe specimens in an atmosphere containing water vapor up to 20 torr (about 2.7 kPa), which is the saturation vapor pressure at room temperature. This permits the imaging of structural features at elevated relative humidities and moisture contents. Although the pressure in the specimen chamber is usually maintained with water vapor, almost any gas (such as nitrogen, oxygen, argon, methane) which neutralizes charge build-up across the specimen surface can be used.

ADVANTAGES OF ESEM MICROSCOPY

In a conventional SEM, the vacuum system is maintained at high vacuum to avoid scattering the electron beam as it passes through high pressure gas. In an ESEM, however,

(a)

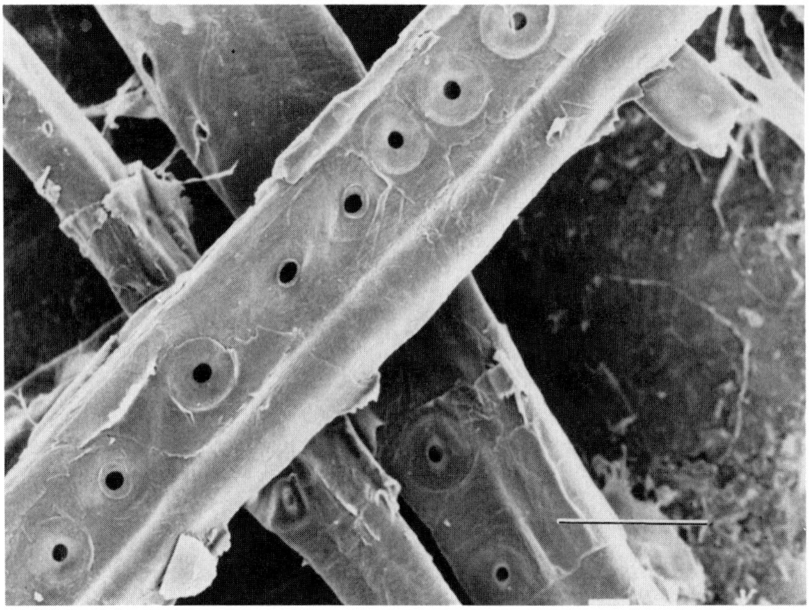

(b)

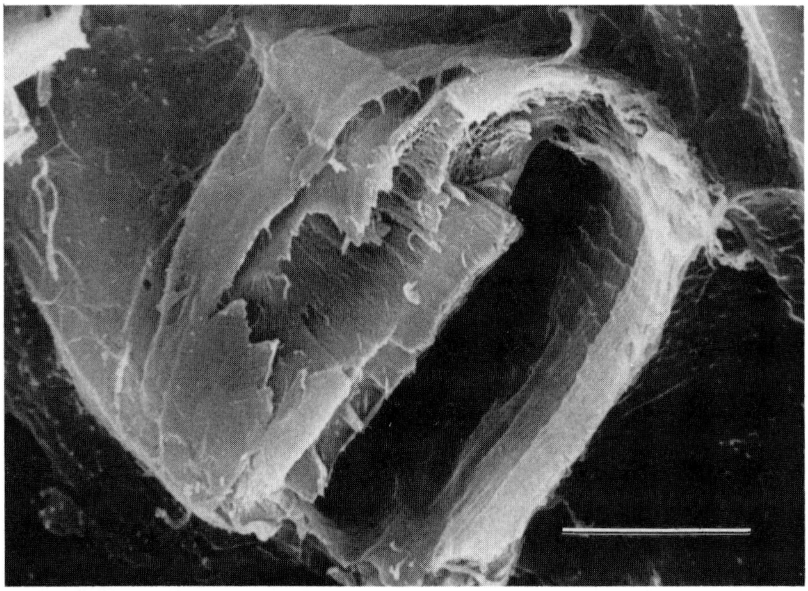

Figure 16. Pulp samples analyzed in the Cryo-SEM: (a) Hydrated TMP fibers showing extent of fiber wall damage and selective removal in certain regions of either the P layer or the S_1 layer (bar equals 40 μm); (b) Laboratory-prepared (to avoid introducing mechanical changes to the fiber wall structure) bleached chemical pulp fiber showing internal delamination of the concentric fiber wall lamellae giving it the appearance of a beaten fiber (bar equals 10 μm).

different parts of the microscope are maintained at different pressures. The electron gun itself still requires high vacuum, and the vacuum is gradually lowered towards the specimen chamber using a series of pressure limiting apertures (Figure 17). This is made possible by a differential vacuum pumping system, which maintains the pressure gradient accurately and prevents any contamination from reaching the high vacuum gun.[29] The pressure in the electron gun is reduced to about 10^{-5} Pa by means of an ion pump. The condenser and objective lenses, however, only require pressures less than 10^{-2} Pa, which are maintained by two diffusion pumps. Two pressure-limiting apertures separate the objective lens vacuum system from the specimen chamber. This pressure difference is maintained by a mechanical rotary pump.[30] The distances between the two pressure-limiting apertures and the specimen are made small to reduce electron beam scattering; the distance the beam travels at high pressure can be as little as 10 mm, depending on the working distance.

As noted previously, two different types of detectors can be used in the ESEM. The environmental secondary electron detector (ESD) collects ions and secondary electrons. This detector, also called a bullet, is placed co-axially with the final lens and the beam passes through the detector. The back scattered electron detector is a standard light probe, which is operated at low pressures (<2 torr).

The advantage of the ESEM is that no surface-destructive sample preparation is required and specimens can be imaged in their natural state; samples do not need to be dehydrated or sputter coated. Besides freeing the microscopist from artifacts caused by the preparative procedures, the lack of dehydration and coating means that samples can be maintained or archived more easily. Prolonged storage of metal- or carbon-coated specimens, especially in fluctuating temperature and relative humidity conditions, can cause stressing of interfacial regions and cracking of the conductive coatings.

X-ray spectroscopy is covered in another chapter, but it should be noted that the metallic coatings required for imaging insulators like paper in a conventional SEM often limit or conceal data in x-ray spectra. This means that the ESEM is particularly suited to the use of energy dispersive x-ray analysis systems. Insulating samples may be analyzed in the ESEM with higher accelerating voltages without coating the sample, whereby the material is excited to produce x-ray data not observable using lower voltages.[31] X-ray analysis of hydrated samples is most successfully conducted using a cold stage (to lower the saturation vapor pressure for the sample) and a good vacuum. The good vacuum, while not essential, helps to maintain collimation of the electron beam focussed on the sample. The spatial resolution of the x-ray data is improved as a consequence.

A number of accessories are available to use the ESEM to its full extent. Among these are the Peltier thermo-electric stage (temperature range 0 to 100 °C), a separate hot stage (temperature range up to 1500 °C), micromanipulators and microinjectors, and a tensile stage. The Peltier thermo-electric stage can be used to cool and heat samples from 0 to100 °C, allowing the microscopist to dynamically observe the hydration and dehydration of

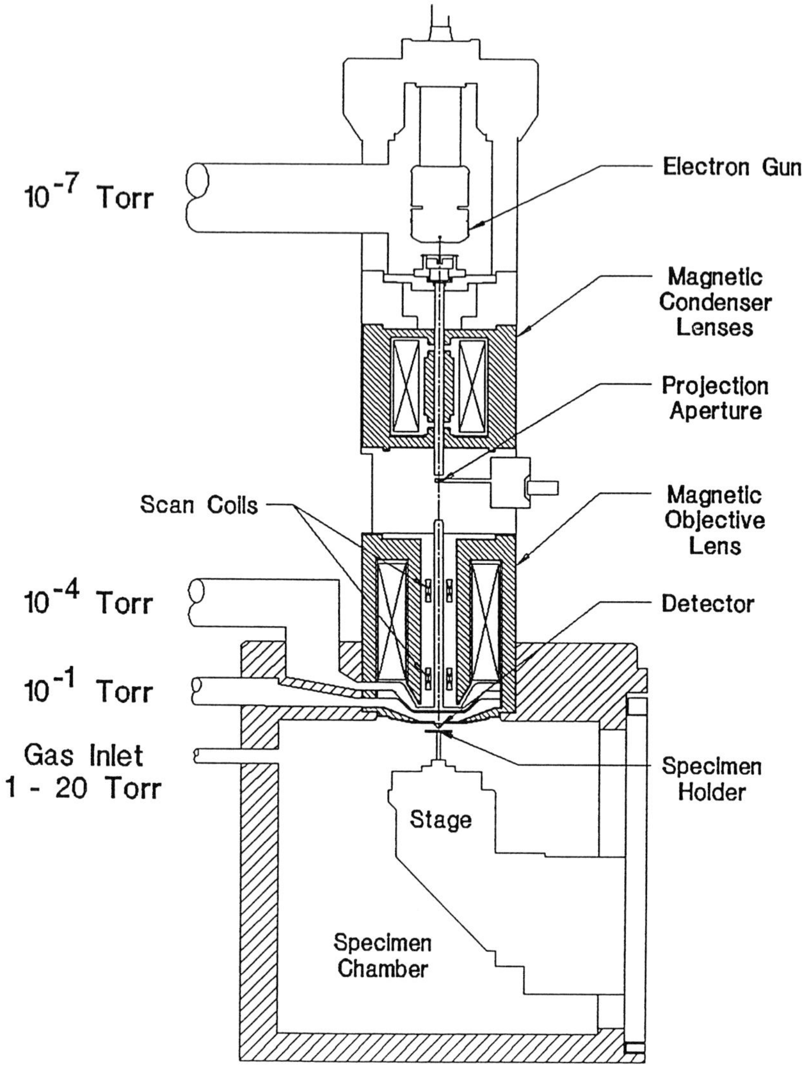

Figure 17. Cross-section of the ESEM electron optical column.

the specimen. As water droplets form and dry, the advancing and retreating contact angles can be recorded on particular features (Figures 18a and b) illustrating the degree of microhomogeneity present. Even volatile-containing inks can be visualized. Other dynamic processes can be observed that were previously possible only in the light microscope at low magnification. Examples of these include fracture testing as well as observation of swelling and shrinkage effects due to wetting and re-drying. The drying of clay coating formulations has also been recorded, although precautions had to be taken to prevent premature drying while the ESEM vacuum chamber was pumped down to the desired vacuum. In addition, the ESEM can be connected to a VCR to record dynamic

(a)

(b)

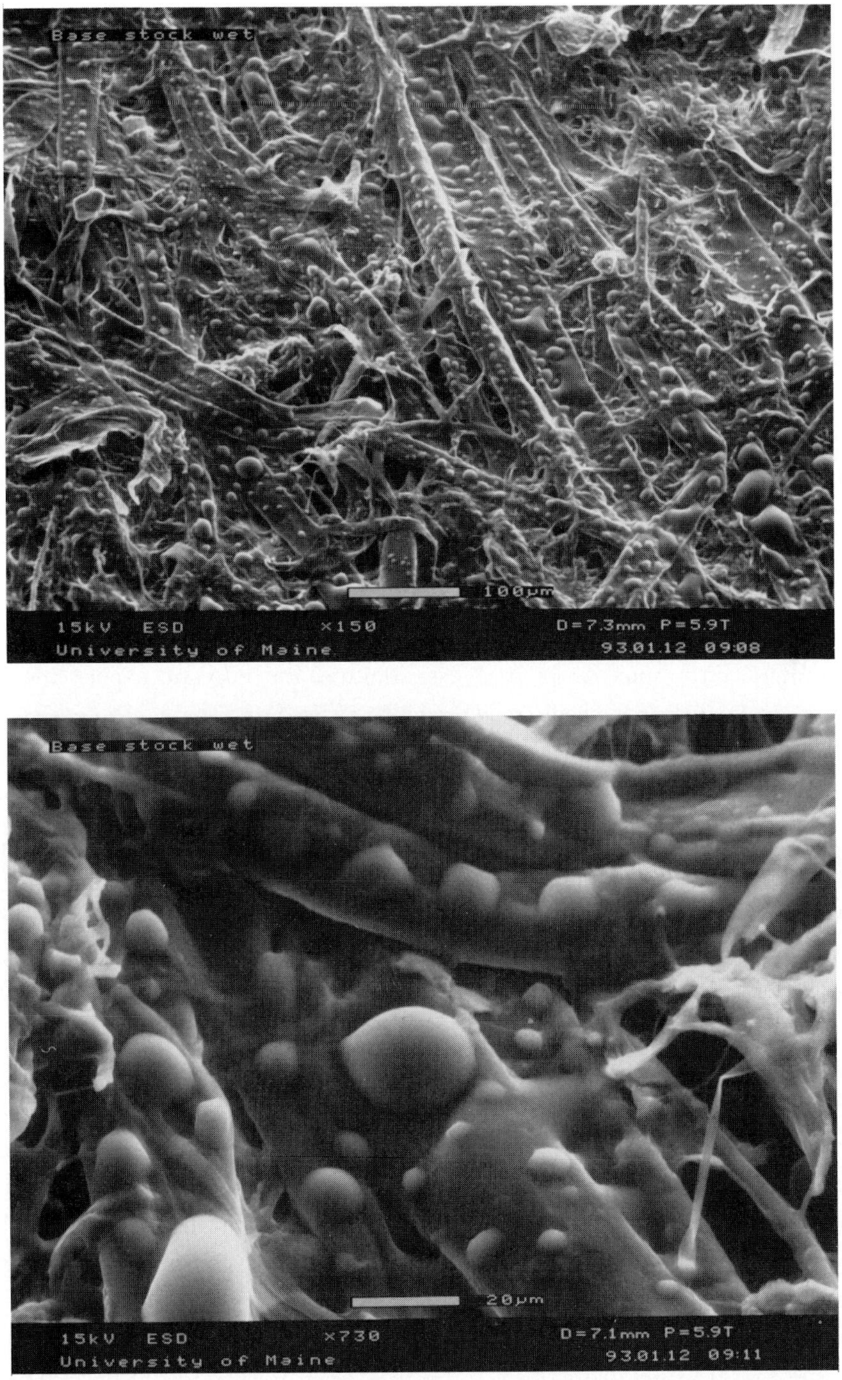

Figure 18. (a) Wet base stock; (b) Same base stock as in (a) at a higher magnification.

processes on a VHS tape. This allows detailed studies of the ESEM images after the session and the use of image analysis.

DRAWBACKS TO ESEM MICROSCOPY

The advantage of the ESEM is that samples can be viewed at any accelerating voltage without coating or other sample preparation. However, like conventional SEM, the resolution of non-conductive samples can be enhanced by coating the sample with a conductive metal (*e.g.*, Au, Pd), especially for observations at high magnifications. Perhaps the main difficulty to be aware of is that the electron beam can sometimes alter uncoated samples. Especially when working with polymers and wet kraft fibers, care should be taken not to burn up the sample. The irradiation damage can be minimized or even avoided by using lower accelerating voltages and higher condenser lens settings. The use of unnecessary moisture or prolonged viewing of the same spot should also be avoided. Wet mechanical pulps do not appear to be affected in our observations to date.[32]

SUMMARY

The scanning electron microscope is an essential tool for pulp and paper science and the development of new paper products. The depth of field obtainable, the ease with which paper samples can be prepared, and the new techniques that allow for the examination of hydrated wood pulp fibers make it one of the best instruments to examine pulp and paper products. The addition of image analysis and image enhancement systems to SEMs has made it possible not only to improve the quality of the micrographs produced but also to carry out a number of quantitative measurements that help in analyzing the specimens.

The emission of a variety of signals by the sample as a result of the impingement of incident electrons permits the scanning electron microscopist to obtain information about the topography (SE signal) of the sample, the atomic number of the elements (or the average atomic number in the case of compounds) (BSE signal) present within the sample and the semi-quantitative chemical analysis (x-ray signal) of the elements constituting the specimen. Other signals that are emitted simultaneously, such as Auger, cathodoluminescence, plasmons and phonons can also be detected with dedicated systems.

Paper samples are relatively easy to prepare for the SEM, but hydrated pulp fibers, inks and pigment/polymer slurries require dedicated systems equipped with a cryo stage (LTSEM) or an environmental chamber (ESEM). These permit the analysis of hydrated specimens and make the dubious sample preparation methods involving chemical substitutions used to date superfluous. Another advantage of the ESEM is the fact that fully hydrated pulp fibers, inks and coating pigment slurries can be studied while the relative humidity conditions are dynamically changed.

ACKNOWLEDGEMENTS

Most of the micrographs were taken at PAPRICAN, and we appreciate the assistance and cooperation of that organization. The authors would also like to thank the University of Maine for the use of the ESEM while P. Forsberg was in residence there as a Visiting Scientist. Finally, we would like to thank ElectroScan for providing the cross-section of the ESEM column shown in Figure 17.

REFERENCES

1. Oatley, C.W. and Everhart, T.E. 1952. *J. of Electron Microscopy* No. 2.
2. Smith, K.C.A. 1956. Ph.D. Dissertation, University of Cambridge, Cambridge, U.K.
3. Humphreys, W.H., Spurlock, B.0. and Jackson, J.S. Principles and Techniques in SEM Biological Applications. Volume No. 6. Hayat, M.A., ed. 1977. Van Nostrand, Reinhold, New York.
4. Watt, I.M. 1985. The Principles and Practice of Electron Microscopy. Cambridge University Press. Cambridge, New York.
5. Goldstein, J.I., Newbury, D.E., Echlin, P., Joy, D.C., Romig,Jr., A.D., Lyman, C.E., Fiori, C. and Lifshin, E. 1992. Scanning Electron Microscopy and X-Ray Microanalysis. Plenum Press, New York and London.
6. Evans, R.D. 1955. The Atomic Nucleus. McGraw-Hill, New York.
7. Heinrich, K.F.J. 1966. X-ray absorption uncertainty. *In*: The Electron Microprobe. McKinley, T.D., Heinrich, K.F.J. and Wintry, D.B., eds. Wiley, New York. 1966. pp. 296–377.
8. Bishop, H. 1966. Some electron backscattering measurements for solid targets. *In*: Proc. 4th International Conference of X-Ray Optics and Microanalysis. Castaing, R., Deschamps, P. and Philibert, J., eds. Hermann, Paris.
9. Heinrich, K.F.J. and Newbury, D.E., eds. 1991. Electron Probe Quantification. Plenum Press, New York.
10. Everhart, T.E., Herzog, R.F., Chang, M.S. and DeVore, W.J. 1972. Electron energy dissipation measurements in solids. *In:* Proc. 6th Intl. Conf. on X-Ray Optics and Microanalysis. Shinoda, G.,Kohra, K. and Ichinokawa, T., eds. University of Tokyo Press, Tokyo.
11. Peters, K.-R. 1986. Metal deposition by high-energy sputtering for high magnification electron microscopy. *In*: Advanced Techniques in Biological Electron Microscopy. Koethler, J.K., ed. Springer-Verlag, Berlin.
12. Echlin, P., ed. 1984. Analysis of Biological and Organic Surfaces. Wiley, New York.

13. Parsons, E., Dole, B., Hall, D.J. and Thomas, W.D.E. 1974. A comparative survey of the techniques for preparing plant surfaces for the scanning electron microscope. *J. Microscopy* 101(1):59–75.
14. Dwivedi, A.K. and Ahmad, K.J. 1985. Scanning electron microscopy of fresh, uncoated plant parts. *Micon. and Microscopica Acta* 16(1):55–57.
15. Echlin, P. 1992. Low Temperature Microscopy and Analysis. Plenum Press, New York.
16. Parham, R.A. 1975. Critical point drying for fiber microscopy. *Tappi J.* 58(3): 138–140.
17. Sachs, I.B. 1985. Preserving and recovering pulp fibrils subsequent to drying. *Paper Tech. and Ind.* 26(1):38–41.
18. Sachs, I.B. 1986. Retaining raised fibrils and microfibrils on fiber surfaces. *Tappi J.* 69(11):124–127.
19. Howard, R.C. and Sheffield, E. 1987. The wet structure of pulp and paper examined in cryo-SEM. *Paper Technology and Industry* 27(2):425–427.
20. TAPPI. 1994. Standard T 402 om–93, Standard conditioning and testing atmospheres for paper, board, pulp handsheets and related products. TAPPI Press, Atlanta, Georgia.
21. Beckett, A., and Read, N.D. 1986. Low temperature scanning electron microscopy. *In*: Ultrastructure Techniques for Microorganisms. Aldrich, H.C. and Todd, W.J., eds. Plenum Press, New York.
22. Sargent, J.A. 1988. Low temperature scanning electron microscopy: Advantages and applications. *Scanning Microscopy* 2:835–849.
23. Sheehan, J.G. 1992. A conduction cooled stage for high resolution cryo-SEM. Proc. 5Oth EMSA Annual Meeting. San Francisco Press, page 53.
24. Moss, P., Howard, R.C. and Sheffield, E. 1989. Artefacts arising during preparation of hydrated pulp samples for low temperature SEM. *J. Microscopy* 156:343–351.
25. Moss, P. 1990. A Study of the Frozen Hydrated Structure of Pulp. Ph.D. thesis, UMIST, Manchester, U.K.
26. de Silveira, G. 1995. The Ultrastructure of Pulp Fibres Studied by Low Temperature Scanning Electron Microscopy. M.Sc.A. thesis, École Polytechnique, Montreal, Quebec, Canada..
27. Danilatos, G.D., Robinson, V. N. E. 1979. Principles of scanning electron microscopy at high specimen pressures. *Scanning* 2:72–82.
28. Ruggiero, M., ElectroScan, personal communication 12/15/93.
29. Sujata, K., Jennings, H. M. 1991. Advances in scanning electron microscopy. *MRS Bulletin* XVI (3):41–45.
30. Sheehan, J. G., Scriven, L. E. 1991. Assessment of environmental scanning electron microscopy for coating research. TAPPI Coating Conference Proceedings, pp. 377–383.

32. Forsberg, P. and P. Lepoutre. 1994. ESEM examination of paper in high moisture environments: Surface structural changes and electron beam damage. *Scanning Microscopy International* 8(1):31–34.

3

Mechanical and Physical Properties of Paper Surfaces

John F. Waterhouse

Institute of Paper Science and Technology, Atlanta, Georgia, U.S.A.

INTRODUCTION

Paper, although ubiquitous, is a complex material. We can classify paper as a basic network of self bonding cellulosic fibers; the chemical and physical characteristics of its surface are controlled by the raw materials, papermaking and converting processes used to produce it. Paper sometimes contains non-cellulosic fibers such as non-wood and synthetic fibers, chemical additives, fillers, and bonding agents. Other chemical additives used in the papermaking process, *e.g.*, formation, drainage, and retention aids or coating materials may also change the physical and chemical characteristics of the paper's surface. With the greater use of recycled fibers we may expect additional changes in surface chemistry and structure.

The uniformity and structure of paper is highly dependent on the spatial arrangement of its fibers, fines and other components. Sheet formation (measured in terms of its mass density or small scale grammage distribution) is one measure of its uniformity. If we use the coefficient of variation of mass density CV(W) that has been suggested as a universal index of formation by Dodson,[1] it can be shown for an ideal random network of fibers that CV(W) is dependent on fiber length, fiber coarseness, and average grammage. Therefore the shorter and less coarse fibers, such as hardwoods, generally have better forming characteristics than softwood fibers and thus produce papers with superior formation characteristics.[2] Other properties which may vary on a small scale (*e.g.*, in the range of 0.1 mm to 100 mm) include porosity, surface roughness, surface wettability, and surface strength. Many of these properties are clearly related to the mass density distribution, but little work has been done to establish these links.

Paper is a highly anisotropic planar material which we often view and treat as a continuum, although the porous nature of paper is vital in many end-use applications. There are two major factors, namely fiber orientation and drying restraint, controlling

0-8493-8992-5/95/$0.00+$.50

the anisotropy of paper. This can be measured, for example, by the elastic properties. Paper fibers lie mainly in the plane of the sheet; the fibers themselves are highly anisotropic with their elastic constants being much higher along the axis of the fiber than those normal to the fiber axis. Thus, the elastic constants in the thickness direction of paper are generally much lower than those measured in the plane of the sheet. The in-plane elastic anisotropy is mainly controlled by fiber orientation, but bonding and the effective machine and cross-machine direction restraints on the sheet during drying will also play a significant role. Changes in drying restraint will usually have a negligible effect on fiber orientation, but the effective axial modulus of the fiber can be changed quite dramatically, and therefore the anisotropy of the sheet will be affected commensurately.

In addition to structural differences in the in-plane and out-of-plane directions of paper, there are also differences from layer to layer in the thickness or z-direction when paper is viewed as a laminar structure.[3] Clearly, the spatial arrangement of the fibers and other principal structural components which takes place at the wet end of the papermachine can influence the surface structure of paper. Furthermore, the surface structure can be modified by other unit processes of papermaking such as wet pressing, the size press, and calendering and converting operations (*e.g.*, coating and supercalendering). The surface properties of paper and their relationship to paper structure has been dealt with in some detail by Bristow.[4]

The surfaces of paper are commonly referred to as the wire side and felt or top side. Traditionally these sides have had very different characteristics, but, with the recent emphasis on quality, particularly print quality, more effort is being expended by the papermaker to minimize these differences. Felt and wire side differences may be due to non-uniformities in fiber orientation, fines, filler and additive distributions. Surface structure is also affected by the forming fabric, wet press, breaker stack and dryer felt clothing. Subsequent converting operations such as calendering (hard nip, soft nip, and supercalendering) are used to improve the surface characteristics of paper while maximizing bulk without adversely effecting mechanical properties.

COATED PAPERS

Improvements in surface structure and properties can be achieved by surface coatings applied, for example, by roll, air knife, or blade methods. Aqueous pigment coatings are usually intended to improve the printing and writing characteristics of paper (optical) properties. However, the application of aqueous coatings may result in an increased surface roughness.[5] This can be reduced by supercalendering, but surface roughness may recur during subsequent printing operations, giving rise (in some situations) to a variety of mottle problems.

The average pore size in a pigment coating is an order of magnitude lower than the base paper, (*e.g.*, 0.2 μm *vs* 3 μm). In spite of the smaller pore size, in-plane and thickness direction non-uniformities in both coating structure and wetting behavior *(e.g.,* "binder migration") can give rise to poor printing characteristics. Common printing concerns, some of which can apply to both uncoated and coated papers, include print mottle, back trap mottle, half tone dot and solid color uniformity, and image quality of non-impact processes such as ink jet printing.

Functional coatings are another category of coatings used to control specific product properties such as friction, resistance to blocking, surface charge, abrasion resistance, lubrication, improvement of adhesion, surface strength, etc.

MECHANICAL PROPERTIES OF PAPER AND BOARD SURFACES

SURFACE STRENGTH

Surface failure may range from incipient picking to total disruption of the surface layers of paper and board, and may be important in both the dry and wet states. Examples of different types of surface failure and associated terminology, as applied to printing related surface strength problems, are given in a recently updated TAPPI Monograph.[6]

Generally, we expect surface strength to be dependent on both in-plane and out-of-plane deformation behavior (which can also vary from layer to layer), as well as sheet structure. It is well known that surface strength as judged from simple Scotch tape peel tests can be directional, *i.e.*, there may be little fiber pull in the direction the paper is made, whereas, if the tape is pulled in the opposite direction surface failure will be very evident. Surface strength is also dependent on the level of stress concentration at the surface which complicates the problem of an appropriate measure of surface strength for a given application.

Failure of the surface or surface layers implies that applied surface stresses are exceeding the surface strength of the paper when the appropriate stress concentration factors are accounted for. Stresses can be applied to the surface in a number of ways including: adhesive layers, ink (especially in offset printing where tacky inks are used) and wax films, and friction forces. In the case of adhesives and ink films, a simplistic view is to say that either the adhesive layer fails (adhesive failure) or the paper fails (cohesive or adherend failure). A change in surface chemistry which affects adhesion can give misleading results with some measurements of surface strength.

The following are important steps involved in bond formation that have special relevance to paper and board: *i*) wetting of substrate by adhesive; *ii*) penetration of adhesive/solvent into substrate; and *iii*) solidification and curing of adhesive. In their studies of bond formation and strength development Lepoutre and co-workers[7] have shown how solvent and aqueous based adhesives solidify. The cohesive strengths of both the adhesive and adherend develop at different rates. One therefore has to be careful in interpreting adhesive strength pull tests, which are commonly used to assess the ultimate strength of an adhesive joint (*e.g.*, lap joints, manufacturer's joint in corrugated containers, and single face and double face bonding in corrugated board manufacturing). This is particularly so in the case of aqueous adhesives, which can, according to Lepoutre,[8] create "weak boundary layers" as water is transported away from the adhesive to the adherend by capillary forces or diffusion. The surface layer of the paper can be reinforced by adhesive penetration, so, depending on the level of stress concentration, the weak boundary layer may be responsible for adhesive joint failure.

Before considering specific tests for the measurement of surface strength let us next consider elastic properties. These may be helpful in assessing surface strength non-destructively, although there are obvious pitfalls in relating surface properties to bulk properties. Mechanical[7–12] and ultrasonic tests[13] have been developed for determining the out-of-plane longitudinal and shear moduli of paper and board. Surface and bulk compressibility are considered to be important in roll building and printing.

Improvements in fiber and bond properties by refining and wet pressing are two ways to improve the overall elastic and failure properties of paper. A concomitant improvement in surface strength would also be expected. Stratton[14] and Waterhouse[15] have found bond strength differences in the longitudinal and shear failure modes.

MEASUREMENT OF SURFACE STRENGTH

We have already noted that surface or interface failure is important in adhesion and print-related problems. Tests that have been developed to measure surface strength are somewhat contrived, since it is not a trivial matter to simulate the conditions under which paper fails in real-world situations.

Dennison Wax Pick: The Dennison wax pick test was one of early tests for measuring surface strength as it might relate to offset printing and picking. The procedure is relatively straightforward and quite reproducible for a given operator. A series of waxes with different tacks numbered from 2A to 26A is used. A small oil lamp is used to soften and melt the surface of the wax, which is then brought into contact with the paper to be tested. After 15 seconds the wax pick is removed from the surface in a vertical direction with the aid of a small wooden jig to keep the paper flat, while the pick is being removed. Before incipient or total failure of the surface there is a clean separation of the wax from the surface without surface disruption. In the next phase there is a slight disruption of the surface in the form of picking, and if this is the desired end point, then the wax pick number is recorded. If the test is continued to a higher wax pick number then there will eventually be total disruption or failure of the surface. With care results are reproducible to a specific wax pick number for an experienced operator.

The question is often asked: Is wax pick is a good measure of surface failure which can occur in an offset press? The general answer is *maybe.* It may work well in one situation but not in another. This is not surprising when one considers all the variables involved and our incomplete understanding of the process. In measuring the surface strength of coated papers the type of pigment binder can be important. For example, the indicated relative surface strength of a coating with a starch binder may be quite different from one with an SBR (styrene butadiene rubber) binder. This is, in part, due to differences in how the wax(es) wet the two polymers.

Viscosity – Velocity Product Test (VVP): Measurement of surface strength and adhesion was dealt with by Wink and co-workers in a series of papers published from 1946 to 1957.[16–19] Their research culminated in a device for measuring surface strength, where failure ranged from incipient picking to complete disruption of the surface, and the internal strength (bond strength) of the sheet.

A schematic of the device is shown in Figure 1. The basic principle employed is film splitting in a nip during an accelerating flow situation. As the paper–film combination accelerates through the nip, film splitting can change to surface failure of the paper or board which it is in contact with. We shall see that any particular phase of surface failure can be associated with the product of viscosity and velocity.

Both normal stress and shear stress are proportional to the product of viscosity and velocity. As an illustration of this consider Couette flow between two smooth parallel surfaces a distance h apart, and moving at velocity U relative to each other. Then the shear stress $\tau = \mu U/h$ (*i.e.*, the shear stress for a fixed film thickness) is proportional to the product of viscosity (μ) times velocity (VVP).

As the magnitude of these stresses is also dependent on film thickness, it is important to ensure that the film thickness is constant. A film applicator–doctoring device was developed to ensure that this requirement is satisfied, but variations in surface roughnessstill have to be corrected for as the effective film thickness undergoing shear is less than the total film thickness. When surface failure (incipient picking through to blister and complete failure) is of interest, the surface of the sample is not pre-filmed.

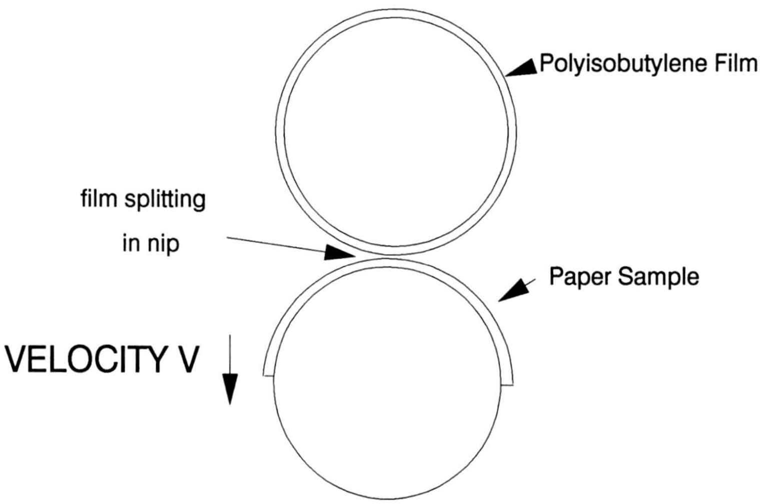

Figure 1. Schematic of velocity–viscosity product test.

However, when a measurement of bond strength is required[18] the surface is pre-filmed. The applied viscous film should wet the surface and be Newtonian over the anticipated shear rate range. Polyisobutylene closely approximates this requirement.

As already stated, the product of viscosity and velocity referred to as VVP (when a pre-film is used values are denoted by VVP') is used as a measure of surface strength and bond strength. For VVP to be a material property it should be independent of the following test variables which are summarized below together with the main findings from Wink *et al.*:[18,19]

i) contamination – contamination is important and can affect wetting of the surface;

ii) backing of the paper sample – the backing material can influence the nip width, but its effect is small;

iii) nip load – nip load can also influence nip width and have a tendency to extrude out the film at high film thicknesses;

iv) viscous film thickness – VVP is relatively insensitive to film thickness in the range of 0.0006" to 0.0018" (0.015 to 0.046 mm) for smooth samples. However, without pre-filming, the rate of change of VVP with film thickness can vary and is attributed to variations in surface roughness. Surface roughness and poor coverage are evident in the low film thickness range where VVP rises rapidly. At high film thickness VVP rises again (attributed to a loss in shear stress);

v) pre-filming – pre-films were applied by film splitting and doctoring in an attempt to compensate for incomplete coverage due to roughness and wetting. It was found that pre-filming by film splitting was generally unsatisfactory since a "rough" film surface was generated, resulting in higher VVP' values than those for the pre-films which were applied by doctoring;

vi) penetration – penetration appears to have a negligible effect on VVP, but the time between filming and testing should be as short as possible to minimize this effect;

vii) film thickness, viscosity, wetting, and smoothness – VVP values were largely unaffected by the test film thickness, and the viscosity of the pre-film, which was very much higher than the test film viscosity. The data in Table I illustrate the effects of pre-filming. The no pre-film condition (column 3) is a 0.8 mil (0.02 mm) test film. The divided pre-film condition (column 4) is produced by splitting a 0.8 mil film, *i.e.*, 0.4 mil is the test film and 0.4 mil is applied to the surface of the paper. This condition is supposed to improve wetting, but the film thickness is insufficient to overcome surface roughness. In the penultimate column, both wetting and roughness effects are eliminated by applying a relatively thick film (*i.e.*, 1.8 mil, high viscosity [7.2 kilopoise, or 720 x 10^{-3} newton-sec/m^2]) to the surface of the paper. The test film is 0.8 mil.

Table I
The Effects of Pre-Filming on VVP Measurements[19]

Paper Type	Smoothness, Chapman %	VVP: No Pre-Film	VVP′: Divided Film	VVP′: With Pre-Film	R
100% Rag Bond	10	248	242	5.4	45
Newsprint	18	140	104	2.6	40
Coated Paper	24	34	33	2.8	12
Film Coated #2	42	24	21	2.8	7.5
Cast Coated #1	57	282	16	5.0	3.2
Cast Coated #2	63	208	13	5.6	2.3
Cast Coated #3	71	199	14	5.7	2.5
Cast Coated #4	77	83	4.5	5.4	0.8
Cast Coated #5	81	80	4.4	3.4	1.3

The VVP values obtained with film splitting are considerably lower than those without prefilm, particularly for the cast coated films. The last column is the ratio R obtained by dividing the fourth column by the fifth column. As noted by Wink and co-authors[19] a good correlation is found between R and Chapman smoothness.

In the first paper in this series,[16] the product VVP was justified on the basis of dimensional analysis. However, it arises in solutions of film splitting and forward roll coating processes. One solution of the forward roll coating process[20] shows that the pressure rises prior to mid nip after which it drops and falls to a maximum negative pressure just downstream of mid nip. The maximum negative pressure at this point is given by:

$$p = P_{\max} \frac{(\mu V R^{0.5})}{H_o^{1.5}} \tag{1}$$

where μ, V, R, and H_0 are respectively the viscosity, velocity, roll radius, and mid-nip film thickness. For a Newtonian fluid $P_{max} = -1.2$.

Scott Bond Test: The Scott bond test is shown diagrammatically in Figure 2. In this test a “shoe” is attached to the surface of the paper or board to be tested using a double-sided adhesive tape. The pendulum is released and the “shoe”, a right angle piece of steel, is impacted by it. The geometry is such that the layers of the test sample experience both normal and shear stresses. Now, since the normal or z–direction failure stress is generally lower than the shear failure stress,[14,15] this is expected to be the predominant mode of

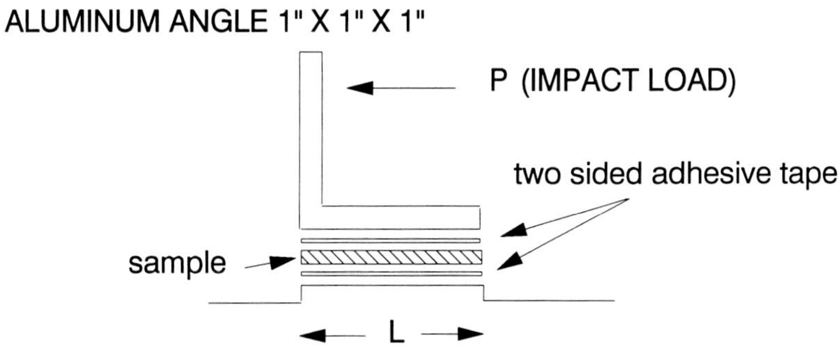

Figure 2. Schematic of Scott bond tester.

failure although the stress situation in this case is more complex. When the pendulum strikes the "shoe" a shear force and bending moment are resisted by the adhesive layer and sample; in effect, the shoe wants to both slide and rotate. The bending moment results in a peeling force and there is progressive failure from the point of maximum bending moment (stress) to the fulcrum.

Under impact conditions the effective viscosity–velocity product will be high, and because failure is by peeling, papers with a high z–direction strength can be tested. This is convenient since no time-consuming methods of preparation are required (*e.g.* as used in the ZDT test[12]). Because of the peeling-type failure associated with the Scott bond test, a poor correlation might be expected between it and the z–direction strength test.

Using a simplified analysis based on secant modulus, it can be shown with reference to Figure 2 that:

$$\sigma_z \max = \left(\frac{3P}{L} \right) \tag{2}$$

where P is the impact load and L is the distance from the impact point to the fulcrum and also the length of the side bonded to the paper. We see that the maximum stress at the initiation of failure is three times the average stress P/L per unit width. However, the instrument does not measure the impact failure load, but rather the amount of (kinetic) energy absorbed by the impact and the rupture process.

Abrasion Testers: Other examples of surface failure include scuffing, resistance to abrasion, surface durability etc. In sheet-fed processes the level of interaction between the surfaces might be such as to cause partial or total disruption of the surface of the sheet as one sheet slides over another. This is not an easy situation to analyze, nor is it easy to measure a meaningful related property. Abrasion testers are used to measure the dry or wet resistance of a surface to abrasion, *i.e.*, a visual assessment of surface damage may be made after so many cycles (or revolutions), or the loss of weight might be used to quantify the damage (see TAPPI T 476).[21]

In somewhat different circumstances paper itself may be abrasive, *e.g.*, when being guillotined or cut or dragged over working surfaces in converting machinery. The abrasive quality of paper can be assessed using an industrial sewing machine. The test consists of measuring the weight loss of metal needles made from alloys of interest after typically 10,000 penetrations. In this case it is mainly the type and amount of filler which contributes to paper's abrasive character.

Linting and Dusting: Linting and dusting can be serious problems in converting operations, and significant effort has been devoted to understanding this problem, as well as measuring a paper's propensity for linting and dusting. Lint and dust can be mainly regarded as cellulose debris and, if present in sufficient quantity, can rapidly lead to poor print quality, web breaks, and significant downtime of the printing press while the press is being washed up.

PHYSICAL PROPERTIES OF PAPER SURFACES

SMOOTHNESS AND ROUGHNESS

Smoothness and roughness are important characteristics of a paper's surface, especially with regard to its printing performance. Roughness can be defined and measured in a variety of ways; more likely than not it may exhibit fractal qualities. As with basis weight or grammage, we may be interested in both large- and small-scale variations in surface topology where each is related to different factors. It should also be noted that surface roughness may have an effect on FT-IR measurements due to changes in scattering at the surface.

Caliper Methods: Figure 3 shows the variation of hard platen caliper (a caliper measurement which relies on contact with the outermost surface or high spots of the paper) with grammage for paper made under identical papermaking conditions. We note that the variation is linear, and that the line does not pass through the origin. The intercept on the ordinate is identified as "surface roughness". Clearly, if there were no roughness the line would pass through the origin.

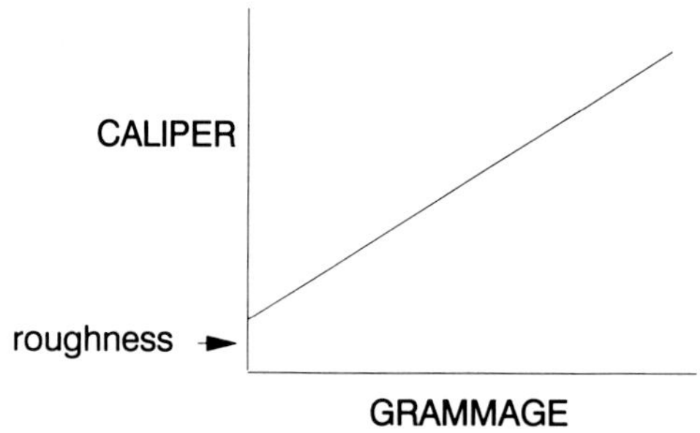

Figure 3. Variation of caliper with grammage.

A linear variation, as shown in Figure 3, implies that "roughness" is invariant with grammage. Factors which contribute to sheet roughness include furnish composition and the surfaces which are impressed upon the web during forming, consolidation, and drying. Therefore, the variation shown in Figure 3 might be expected from handsheets which have been made under constant forming, pressing, and drying conditions. If the handsheets have been plate-dried then they will have one very smooth surface, since during drying the surface of the plate will be replicated with the aid of surface tension forces. A similar effect is found on the paper machine, *e.g.*, smoothing presses, Yankee dryers. Caliper variations and sheet smoothness can generally be improved by calendering; this might involve some combination of hard nip calendering,[22] soft nip calendering, or supercalendering.[23]

A simple caliper measurement, which is largely independent of surface roughness, has been developed by Baum and Wink[24] and is known as the soft platen caliper technique. The technique was originally developed to enable longitudinal out-of-plane ultrasonic velocity measurements to be made by providing a suitable coupling medium and a meaningful caliper measurement.[13]

The difference between the hard and soft platen calipers (HP – SP) can be used as a measure of surface roughness, or can be expressed as percentage roughness, *i.e.*, [(HP – SP) ÷ SP] x 100%. In the development of the soft platen caliper technique Baum and Wink[24] reported good agreement between this technique and other methods (including mercury displacement, bending stiffness and stylus profilometry) for determining a "core" or true caliper of paper.

Liquid Transfer: The Bristow wheel[25,26] is another technique that can be used to measure surface roughness under circumstances where liquid transfer is involved (*e.g.*, printing). A schematic of the method is shown in Figure 4. The technique involves the transfer of liquid to a paper surface from a small reservoir or "headbox" having an opening of length L. The speed of the wheel V is varied, and the amount of liquid transferred into the paper by sorption in time $t = L/V$ is measured. We note that there is generally a positive intercept when the sorption curves are extrapolated back to zero time which represents the amount of liquid transferred to the surface voids as a consequence of surface roughness.

Walker and Fetsko[27] and Walker[28] have analyzed the more complicated ink transfer process relevant to printing, particularly the offset process. The process is outlined in Figure 5. Generally, ink transfer involves film splitting and immobilization.

The amount transferred y is also dependent on the smoothness of the paper. In an ideal situation involving a non-absorbent smooth surface, ink film split is assumed to be half of the amount of ink x applied to the printing plate. In reality, the film split fraction f is dependent on the characteristics of the paper surface, since, just prior to splitting, a portion b of the ink in contact with the paper is immobilized. It should be noted that a

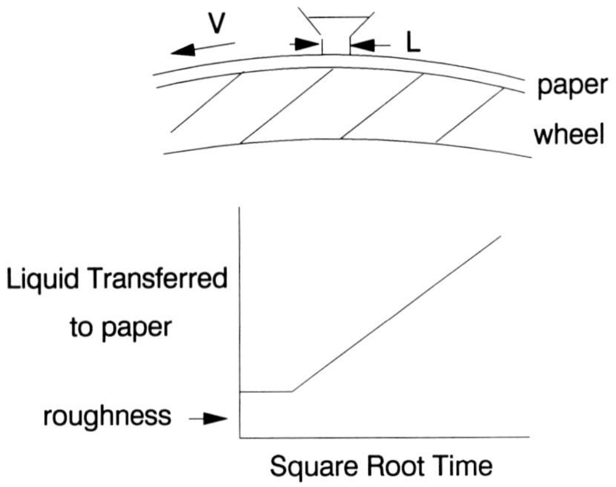

Figure 4. Schematic of Bristow wheel.

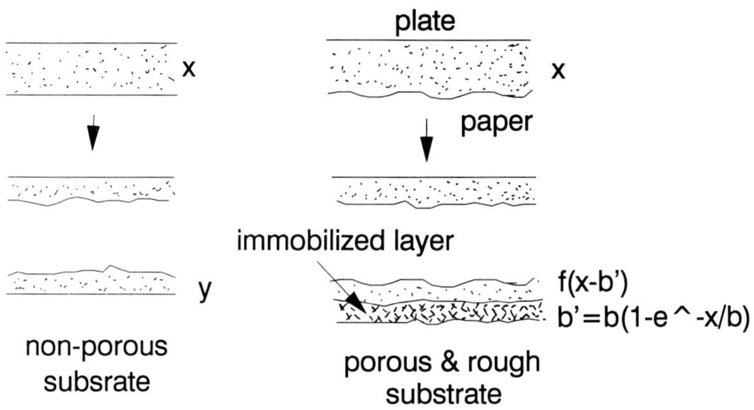

Figure 5. Ink transfer process.

printing ink consists of a vehicle, pigment, and other additives, and preferential absorption of the vehicle into the paper is responsible for immobilization of part of the ink film transferred to the paper. Walker and Fetsko also accounted for surface smoothness which effectively limits ink transfer by incomplete contact of the ink with the surface of the paper. This is accounted for by a paper smoothness parameter k and an exponential function involving the amount of ink applied to the plate x, *i.e.*, $(1 - e^{-kx})$. The extent to which the ink film is immobilized will also depend on surface smoothness and the amount of ink applied to the plate. The Walker–Fetsko (WF) ink transfer equation taking into account these factors is given as follows:

$$y = (1 - e^{-kx})[b(1 - e^{-x/b})(1 - f) + fx] \tag{3}$$

For relatively smooth papers this equation has a well defined asymptote for large values of x given by:

$$y = b(1 - f) + fx \tag{4}$$

Therefore, the parameters b and f can be obtained from the intercept and slope of the line respectively. Using the values of b and f and small values of x and y the above equation can be solved for k. Since this is not a very good procedure for uncoated papers Walker[28] has suggested better procedures for obtaining k. Smoother papers have larger values of k.

Other developments and modifications of the WF equation have occurred, and more recently Mangin[29] has developed a procedure for calculating the effective surface roughness R_g under actual printing conditions using the ink coverage factor embodied in a modified form of the Walker–Fetsko equation. He also showed that the distribution of surface pores is similar in form to the ink coverage factor equation. His equation for printing surface roughness is given below:

$$R_g = \left[\frac{6(1 - F_o)}{k^{3\Gamma}}\right]^{0.33} \tag{5}$$

The three parameters k, Γ and F_0 describing the pore shape function of the paper in the printing nip are calculated from the above ink transfer equations where the ink coverage factor $(1 - e^{-kx})$ is replaced by the more complex factor:

$$[1 - (1 - F_o)\, e^{-(kx)\Gamma}] \tag{6}$$

Evaluation of these three parameters was performed by Mangin[29] under printing conditions where both press pressure and speed were varied. As expected, an increase in pressure reduced printing surface roughness. Furthermore, the correlation with other roughness measurements (air leak method) was only good for a fixed printing pressure. For uncalendered samples the printing surface roughness was not as great as expected from Parker Print Surf (PPS) measurements due to effective calendering within the printing press nip.

Paper surface roughness compressibility K' has been defined by Mangin[29] following the work of Bristow[4,30] as follows:

$$K' = \frac{-dR}{d(\log P)} \tag{7}$$

which makes it effectively independent of press pressure. However, it does vary with press speed and only effectively correlates with PPS measurements for a fixed press speed.

As already noted, one difficulty associated with caliper and roughness measurements is that paper and board are compressible and viscoelastic. Both hard and soft caliper measurements are usually made at 50 kPa. We expect that paper will appear to be smoother as the compressive stress used to measure caliper is increased (this is the major difficulty with stylus measurement techniques). In fact, the Chapman smoothness method for measuring the degree of surface roughness under pressure relies on the change in optical contact between a glass plate and the paper's surface.

Graphite Tester: Differences in paper compressibility, conformability, and formation are readily seen with Beri's graphite tester.[31] In this test a print is made with a graphite bar which is drawn across the surface of the paper under a uniform load. When the paper is supported on a hard backing material the graphite print is similar to a beta radiograph[32] used to measure the mass density variation in paper, and therefore is considered to be directly related to the mass density formation of the sheet. When the backing material is relatively soft and conformable, *e.g.*, an offset blanket, the appearance of the graphite print alters dramatically and is much more uniform as illustrated in Figure 6. The graphite print can be readily analyzed using image analysis techniques. This method is similar in principle to that developed by Mangin[29] in as much as the effective roughness of the sheet is obtained under conditions as close as possible to that which the sheet undergoes while being printed.

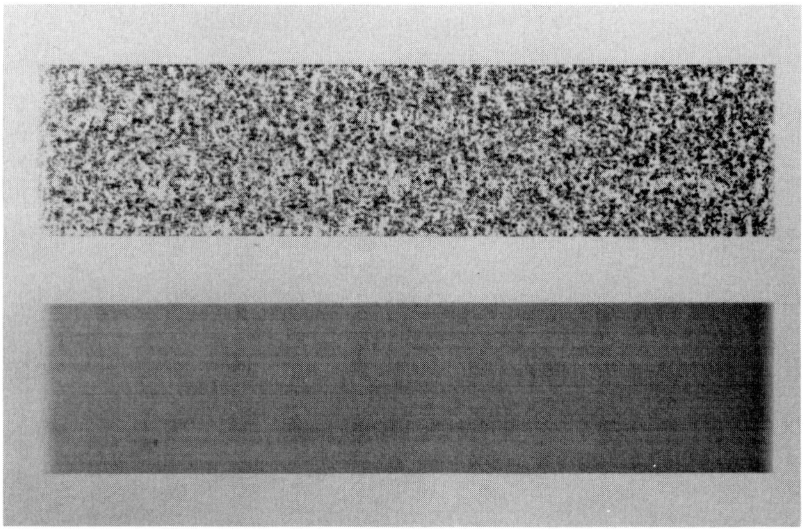

Figure 6. Graphite prints made with hard and soft backing. Hard backing print (top) is much less uniform.

With the ink transfer method, the effects of the fountain solution on roughness could, presumably, be accounted for, while this would not be possible with the graphite method.

As sheet formation deteriorates so might sheet smoothness. In fact it has been found that larger scale roughness is associated with formation whereas smaller scale roughness is largely independent of formation.

Air Leak Methods: The air leak principle is embodied in a number of instruments for measuring surface roughness including the Bekk, Sheffield, Bendtson, and Parker Print Surf. The measuring heads of these instruments are shown schematically in Figure 7. The more recent of these instruments is the Parker Print Surf[33] and is considered to be the most appropriate for measuring printing related roughness, although this must be seen in the light of Mangin's research which has already been discussed. The land of the Parker Print Surf is much smaller than the other instruments, *i.e.*, of the order of a half tone dot (50 μm), and the pressure and backing material can be changed according to the type of printing process for which the paper is intended.

FRICTION

Friction is of major importance in many engineering applications, including wear and energy consumption. Friction is also of importance in many converting and end use applications of paper including roll building, sheet-fed processes, corrugating, guillotining and cutting operations. The conversion from acid to alkaline papermaking has resulted in slipperier surfaces, and as a consequence roll and related converting problems have been reported.[34]

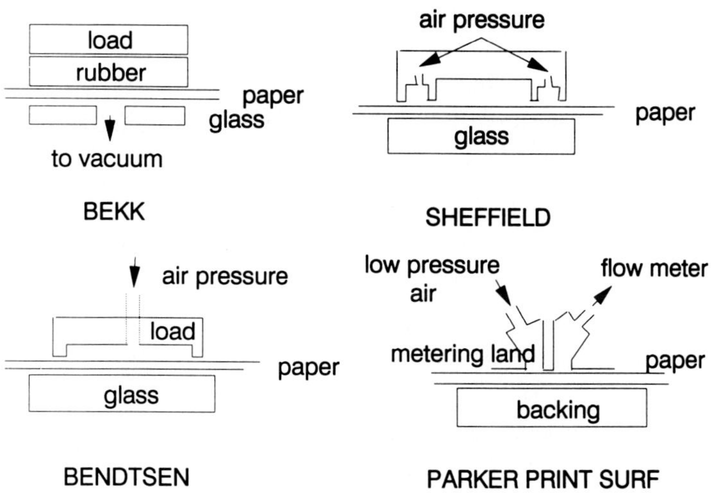

Figure 7. Air leak-type smoothness testers.

Roll quality is dependent on the radial and tangential stress distribution within the roll. The initial stress distribution is independent of the friction characteristics of the paper. However, if the friction level is not high enough, then there will be a change in stress distribution due to slippage resulting in roll defects, *e.g.*, telescoping. In contrast, too high a coefficient of friction in sheet-fed processes can result in jamming and surface damage.

Corrugating or flute formation in medium is another converting operation where friction is important.[35] In hot corrugating the coefficient of friction of the medium is significantly reduced, allowing the medium to be corrugated without flute fracture as well as minimizing highs and lows. The friction coefficient (μ) of the medium at room temperature can vary widely, but if sufficiently low, *i.e.*, $\mu < 0.3$, then the medium could be cold-corrugated.[36] This can be achieved by applying a very thin film of wax such that it does not interfere with the single face bond between the liner and the medium. However, it should be noted that cold-corrugating has not yet been commercialized.

The coefficient of friction μ is defined as follows:

$$\mu = F / W \tag{8}$$

where F is applied force parallel to the plane on which the body is resting and W is the force normal to this plane. This is known as Amontons's law and is independent of the size of the body. Of general interest is the friction behavior of paper to paper, as well as paper and other materials, *e.g.*, metals, plastics.

The coefficient of friction may be measured statically or dynamically (kinetic) for paper and board using TAPPI-recommended procedures[21] T 548 pm–90 (inclined plane method), T549 pm–90 (horizontal plane) for uncoated printing and writing papers; and T 816 om–92 (inclined plane method) and T 816 om–92 (horizontal plane) for corrugated and solid fiberboard respectively.

The static coefficient of friction is calculated from the force required to just move the body, while the kinetic coefficient of friction is calculated using the force just required to keep the body in steady motion along the plane. The static coefficient of friction may be measured using either the horizontal or the inclined plane method. Using the latter, the coefficient of friction is conveniently calculated from the tangent of the angle of incline at which the body just begins to move. In the case of paper, the sample to be tested is attached to a sled, and the coefficient of friction is determined from the tangent of the angle at which the sled just begins to move.

Variables Affecting the Friction Behavior of Paper: An interesting review of paper-to-paper and paper-to-metal friction has recently been made by Back.[37] In terms of mechanisms both surface chemistry and mechanical factors are important. According to Tabor,[38] the main factors involved are: *i*) the true contact area; *ii*) the strength of the interfacial bond; and *iii*) deformation behavior when interfacial bonds are broken during sticking. For paper this implies that the surface structure, *i.e.*, its roughness and roughness anisotropy, may be important. Differences in coefficient friction are therefore expected between the machine and cross machine directions, wire and felt sides of the sheet. Calendering and moisture content are also expected to change the coefficient of friction. Since paper and coatings are viscoelastic and their behavior is time-dependent, friction behavior involving stick–slip phenomena may be important.

The importance of surface chemistry has already been mentioned. The presence of soaps and fatty acids in paper and board and their migration to the surface can greatly affect the coefficient of friction. Inoue and co-workers[39] and Gurnagul and co-workers[40] have examined the role of surface chemistry in more detail. The latter showed that the coefficient of friction linearly increases with the surface oxygen/carbon ratio (measured by ESCA). Oxidation of the surface by corona discharge also increased the coefficient of friction. Inoue *et al.'s*[39] study of linerboard static friction determined that the coefficient of friction increased with surface free energy, from which they concluded that adhesion at the linerboard–linerboard interface is important. Colloidal silica was found to increase surface free energy and, therefore, the friction coefficient. Inoue *et al.* speculated that contamination may be responsible for low coefficients of friction in a mill environment. Withiam[41] in a study on the effect of fillers on paper friction concluded that sheet friction increases as pigment surface area and porosity increase.

REFERENCES

1. Dodson, C.T.J. 1990. A universal law of formation. *J. Pulp and Paper Sci.* 16(4): J136.
2. Waterhouse, J.F. 1993. The effect of some papermaking variables on formation. *Tappi J.* 76(9):129.
3. Waterhouse, J.F. Stera, and Brennan, D. 1987. Z–Direction variation of internal stress and properties of paper. *J. Pulp and Paper Sci.* 13(1):J33–J37.
4. Bristow, J. A. The paper surface in relation to the network. *In*: Paper – Structure and Properties. Bristow, J.A. and Kolseth, P., eds. Marcel Dekker, Inc. New York. 1986.
5. Skowronski, J. and Lepoutre, P. 1985. Water–paper interaction during paper coating – Changes in paper structure. *Tappi J.* 68(11):98–102.
6. Evanoff, P.C., Gerlach, W. and Lyne, M.B., eds. 1993. Surface Strength Terminology. Second edition revised by J. Lind. TAPPI Press.
7. Lepoutre, P. Huynh, K., and Robertson, A.A. 1980. The setting of water soluble adhesives on paper – The effect of adhesive thickness and pressure. *Svensk Papperstidning* 83(18):520–24.
8. Lepoutre, P. 1986. Adhesion in paper products. *Svensk Papperstidning* (1):20–29.

9. Fellers, C. 1977. Procedure for measuring the interlaminar shear properties of paper. *Svensk Papperstidning* 80(3):89.
10. Byrd, V.L., Setterholm, V.C. and Wichmann, J.F. 1975. A method for measuring the interlaminar shear properties of paper. *Tappi J.* 58(10):132.
11. Waterhouse, J.F. 1984. Out-of-plane shear deformation behavior of paper and board. *Tappi J.* 67(6):104–108.
12. Wink, W.A. and Van Eperen, R. 1967. Evaluation of z-direction tensile strength. *Tappi J.* 50(8):393–400.
13. Habeger, C.C. and Wink, W.A. J. 1986. Ultrasonic velocity measurements in the thickness direction of paper. *Appl. Polymer Sci.* 32:4503–40.
14. Stratton, R.A. 1993. Characterization of fibre–fibre bond strength from out-of-plane paper mechanical properties. *J. Pulp and Paper Sci.* 19(1):J6–J12.
15. Waterhouse, J.F. 1991. The failure envelope of paper when subjected to combined out-of-plane stresses. *In*: Proc., Int. Paper Physics Conference, Kona, Hawaii. TAPPI Press, Atlanta, GA.
16. Staff, Institute of Paper Chemistry. 1946. *Paper Trade J.* 123 (18):24,26,28,29. (Oct. 31).
17. Staff, Institute of Paper Chemistry. 1946. *Paper Trade J.* (19):24,26,28,30,32,34. (Nov. 7).
18. Wink, W.A., Clinton, T.J.,Thickens, R.W. and Van den Akker, J.A. 1952. Determination of the bonding strength of paper. IV. Progress report on the development of a new method and instrument. *Tappi* 35(9):181A.
19. Wink, W.A., Shillcox, W.M., Van Eperen, R.H. 1957. Determination of the bonding strength of paper. IV. Progress report on the development of a new method and instrument. *Tappi* 40(7):181A.
20. Middleman, S. 1977. Roll coating. *In*: Fundamentals of Polymer Processing. McGraw–Hill Book Co., New York. pp. 191–196.
21. TAPPI Standards, 1994 – 1995. TAPPI Press, Atlanta, GA.
22. Crotogino, R.H. 1981. Machine calendering–recent advances in theory and practice. *Pulp and Paper Mag. Canada* Trans. Dec TR-75–87.
23. Peel, J.D. 1989. Recent developments in the technology and understanding of the calendering process. *In*: Fundamentals of Papermaking, Trans. 9th Fundamental Res. Symp. Cambridge, Vol. 2. Baker, C.F. and Punton, W., eds. p. 951.
24. Wink, W.A. and Baum, G.A. A rubber platen caliper gauge – A new concept in measuring paper thickness. 1983. *Tappi J.* 66(9):131.
25. Bristow, J.A. 1965. Water vapor transmission rates through paper and board. *Svensk Papperstidning* 68(23):834.
26. Lyne, M.B., Aspler, J.S. 1982. Wetting and the sorption of water under dynamic conditions. *Tappi J.* 65(12)98–101.

27. Walker, W.C. and Fetsko, J.M. 1955. Measurements of ink transfer in the printing of coated papers. *American Ink Maker* 33(12):38,42,44,69,71.
28. Walker, W.C. 1981. Determination of Ink Transfer Parameters. *Tappi J.* 64(5):71.
29. Mangin, P.J. and Geoffrey P. Printing roughness and compressibility: A novel approach based on ink transfer. *In*: Fundamentals of Papermaking, Trans. 9th Fundamental Res. Symp. Cambridge, Vol. 2. Baker, C.F. and Punton, W., eds. p. 951.
30. Bristow, J.A. 1982. The surface compressibility of paper. *Svensk Papperstidning* R127–R130.
31. Beri, Y. 1986. Graphite tester, conceptual development. *American Inkmaker*, pp. 29–37, July.
32. Tomimasu, H., Kim, D., Suk, M., and Luner, P. 1991. Comparison of four paper imaging techniques: ß–radiography, electrography, light transmission, and soft x–radiography. *Tappi J.* 74(7):165–176.
33. Parker, J.R. 1981. The measurement of printing roughness. *Tappi J.* 64(12):56.
34. Roisum, D.R. 1994. Interlayer slippage. *In*: The Mechanics of Winding. TAPPI Press, Atlanta, GA.
35. Whitsitt, W.J. 1987. Runnability and corrugating medium properties. *Tappi J.* 70(10):99–103.
36. Sprague, C.H. 1979. Cold corrugation – Development and commercial application. *Tappi J.* 62(6):45–48.
37. Back, E.L. 1991. Paper to paper and paper to metal friction. *In*: Proc., 1991 Int. Paper Physics Conference, Kona, Hawaii, Vol. 2. TAPPI Press, Atlanta, GA.
38. Tabor, D. J. 1981. *Lubrication Tech.* 103(4):169.
39. Inoue, M., Gurnagul, N. and Aroca, P. 1990. Static friction of properties of liner-board. *Tappi J.* 73(12):81–85.
40. Gurnagul, N., Ouchi, M.D., Dunlop-Jones, N., Sparkes, D.G., and Wearing, J.T. 1992. Factors affecting the coefficient of friction of paper. *J. Appl. Polymer Sci.* 46(5):805–814.
41. Withiam, M.C. 1991. Effect of fillers on paper friction properties. *In*: Proc., 1991 TAPPI Int. Papermakers Conference. TAPPI Press, Atlanta, GA. pp. 341–347.

4

The Surface Chemistry of Paper: Its Relationship to Printability and Other Paper Technologies

Frank M. Etzler and James J. Conners

Boehringer-Ingelheim Pharmaceuticals
Ridgefield, Connecticut, U.S.A.

INTRODUCTION

Little attention has been given to the surface chemistry of paper and its effect on converting operations. Two of the most important converting operations are gluing and printing. Here, our discussion will focus primarily on printing issues; however, the topics addressed are easily applied to gluing. Both of these operations, as will be seen from the discussion below, are quite closely related from a surface chemical point of view.

Environmental concerns have recently forced industry to turn away from traditional oil- and solvent-based ink formulations to so-called aqueous-based formulations. Traditional formulations use nonpolar materials which have low surface tensions. Such liquids spread over, penetrate into, and adhere to a wide range of materials. Because the formulations are essentially hydrocarbon-based, they have very similar surface chemistries and thus behave alike with regard to wetting behavior. Again, traditional ink formulations are tolerant to a wide variety of surface chemistries due to their low surface tension and due to the lack of polar interactions. In contrast, so-called aqueous based formulations are much more sensitive to the surface chemistry of paper as they have both higher surface tensions and interact with the paper through polar interactions which, as illustrated below, may be repulsive in nature.

In this chapter we discuss those physico-chemical concepts required for better understanding of the interaction of fluids with paper and show with experimental data how theory may be used to address various technological issues of interest to the paper and ink maker.

0-8493-8992-5/95/$0.00+$.50

FUNDAMENTALS

SURFACE FREE ENERGY AND CONTACT ANGLE

A more detailed discussion of the topics discussed in this section can be found in Adamson.[1] Here we provide a brief introduction. The differential energy expression, derived from the first and second law of thermodynamics, which describes the Gibbs free energy of a system is:

$$dG = -S\,dT + v\,Dp + \sigma\,dA + \sum_i \mu_i\,dn_i \tag{1}$$

Here G is the Gibbs free energy; S, the entropy; σ, the surface tension; A is area; and μ_i, the chemical potential of the ith component. P, V, and T have their usual thermodynamic meaning.

Surface tension is defined as:

$$\sigma \equiv \left(\frac{\delta G}{\delta A}\right)_{T,P,n_i} \tag{2}$$

and the surface free energy, g^s, is:

$$g^s = \sigma + \sum_i \Gamma_i \mu_i \tag{3}$$

Here Γ_i is the Gibbs surface excess. For positive values of the surface excess, the component in question is preferentially selected into the interfacial zone. Bulk and surface concentrations are thus not generally equal. A more detailed discussion of surface excess is given by Adamson.[1] The term surface tension is often reserved for the air-material interface and interfacial tension for interfaces of other types.

As dictated by thermodynamics, processes such as spreading, penetration, and adhesion require ΔG to be negative if these processes are to be spontaneous. Equation (1) and those equations which directly follow from it indicate that interfacial tension is a critical parameter for determining the spontaneity of the above processes.

The principal equation which determines the extent of spreading of a liquid over a solid is the Young equation.

$$\gamma_{SV} - \gamma_{SL} = \gamma_{LV}\cos(\theta) \tag{4}$$

Here γ is the interfacial free energy (surface tension), the subscripts SV, SL and LV refer to the solid–vapor, solid–liquid and liquid–vapor interfaces respectively, and θ is the contact angle.

When a drop of liquid is placed on a solid material, the drop spreads until the interfacial forces are balanced. At this point, the Young equation is obeyed. The contact angle, θ, is defined by the intersection of the liquid–vapor and solid–vapor interfaces. Contact angle is related to thermodynamic quantities through the Young equation.

SPREADING AND SURFACE FREE ENERGY

The free energy associated with spreading of a liquid over a surface is given by the expression below:

$$dG = \frac{\partial G}{\partial A_1} dA_1 + \frac{\partial G}{\partial A_2} dA_2 + \frac{\partial G}{\partial A_{12}} dA_{12} \tag{5}$$

where

$$dA_2 = -dA_1 = dA_{12} \quad and \quad \gamma_i = \frac{\partial G}{\partial A_i}$$

The coefficient, $-\partial G/\partial A_2$, gives the free energy change for the spreading of liquid *2* over surface *1*. The spreading coefficient, $S_{2/1}$, for spreading liquid *2* over material *1* is given by the expression below.

$$S_{2/1} = \gamma_1 - \gamma_2 - \gamma_{12} \tag{6}$$

If $S_{2/1}$ is positive, then the free energy change is negative and spreading is spontaneous. Spreading is thus controlled by the surface chemistry of the materials in question.

CAPILLARY PENETRATION AND SURFACE FREE ENERGY

A pressure gradient exists across the meniscus of a liquid contained in a cylindrical capillary. The pressure difference is given by the Kelvin equation.

$$\Delta P = \frac{2\gamma_{LV} \cos(\theta)}{r} \tag{7}$$

Combining the above relation with the Young equation yields:

$$\Delta P = \left(\frac{2(\gamma_{SV} - \gamma_{SL})}{r} \right) \tag{8}$$

Note that ΔP can be maximized by minimizing γ_{SL}. The pressure gradient given by the Kelvin equation serves as the driving force for liquid penetration into porous media such as paper. Washburn[2] has developed a well known equation for the rate of liquid penetration into porous media. The equation has been the subject of discussion in many works over the last several decades. The Washburn equation for liquid penetration into cylindrical capillaries where gravity effects can be ignored is:

$$\frac{dl}{dt} = \frac{r \gamma_{LV} \cos(\theta)}{4 \eta l} \tag{9}$$

Here η is the liquid viscosity and l is the depth of penetration. For $\theta > 90°$, dl/dt is negative indicating that liquid does not spontaneously penetrate into small capillaries. For maximal liquid penetration, good wetting ($\theta \rightarrow 0°$, $\cos(\theta) \rightarrow 1$) must be achieved.

ADHESION AND SURFACE FREE ENERGY

Surface free energies of materials also determine their adhesive properties.[1,3] The ideal work of adhesion between surfaces *1* and *2* is given below.

$$W_{12} = \gamma_1 + \gamma_2 - \gamma_{12} \tag{10}$$

Combining the above equation with Young's equation results in the following expression.

$$W_{12} = \gamma_2 [1 + \cos(\theta)] \tag{11}$$

The ideal work of adhesion applies to adhesion between atomically smooth surfaces. Real surfaces are rarely atomically smooth; thus, W_{12} is much larger than empirically determined adhesion strengths. The relationships discussed here are nonetheless important. These issues are:

- Good wetting is important for good adhesion.
- Contact angles less than 90° are required for good adhesion.

Real adhesion differs from ideal adhesion. This difference results from roughness of the surface. Surface roughness on a dimension greater than atomic sizes results in reduced area of contact and hence reduced adhesion strength. Excessive capillary penetration (usually caused by too low a viscosity) of an adhesive may result in weak adhesion as the adhesive is absorbed into the substrate and away from the bonding interface.

For a liquid adhesive to achieve the maximum area of contact, the liquid must wet the surface (zero contact angle). Adhesive strength depends on:

- Compatible surface energies such that zero contact angle is achieved.
- Adequate contact of adhesive and substrate: liquid adhesives must flow into crevices to maximize area of contact (see also Yasuda *et al.*[4]).
- Cohesive energy of the adhesive must be adequate.

ESTIMATION OF CONTACT ANGLE, INTERFACIAL ENERGIES, AND ADHESION STRENGTHS

From the above discussion it can be seen that physical processes involved in printing, namely spreading, penetration and adhesion, are related to the surface chemistry of the paper and ink. More specifically, the magnitude of γ_{12}, the interfacial tension between paper and ink, must be known.

Zisman Approach: The calculation of solid–liquid interfacial tensions has been a topic of interest for a number of decades. The early work of Zisman and co-workers (*e.g.*, Zisman[5]) is notable. Zisman noted for a homologous series of series of liquids that cos(θ) had a nearly linear dependence with liquid surface tension. The proposed function is:

$$\cos(\theta) = 1 - \beta(\gamma_L - \gamma_c) \tag{12}$$

The intercept at cos(θ)=1 is referred to as the Zisman critical surface tension. The critical surface tension has been considered as an approximation of the solid surface tension as many homologous series of liquids give approximately the same value. As β = 0.03–0.04 in many instances, the Zisman equation provides a method for estimating contact angles.

Evaluating the Young equation (Equation (4)) at cos (θ)=1 gives:

$$\gamma_c = \gamma_{SV} - \gamma_{SL} \tag{13}$$

Thus, γ_c is not the solid–vapor interfacial tension as $\gamma_{SL} \neq 0$.

The use of so-called "dyne solutions" for monitoring the surface tension of paper is based on the arguments of Zisman. In this approach, the alcohol–water or lactic acid–water solutions are used to determine the Zisman critical surface tension. If one is using inks which are indeed essentially alcohol–water mixtures, this approach gives useful information. The Zisman approach, however, does not easily give information on the details of Lewis acid–base interactions at the fluid–solid interface. The so-called "dyne solution" may thus be inadequate to describe the wetting behavior in a particular system.

Good–Girifalco–van Oss Approach: Good and co-workers[6,7] have offered another approach for estimation of contact angles. Although this approach was first suggested more than 30 years ago, it remains as a forefront topic in surface and colloid science.

The calculation of interfacial tension is based on the axiom which states that the total surface tension can be expressed as follows:

$$\gamma_T = \sum_i \gamma_i \tag{14}$$

where the γ_i result from specific classes of intermolecular forces. Separation of the surface energy components is possible using contact angle measurements employing appropriate probe liquids.

Good and Girifalco[6] more than thirty years ago suggested division of surface energy into polar and non-polar (or dispersion) components. Thus

$$\gamma_T = \gamma^d + \gamma^p \tag{15}$$

Here γ^d is the surface free energy resulting from London dispersion forces. London forces result from the polarizability of electron orbitals and are common to all intermolecular interactions. The term γ^p is the surface free energy resulting from Lewis acid–base (polar) interactions. According to Good–Girifalco theory

$$\gamma_{12} = \gamma_1 + \gamma_2 - 2\sqrt{\gamma_1^d \cdot \gamma_2^d} - 2\sqrt{\gamma_1^p \cdot \gamma_2^p} \tag{16}$$

More recently, van Oss[7] and Fowkes *et al.*[8,9] have suggested separation of the polar component into separate Lewis acid and Lewis base terms. Thus

$$\gamma^p = 2\sqrt{\gamma^+ \cdot \gamma^-} \tag{17}$$

According to the above model the polar contribution to the interfacial tension between the two surfaces is given by the following:

$$\gamma_{12}^p = \gamma_{12}^{AB} = 2\left[\sqrt{\gamma_1^+ \cdot \gamma_2^+}\right] \cdot \left[\sqrt{\gamma_1^- \cdot \gamma_2^-}\right] \tag{18}$$

Good's original equation can be combined with the Young equation to calculate contact angles. The resulting equation is

$$1 + \cos(\theta) = \left(\frac{1}{\gamma_1}\right) \cdot \left[2\sqrt{\gamma_1^d \cdot \gamma_2^d} + 2\sqrt{\gamma_1^p \cdot \gamma_2^p}\right] \tag{19}$$

Accounting for Lewis acid–base interactions the above equation becomes:

$$1 + \cos(\theta) = \left(\frac{1}{\gamma_1}\right) \cdot \left[2\sqrt{\gamma_1^d \cdot \gamma_2^d} + 2\left\{\sqrt{\gamma_1^+} - \sqrt{\gamma_2^+}\right\}\left\{\sqrt{\gamma_1^-} - \sqrt{\gamma_2^-}\right\}\right] \tag{20}$$

Complete wetting occurs when $\theta = 0°$ or $1 + \cos(\theta) = 2$. Non-wetting occurs when $\theta > 90°$ or $1 + \cos(\theta) < 1$. Note that Equation (19) differs significantly from Equation (20).

When the Lewis acid–base terms are not explicitly separated, polar interactions always decrease the contact angle and thus improve wetting. In contrast, when the Lewis acid and base terms are separated, one finds that polar interactions may either increase or decrease the contact angle. Lower contact angles are observed between surfaces and liquids which have opposite acid–base chemistries.

Practical Considerations: The calculations by Good,[6] Fowkes[8,9] and van Oss[7] are of considerable practical importance to the topic discussed here. It can be seen that a sizing test which only uses a single contact angle measurement (usually water) is inadequate for characterization of the surface chemistry as such a measurement does not clearly or uniquely describe the acid–base chemistry between ink and paper. Furthermore, simple tests which use so called "dyne solutions" and necessarily employ Zisman's approach are also generally inadequate for description of wetting a polar surface such as paper with polar or aqueous based inks, as an understanding of Lewis acid–base chemistry is critical for the understanding of wetting behavior. Zisman's approach, however, may be adequate for description of wetting of papers by inks which largely interact with the paper via London forces or for occasions when the selected solutions closely match the significant surface chemistry of the ink formulation.

Van Oss[7] has noted that $\gamma^- >> \gamma^+$ for a large number of polymers. This observation suggests that very careful attention to the acid–base surface chemistry of aqueous based ink formulations will be necessary to avoid non-wetting through repulsive polar interactions. The present authors are unaware of any studies which address this issue.

THE EFFECT OF SURFACE MODIFICATION ON WETTABILITY

Whitesides[10,11] has studied the effects of particular surface groups on the wettability of polyethylene films. Such films were chosen, in part, for their relative ease of modification. As discussed previously, the detailed surface chemistry of a material can dramatically affect the ability of a particular liquid to wet a given solid. For example, the contact angle of an aqueous solution on polyethylene is quite high ($\theta = 103°$) as polar liquids do not wet nonpolar solids. In contrast, functionalization of polyethylene with surface carboxylic acid groups allows the aqueous solution to wet the modified polyethylene. It is important to note that when only γ^p and γ^d are considered, the calculation of the contact angle and interfacial free energy by the method of Good and Girifalco gives incorrect results. It is necessary to consider specific Lewis acid–base interactions between the liquids and the solid substrate. Therefore, the wetting of the carboxylic acid functionalized polyethylene by aqueous solutions is, in fact, pH-dependent.

Whitesides and coworkers[10,11] have examined the effect of specific surface chemical modifications of polyethylene films on the wettability of these films by aqueous solutions. Acidic, basic, polar, and nonpolar groups have been incorporated on the surface of the polyethylene to explore the interactions between probe solution and surface. As expected, if the group incorporated onto the surface is less polar than the methylene group (*e.g.*, fluorinated ether), the contact angle of water on the surface will increase dramatically (from 103° to approximately 140°). In contrast, increasing the polarity of the surface with the presence of primary and secondary alcohol groups at the surface decreases the contact angle of an aqueous solution from 103° to approximately 60°. In this case, the contact angle is independent of pH because the alcoholic moieties at the surface are not ionizable. The contact angle of water is pH-dependent when ionizable groups, such as amines or carboxylic acids, are incorporated onto the surface of the polyethylene. Interestingly, the contact angle of the aqueous solution is always less than that on the surface of nonfunctionalized polyethylene due to the increased polarity of the group even in the unionized form.

The chemical derivatization methods for polyethylene used by Whitesides and coworkers[10,11] do not appear to be suitable for industrial use although the lessons they teach are valuable. Surfaces of industrial materials, such as paper and polymers, can be modified in a continuous process using corona and plasma discharge techniques.[12,13,14] These techniques have been used to change the surface characteristics of paper and polymeric materials (*e.g.*, polyethylene) to improve the adhesion between these two materials. The adhesion between these two materials is not very good without corona pretreatment as the polyethylene surface tends to be quite hydrophobic (contains methylene groups), while the cellulosic material has a polar surface. Corona treatment of these materials results in the oxidation of the surface which in turn increases the polar component of the surface free energy. As polyethylene is initially of very low polarity, corona treatment of the polyethylene to increase its surface polarity is more dramatic than is functionalization of the cellulose.

Plasma treatment of a surface can be used within limits to customize the surface chemistry of materials. Depending on the gas used the surface can be made more polar or less polar. Also acidic or basic functional groups can be incorporated onto the surface using these techniques. For example, an air or oxygen plasma reacting with polyethylene oxidizes the surface, resulting in the presence of carbonyl, alcohol, carboxylic acid, and other oxygen containing groups. The presence of these groups significantly alters the wetting properties of polyethylene by polar liquids such as water.

The oxidation of polyethylene surfaces by plasma or corona discharge may improve the adhesive qualities of polyethylene due to the change in surface chemical interactions. In fact, this change in surface chemistry due to the corona or plasma treatments makes it possible to print directly on to the surface of polyethylene.

The experiments by Whitesides and co-workers[10,11] show clearly the relation between the details of surface chemistry to the wettability of surfaces. The surface modification techniques introduced by Whitesides as well as corona and plasma treatment suggest experimental methods for investigation of the effect of surface chemistry on printing and other industrially important processes. Again, it remains an important task to choose furnishes and ink formulations which give optimal performance.

MEASUREMENT TECHNIQUES

MEASUREMENT OF SURFACE CHEMISTRY BY XPS

Of the many spectroscopic techniques developed for the analysis of surfaces, one of the more suitable techniques for studying the surface of paper and other non-metallic materials is XPS (X-ray Photoelectron Spectroscopy) which is also referred to as ESCA (Electron Spectroscopy for Chemical Analysis) (see Adamson[1]). XPS uses monoenergetic x-radiation to eject inner shell electrons (*e.g.*, 1s, 2s, 2p). The energy of the ejected electrons is measured by an electron spectrometer. The energy of the electrons ejected from the *j*th orbital, E_j is given by the Planck equation.

$$E_j = h\upsilon_0 - E_i \tag{21}$$

where $h\,\upsilon_0$ is the energy of the x-radiation and E_i is the energy of the *i*th orbital. Because x-rays can penetrate deep into materials, the x-ray beam may strike the surface at a grazing angle so that only surface atoms are studied.

XPS is more sensitive than Auger spectroscopy. Both the identity and oxidation state of the elements may be determined. When a given element exists in more than one oxidation state, XPS can be used to determine the respective amounts of each state.

Whitesides[10,11] has shown that the specific chemical nature of the interface affects the measured contact angle. For instance, ionization of surface groups which may behave as acids or bases produce contact angles with aqueous solutions which are strongly pH dependent. XPS provides a way to understand how specific surface groups affect contact angle. XPS is discussed further in Chapter 11 in this volume.

MEASUREMENT OF SURFACE FREE ENERGY AND CONTACT ANGLE

Several methods for measuring surface tension are well known. With regard to the present work, only a few of these methods are likely to be suitable for learning about the surface free energy of paper. Furthermore, only a few methods exist for measuring the surface tension components. Here we outline a few methods which may be suitable for measuring some of the relevant quantities for inks and papers.

The maximum bubble pressure method (see Adamson[1]) can be used to determine the surface tension of inks. In this method, a bubble of inert gas is made at the orifice of a capillary. Under ideal conditions, the maximum pressure occurs where the radius of the bubble is equal to the outside diameter of the capillary. In practice a correction factor must be applied to the calculation to compensate for the asphericity of the drop. This method can be used to determine the surface tension of fluid inks but cannot be used to determine polar and non-polar contributions to the surface tension. As bubbles may be formed at various rates, the dynamic surface tensions of inks may be measured by this method. Aqueous surfactant solutions require fairly long periods to reach surface equilibrium. This long equilibrium time results from the time required for large surfactant molecules to diffuse from the bulk solution to the newly created surface. Solutions containing only small molecules reach equilibrium quickly. In general, the surface tension declines exponentially from the surface tension of water to the value of the equilibrium surface tension over time. Dilute surfactant solutions may take as much as a few hours to reach equilibrium. Times of the order of milliseconds have been observed in inks.

The Wilhelmy plate method[1,15] is one of the more useful methods as the instrumentation is relatively inexpensive, experiments can usually be performed quickly and the method is suitable to measure the surface tensions of liquids and solids which can be formed into sheets as is the case for paper. The method can be used to determine both contact angles and surface tension components. No corrections to the "ideal case" are usually required.

A thin plate, such as a microscope cover slide or thin sheet of platinum, will support a meniscus whose weight, W_{tot} - W_{plate}, is given by the following equation:

$$W_{tot} = W_{plate} + \gamma p \tag{22}$$

where γ is the surface tension of a wetting fluid and p is the perimeter of the plate.

The Wilhelmy plate may be implemented in a number of ways. It is possible to use the Wilhelmy plate in a detachment measurement as in the DuNouy ring.[1] The detachment force gives the surface tension without the need for corrections. An alternative method is to raise the liquid until it just touches the plate which is hanging from a balance. The increase in weight is given by the following expression:

$$\Delta W = p\gamma \cos(\theta) \tag{23}$$

Again, p is the perimeter of the plate.

Recently, the Wilhelmy plate apparatus has been configured with a motor driven stage which raises and lowers the stage at a controlled rate. Such instruments are offered commercially by Cahn and Kruss. We have used the Cahn DCA-312.[15] Such instruments allow one to measure surface tension, contact angle, and, by varying the rate of stage movement, dynamic contact angles. The surface tensions of both solids and liquids may be determined.

In this method the force required to immerse or withdraw a solid object of known perimeter is determined by the following relationship:

$$F = p\sigma_L \cos(\theta) + A\,\Delta\rho\, g(x - X_{ZDOI}) \tag{24}$$

Here p is the object perimeter, $\Delta\rho$ is the difference in the density between liquid and vapor, g is the acceleration due to gravity (980 cm/sec^2), x is the position of the plate and x_{ZDOI} is the position of the plate at zero depth of immersion (ZDOI). Evaluating Equation (24) at $x = x_{ZDOI}$ gives

$$F_{ZDOI} = p\sigma_L \cos(\theta) \tag{25}$$

Equation (25) may be used to evaluate σ_L when using a wetting liquid ($\cos(\theta) = 1$) or to evaluate $\cos(\theta)$ if σ_L is known.

For porous media such as paper, some liquid will be retained by the sample upon withdrawal. The force curve must be corrected for the extra force due to the retained liquid when either receding contact angles or σ_L are calculated. If the liquid does not wick at much greater velocity than the advancing meniscus, then the receding force may be corrected as follows. (*n.b.*, the Cahn software makes this correction.) During the receding part of the cycle,

$$F_{rec} = p\sigma_L \cos(\theta) + Ag\,\Delta\rho\,(x - x_{ZDOI}) \ldots$$

$$+ \Delta mg\left[1 - \left\{\frac{x - x_{ZDOI}}{x_{max} - x_{ZDOI}}\right\}\right] \tag{26}$$

where Δm reflects the change in mass of the sample after complete withdrawal of the liquid. Thus

$$F_{rec,ZDOI} = p\sigma_L \cos(\theta) + \Delta mg \tag{27}$$

and

$$F_{rec,ZDOI,corr} = F_{rec,ZDOI} - \Delta mg \tag{28}$$

Using the above relations both receding and advancing contact angles may be determined.

Recently, the Wilhelmy method has been applied to cellulosic materials. Gardner[16] has studied the wettability of various species of wood. Berg and co-workers[17] have studied the wettability of single fibers.

EXPERIMENTAL RESULTS

SURFACE FREE ENERGIES OF PAPER

The surface free energies of various grades of papers were measured via the Wilhelmy plate method on a Cahn DCA-312. A strip of paper 2.54 cm (1 inch) wide was cut into segments of appropriate length. The stage speed was set at 140 μm/sec. The surface free energies were calculated from advancing contact angles using both water and methylene iodide as probe liquids. The results are shown in Figure 1.

It is significant that the total surface free energy of the tested grades varies over a broad range, roughly 25–60 mN/m. This range suggests the need for different ink formulations for various papers. The fractional polarity, (σ^P/σ_T), also varies considerably between grades. The fractional polarity varies from about 0.01 to 0.6. Again, this variation suggests the need for various ink formulations. Papers having a sufficiently large fractional polarity may exhibit poor ink adhesion and/or large contact angles due to polar repulsions. Our experience suggests that substances added to the furnish for sizing and other purposes determine the surface chemistry of paper. We are not aware of a comprehensive and systematic study to investigate the effect of various sizing agents on the surface chemistry of paper.

We have also investigated the surface chemistry of linerboard. In this study the Zisman critical surface tension was determined using isopropanol–water mixtures. The critical surface tension was found to vary from about 24 mN/m to 50 mN/m (see Figure 2). A typical value for the critical surface tension was about 35 mN/m. A significant finding was that the critical surface tension varied considerably between manufacturers and between plants for a given manufacturer. We presume that even at a given site, surface energy variations would be observed over time. It thus appears that for a given

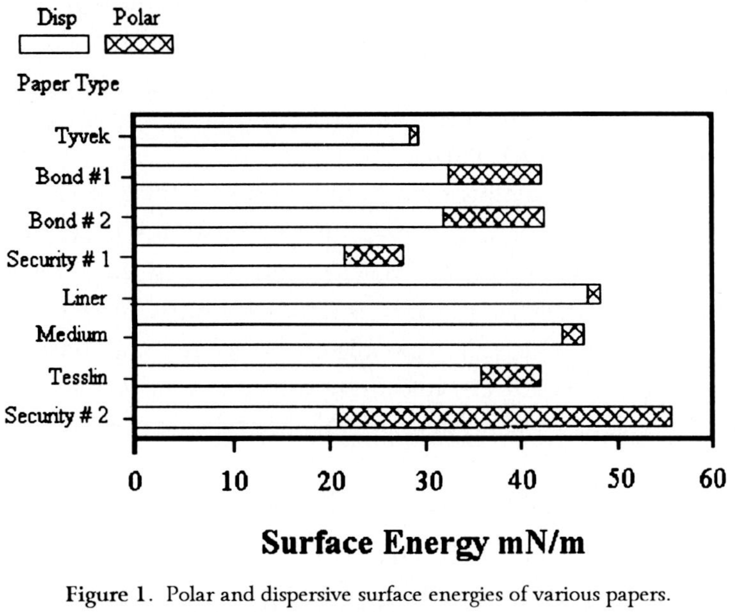

Figure 1. Polar and dispersive surface energies of various papers.

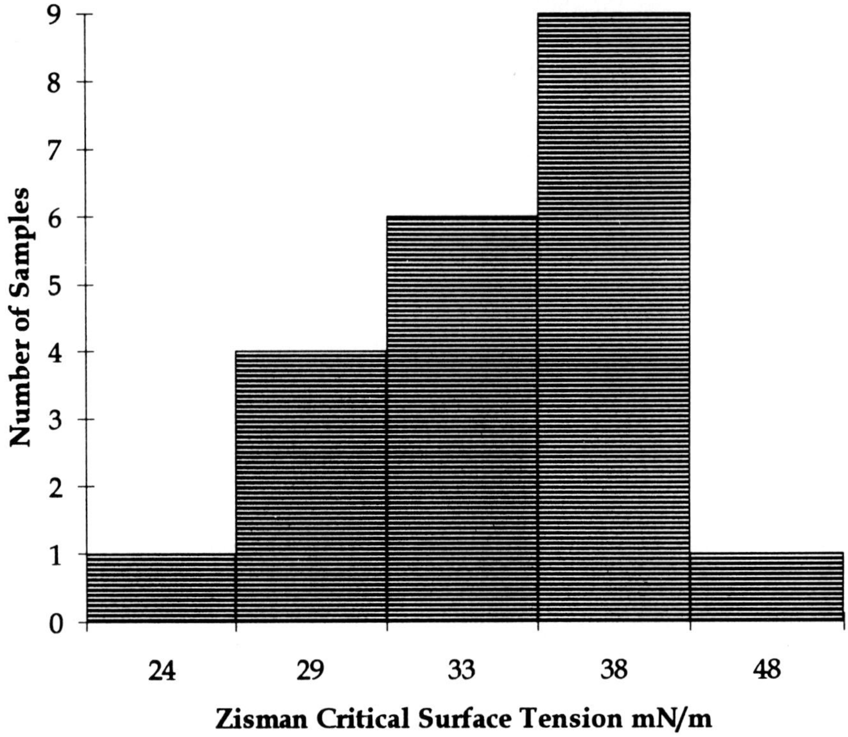

Figure 2. The Zisman critical surface tension of several linerboard samples as determined using isopropanol–water solutions. The linerboard samples were obtained from several companies located throughout the United States.

nominal grade surface energy variations exist. The need to address ink-paper compatibility is again stressed.

XPS was used to correlate the critical surface tension of linerboards to aspects of their surface chemistry (see Figures 3 and 4). The XPS data showed that the surface with the lowest critical surface tensions were generally those with a greater aliphatic character.

ADHESION OF INTAGLIO VARNISHES

Intaglio printing is used for printing security documents. In this process ink is spread on a plate containing an engraving of the image to be printed. Ink is then wiped from the plate, leaving ink only in the grooves of the engraving. Paper is then pressed under great pressure into the engraving. When the press is released a raised image is left. Adhesion of the raised printed image is a critical parameter for determining the durability of the document. In the past, solvent-based inks using linseed oil as a varnish were used. These inks generally give good adhesion. More recently alkyd-based inks have been used with less success.

We have measured the surface tensions of several alkyd varnishes used in intaglio printing. The results are shown in Figure 5. The results were measured using the

Wilhelmy plate method at a stage speed of 6 μm/sec. Polyethylene was used to construct Wilhelmy plates in order to determine the polar and dispersive components.

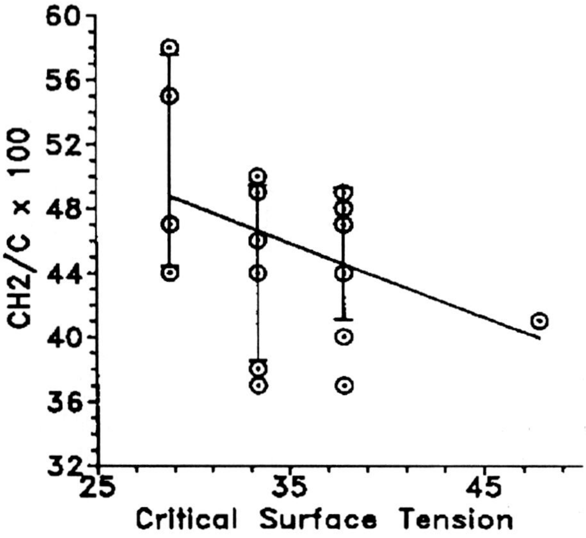

Figure 3. Ratio of methylene carbons to total surface carbon content on linerboard surfaces as determined by ESCA versus Zisman critical surface tension. Note that increased methylene carbons result in a lower critical surface tension. The low methylene content points near 35 mN/m were found to be contaminated with a non-aliphatic material.

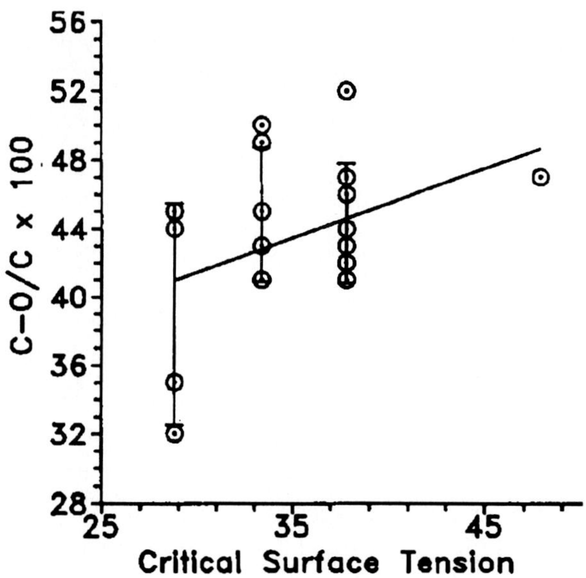

Figure 4. Ratio of carbons attached to oxygen to total surface carbon content of linerboard samples versus Zisman critical surface tension. Samples are the same as those in Figure 3. Note that as oxygen content increases so does the critical surface tension.

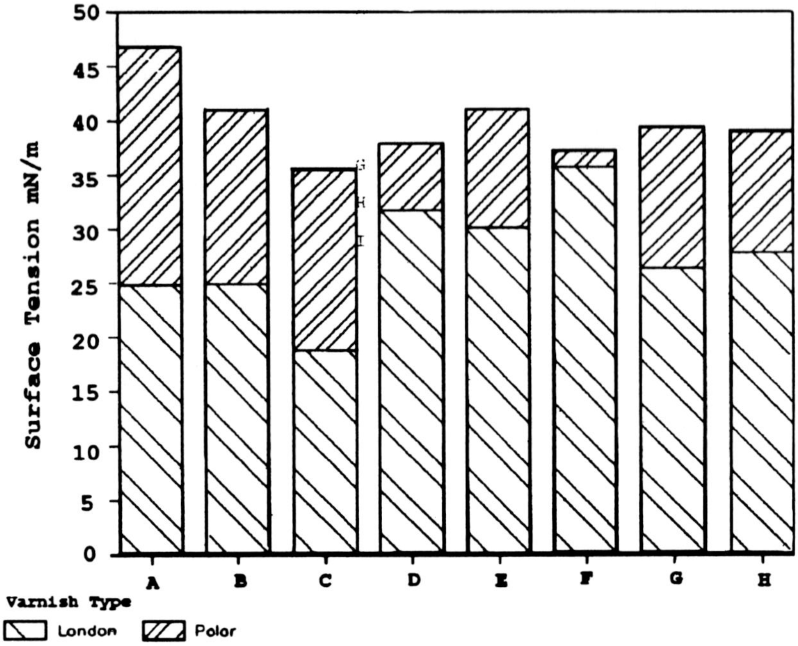

Figure 5. Surface tensions of various alkyd intaglio varnishes.

It is noteworthy that the alkyd varnishes generally have a large polar component to the surface tension. The polar component of the surface tension of linseed oil is essentially zero.

The advancing contact angle of four varnishes on a security paper were measured. The stage speed was 2 μm/sec. The data are shown in Figure 6. The results show that as the polarity of the varnish increases so does the contact angle. Equation (20) suggests that such a result might be caused by base–base or acid–acid interactions at the interface. It is significant that linseed oil which has no polar component also has a zero contact angle. Varnish C is known to exhibit poor adhesion characteristics through mechanical testing.

SURFACE FREE ENERGY AND FLEXOGRAPHIC PRINTING

As indicated above, the range of critical surface tensions for linerboard samples is from 25 – 60 mN/m with an average of 35 mN/m (see Figure 2). We have found the surface tensions of flexographic inks to vary from 28 – 45 mN/m. These values suggest that some inks may be incompatible with some linerboards. Bassimer and Krishnan[18,19] have investigated the surface tension of flexographic inks and their relation to print quality. These authors found that the factors which control print for inks on a given paper are: *i*) total surface tension of the ink; *ii*) dynamic effects on surface tension, and *iii*) polarity of the ink.

The use of surfactants to reduce the surface tension of inks may result in a surface which requires considerable time to reach equilibrium. In general, inks containing

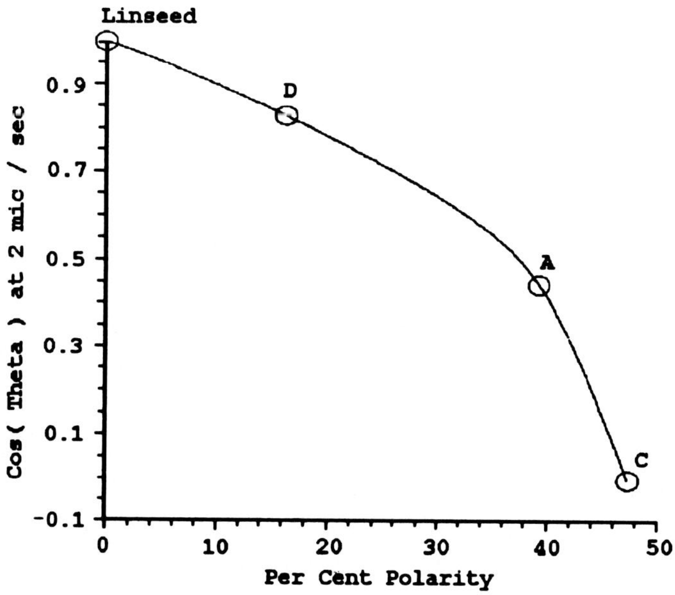

Figure 6. Cosine contact angle of intaglio varnishes on security paper #2 versus surface polarity. Letters correspond to those in Figure 5.

surfactants which have a larger molar masses take longer to reach equilibrium than those containing low molar mass surfactants. For aqueous based inks at short interface lifetimes (fast print speeds), the dynamic surface tension may approach that of water.

As has been shown for the case of intaglio inks, flexographic inks perform best when the polarities of the ink and paper are compatible.

SUMMARY AND CONCLUSIONS

The surface chemistry of paper has been explored by others in the past.[20,21,22,23] Borsch in particular has explored the effect of surface chemistry on adhesion. Triantafillopolous *et al.*[24] have discussed some aspects of acid–base chemistry at the interface and its relation to printability. Sharma[25] and Bassimer[19] have also recently discussed the role of ink and paper surface chemistry in printing. The available data show that the surface chemistry of paper is quite variable. It is hoped that the principles outlined in this paper will be used to improve the printing process and will offer insights to those who are troubleshooting printing problems. Papermakers need to learn how addition of specific components to the furnish affect the surface chemistry so that surface energy may be controlled and adjusted to fit the needs of the end user. These authors are not aware of a systematic study of the effects of sizing chemistry on the overall surface chemistry of paper. The ink formulator must recognize the variability of the surface chemistry of paper and design ink formulations which are both appropriate for the job at hand and which are forgiving to variations in paper surface chemistry.

Of particular import, our data show that the polarity of ink formulations may lead to incompatible surface chemistries. The polar components of surface free energy of ink and paper may result in repulsive forces at the interface. This issue will be very important to those who wish to formulate aqueous-based inks. This effort may require cooperation between ink and paper manufacturers.

NOTE ADDED IN PROOF

Very recently, Banerjee and Etzler[26] have shown that contact angles of a variety of hydrophobic and hydrophilic liquids on polymer surfaces are correlated with surface tension and the interfacial free energy between solid and liquid. They assumed that intimate molecular contact was made between the liquid and the polymer surface. The interfacial free energy was represented by the residual infinite dilution activity coefficient of the liquid in the monomer as computed by the UNIFAC equation.[27] Estimated contact angles were within an average of 7° of measured values.

ACKNOWLEDGEMENTS

This research was, in part, supported by the member companies of the Institute of Paper Science and Technology. The research was also supported by an appointment to the Oak Ridge Faculty Research Program at the Bureau of Engraving and Printing administered by Oak Ridge Associated Universities through an interagency agreement between the U.S. Department of Energy and the U.S. Department of the Treasury. We wish to thank Dr. John Bobalek and Dr. Nabil Radwan for their insights and encouragement. John Collins, Mark Weiss and Julie Evans are also thanked for their assistance. John Moynihan is thanked for preparation of some of the alkyd intaglio varnishes.

REFERENCES

1. Adamson, A.W. 1990. Physical Chemistry of Surfaces. John Wiley & Sons, New York, 5th ed.
2. Washburn, E.W. 1921. Penetration of liquids into capillaries. *Phys. Rev.* Ser. 2, 17:273.
3. Kinloch, A.J. 1987. Adhesion and Adhesives: Science and Technology. Chapman and Hall, New York, pp. 44–46.
4. Yasuda H.; Charlson, E.J.; Charlson, E.M.; Yasuda, T.; Miyama, M. and Okuno, T. 1991. Dynamics of surface property change in response to environmental conditions. *Langmuir* 7:2394–2400.

5. Zisman, W.A. 1964. Relation of equilibrium contact angle to liquid and solid constitution. *In*: Contact Angle, Wettability and Adhesion. Adv. in Chem. Series, Vol. 43, F.W. Fowkes, ed., American Chemical Society, pp. 1–51.
6. Girifalco, L.A. and Good, R.J. 1957. A theory for the estimation of surface and interfacial energies I. Derivation and application to interfacial tension. *J. Phys. Chem.* 61:904.
7. van Oss, C.J.; Chadury, M.K. and Good, R.J. 1987. Monopolar surfaces. *Adv. Coll. Interface Sci.* 28:35–64.
8. Fowkes, F.M. and Mostafa, M.A. 1978. Acid–base interactions in polymer adsorption. *Ind. Eng. Chem. Prod. Res. Dev.* 71:3–7.
9. Fowkes, F.M. 1987. Role of acid–base chemistry interfacial bonding in adhesion. *J. Adhesion Sci.* 1:7–27.
10. Holmes-Farley, S.R.; Reamey, R.H.; McCarthy, T.J.; Deutch, J. and Whitesides, G.M. 1985. Acid–base behavior of carboxylic acid groups covalently attached at the surface of polyethylene. *Langmuir* 1:725–740.
11. Holmes-Farley, S.R., Bain, C.D. and Whitesides, G.M. 1988. Wetting of functionalized polyethylene film having ionizable organic acids and bases at the polymer–water interface. *Langmuir* 4:921–937.
12. Kadash, M.M.; Seefried, C.G., Jr. 1985. Closer characterization of corona-treated PE surfaces. *Plastics Engineering* December 1985.
13. Hansen, H.M.; Finlayson, M.F.; Castille, M.J. and Goins, J.D. 1993. The role of corona discharge treatment in improving polyethylene–aluminium adhesion: An acid–base perspective. *Tappi J.* 76:171–177.
14. Westerlind, B.; Larsson, A. and Rigdahl, M. 1986. Determination of the degree of adhesion in plasma-treated polyethylene/paper laminates. *Int. J. Adhesion and Adhesives Research* pp. 144–46.
15. Domingue, J. 1990. Probing the chemistry of the solid liquid interface. *American Laboratory*, October 1990.
16. Gardner, D.J.; Generalla, N.C.; Gunnells, D.W. and Wolcott, M.P. 1991. Dynamic wettability of wood. *Langmuir* 7:2498–2502.
17. Hodgson, K.T. and Berg, J.C. 1988. Dynamic wettability properties of single wood pulp fibers and their relationship to absorbency. *Wood and Fiber Sci.* 20:3–17.
18. Bassimer, R.W. and Krishnan, R. 1990. Practical applications of surface energy measurements in flexography. *Flexo* July, pp. 31–40
19. Bassimir, R.W. and Krishnan, R. 1991. Surface phenomena in water-based flexo inks for printing on polyethylene films. *In*: Surface Phenomena and Fine Particles in Water-Based Coatings and Printing Technology. Sharma, M.K. and Micale, F.J., eds. Plenum Press, New York. pp. 27–40.
20. Luner, P. and Sandell, M. 1969. The wetting of wood hemicelluloses. *J. Polymer Sci.* Part C, 28:115–142.

21. Swanson, R.E. 1976. The Influence of Molecular Structure of Vapor Phase Chemisorbed Fatty Acids Present in Fractional Monolayer Concentrations on the Wettability of Cellulose Film. Ph.D. thesis, Institute of Paper Chemistry, Appleton, Wisconsin.
22. Borch, J. 1982. Sizing additives affect polymer–paper adhesion. *Tappi J.* 65: 72–73.
23. Aspler, J.S.; Davis, S. and Lyne, M.B. 1987. The surface chemistry of paper in relation to dynamic wetting and sorption of water and lithographic fountain solutions. *J. Pulp and Paper Sci.* 13:J55–J60.
24. Triantafillopoulos, N.; Rosinski, J. and Serfano, J. 1992. The role of ink and paper chemistry in water based publication gravure. *In*: IPGAC Proceedings, TAPPI Press, Atlanta, GA.
25. Sharma, M.K. 1991. Surface phenomenon in coatings and printing technology. *In*: Surface phenomena and fine particles in water-based coatings and printing technology. Sharma, M.K and Micale, F.J., eds. Plenum Press, New York, pp. 1–25.
26. S. Banerjee and F.M. Etzler. 1995. An algorithm for estimating contact angle. *Langmuir*, submitted.
27. Fredenslund, A., Jones, R.L., and Prausnitz, J.M. 1977. Group-contribution estimation of activity coefficients in nonideal liquid mixtures. *A. I. ChE. J.* 21:1086–1099.

5

Brightness Properties of Pulp and Paper

Arthur J. Ragauskas

Institute of Paper Science and Technology
Atlanta, Georgia, U.S.A.

INTRODUCTION

Of the many experimental procedures employed to characterize bleached pulp, paper, and paperboard, optical properties are some of the most important. The optical performance of these products can be characterized with four basic properties: brightness, opacity, gloss, and color. This article reviews the fundamental principles involved in determining the brightness properties of lignocellulosic materials and discusses their relevance in modern papermaking processes.

The production of high-brightness pulps is a sophisticated, technically demanding process, requiring a variety of chemical and/or biochemical procedures.[1, 2] Although the complete spectral properties of pulp can now be readily determined, for historic reasons and for convenience, the optical properties of paper products are frequently determined by measuring "papermaker's brightness." This value, referred to as a brightness value, is determined by measuring the percentage reflectance of light in the blue region of the visible spectrum. Employing a narrow beam of light extending from 380 to 520 nm and centered at 457 nm, testsheets of paper are irradiated and the relative reflectance from the handsheets is measured.[3] The measured brightness values are, therefore, a technical, not colorimetric, evaluation of the visible appearance of paper products. The reflectance of bleached pulp and paper products was shown to be sensitive to this spectral region (see Figure 1) and is in general agreement with visual evaluations of brightness.

BRIGHTNESS TESTING

Although the instrumentation required to measure brightness values was developed in the early 1930s,[4] the test methodology was not standardized until the early 1940s. Van den Akker *et al.*[5] developed a standard TAPPI method to measure pulp brightness values,

0-8493-8992-5/95/$0.00+$.50

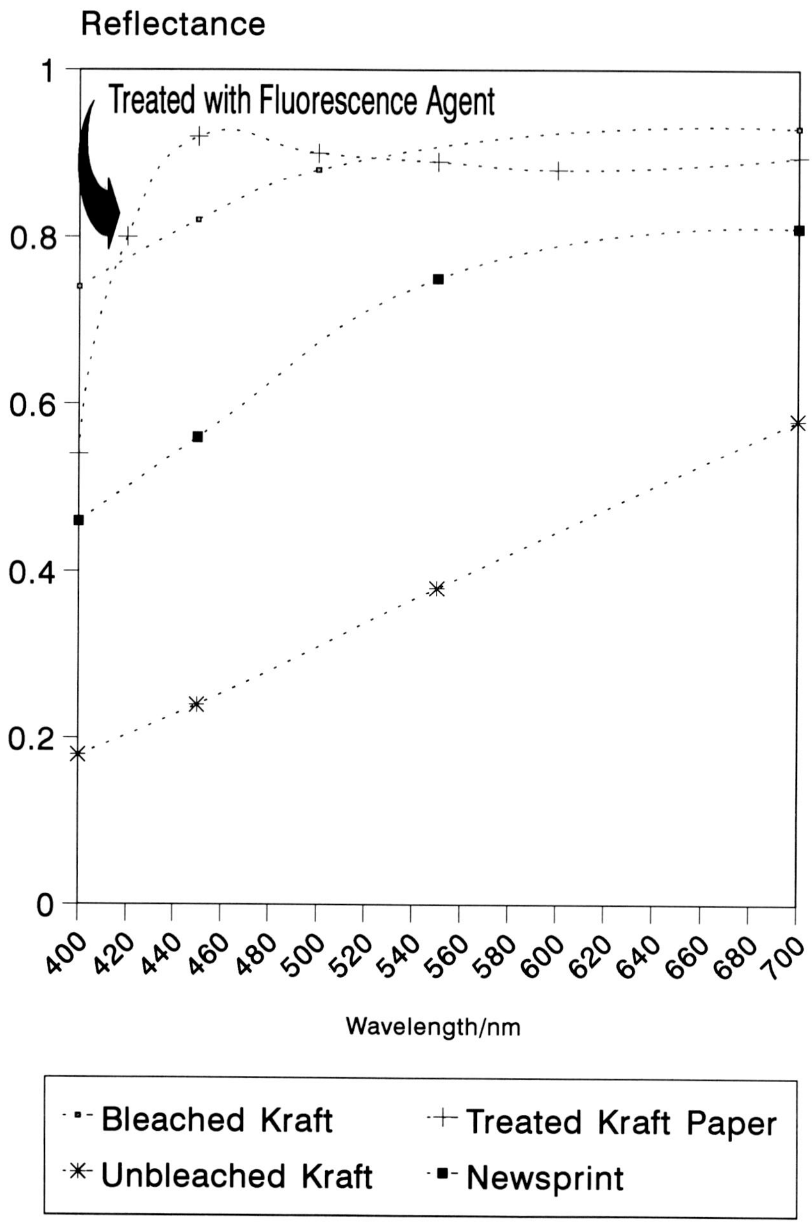

Figure 1. Spectral reflectance of paper.

which with only slight revision, remains actively employed in the pulp and paper industry.[6] The original TAPPI method, also referred to as GE brightness or TAPPI 45°/0° brightness, is dependent upon irradiating standard handsheets with 457 nm light at an incident angle of 45°and measuring the percentage reflectance at 0°(Figure 2). Brightness testing instruments employing this type of directional geometry for measuring brightness are sensitive to the directionality of the fibers in the handsheets. These directionality effects have been employed for determining machine-direction and cross-

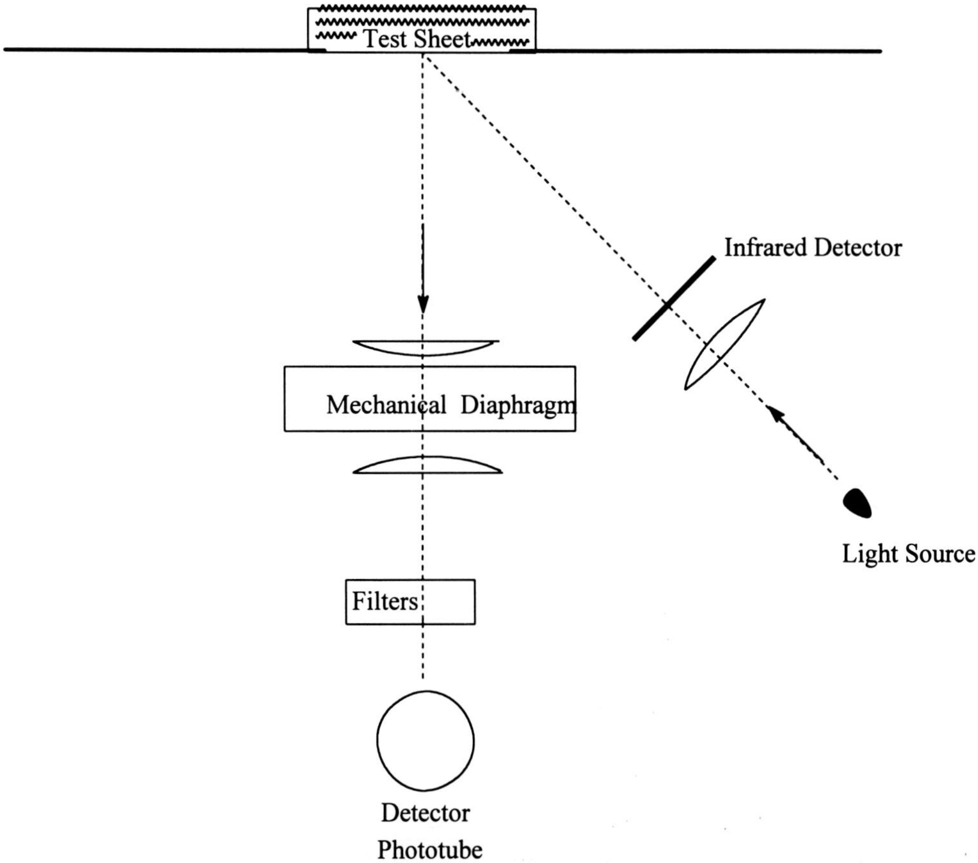

Figure 2. Fundamental design of directional brightness tester.

direction of paper. The machine direction of the paper yields lower brightness readings than the cross machine direction due to decreased scatter in the paper. Likewise, a brightness reading can also be employed to distinguish between the felt and wire sides of paper.

Directional brightness measurements are dependent upon a common standard. For the TAPPI procedure, the standard brightness test is calibrated to 100% reflectance for a magnesium oxide test pellet. Usual calibration of TAPPI brightness testing instruments is accomplished by employing two opal glass standards and five paper standards with a brightness range of 50 – 90%. The exact procedure for calibration and measuring TAPPI brightness is described in TAPPI test method T 452 om–87.[6]

An alternative approach to determining brightness values of paper products employs a diffuse light source centered at 457 nm and measures the reflected light at a 0° viewing angle as shown in Figure 3. This experimental approach is employed in ISO standard 2469, 2470, CPPA E.1, and SCAN P3, C11. In general the diffuse light source is generated employing a $BaSO_4$ integrating sphere, and in this manner, the test sheets are exposed to a substantially greater flux of light. Although initially the "diffuse" brightness

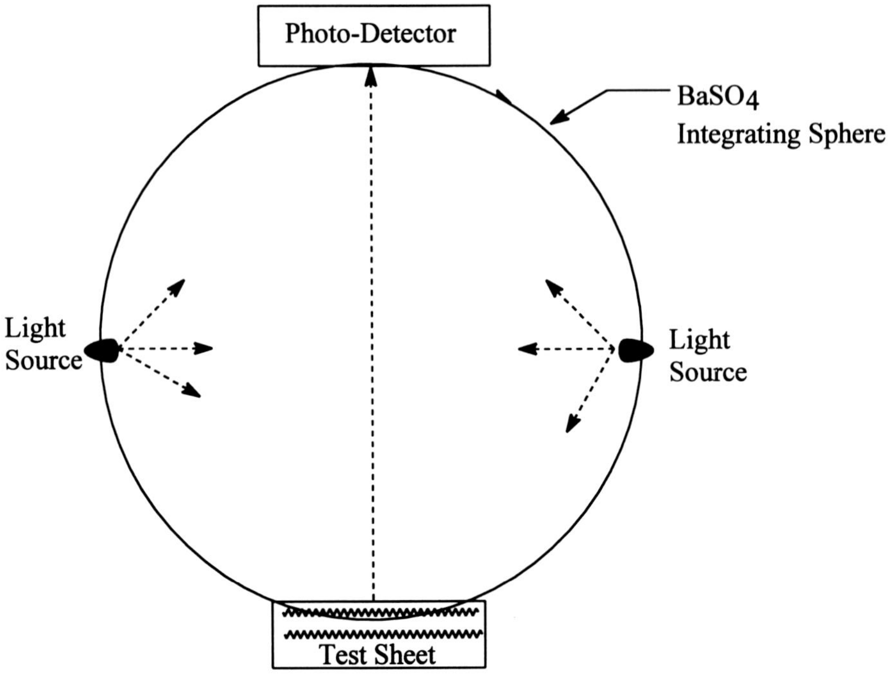

Figure 3. Fundamental design of diffuse brightness tester.

readings were calibrated to 100% reflectance from a magnesium oxide test pellet, the procedure has been changed to absolute reflectance.

Not unexpectedly, brightness measurements recorded with diffuse and directional based-instruments yield different values. Diffuse-based instruments average point to point variations in test sheets and minimize directionality effects which occur with most machine-made or embossed types of papers. Brightness measurements with a directional 45° – 0° instrument are sensitive to these types of directionality effects in the test sheet. As a result, brightness measurements between diffuse and directional instruments can vary between one and five percent depending upon the exact nature of the test sheet. Since these differences are dependent upon the nature of the test sheet being examined, there remains no general overall mathematical relationship which can relate these two types of brightness measurements.

The brightness values of pulp and paper related materials vary greatly dependent upon the final target products. By market standards, fully bleached paper or paperboard must exhibit ISO brightness values of 75% or greater, unbleached pulp materials have ISO brightness values of 45% or less, while paper products which have ISO brightness values of 45 – 75% are referred to as semi-bleached.[7] The goal of almost all bleaching operations is the removal of colored chromophores which are largely lignin-based.[8] In general, this can be accomplished using oxidative lignin-degradation or lignin-retaining processes. Since the manufacture of chemical pulps requires virtual complete removal of lignin, strong oxidizing agents such as ozone and chlorine dioxide are employed. Historically, bleaching of kraft pulps was accomplished with a sequential treatment of chlorine and chlorine dioxide[9], but recent environmental concerns have substantially reduced and/or eliminated the use of chlorine. Modern kraft bleaching operations are frequently based upon chlorine dioxide, and bleaching sequences such as D(EOP)DED or OD(EOP)D can readily yield kraft pulps with ISO brightness values of 85–90%.[10, 11]

The development of high brightness, totally-chlorine free (TCF) bleached pulps has shown continuing improvements. Christensen and Soteland have bleached beech, pine and spruce sulfite pulps to 85–86° ISO brightness values with oxygen and ozone.[12] A final peroxide stage to these pulps afforded 90°+ ISO brightness values. A variety of other combinations have been shown to successfully bleach sulfite pulps to ISO brightness values above 85%.[13] In contrast, kraft pulps are generally more difficult to bleach to high brightness, high strength values without the use of chlorine/chlorine dioxide. By employing bleaching sequences such as OZEP or OZQP, it has been possible to bleach kraft pulps to ISO brightness values of 80–85%.[14] Recent improvements in kraft pulping technology and extended bleaching sequences have further raised the brightness values of TCF kraft pulps into the upper 80 to lower 90% ISO brightness values.[15]

Unlike chemical pulps, bleaching operations for mechanical pulp are targeted at removing lignin chromophores without degrading lignin.[16] Brightness values for these grades of pulp vary greatly dependent upon the exact grade of pulp. Pure mechanical pulps such as thermomechanical pulp (TMP) and stone groundwood (SGW) usually have brightness values in the range of 55–60% ISO brightness, chemi-mechanical pulps in the mid 70% range, and bleached chemi-thermomechanical pulps (BCTMP) vary from 80–86% ISO brightness values.[17]

Brightness values for wood samples have also been shown to vary substantially depending on the nature of the wood species and sample preparation. Typical ISO brightness values can range from approximately 49% ISO brightness for loblolly pine, (*Pinus taeda*) to 29–41% ISO brightness for latewood spruce (*Picea abies*).[18]

SCATTERING AND ABSORPTION COEFFICIENTS

Although directional and diffuse brightness measurements remain the principal method of measuring the brightness of pulp and paper products, many other refined technical methods have been developed. One of the shortcomings of most brightness tests is that these measurements are a combination of two phenomena, the light scattering capabilities of the paper sheet and the light absorbing properties of the test sheet. Since these two properties are usually independent of each other, it is difficult to compare reflectance measurements for pulps produced under differing conditions.[19,20] Furthermore, differences in fiber morphology can also influence the observed values.

The influence of scattering and absorption properties of handsheets can be related to reflectance through the several mathematical models including the Stokes model[21] and the Kubelka–Munk model.[22] The latter approach can be expressed as a simple remission formula, which is as follows:

$$\frac{k}{s} = \frac{(1 - R\infty)^2}{2\ R\infty} \tag{1}$$

where *k*: absorption coefficient
s: scattering coefficient
R∞: measured reflectance from an infinitely thick sheet.

The absorption coefficient at a given wavelength (k_λ) is the product of the Beer–Lambert extinction coefficient (ϵ_λ) and the concentration of absorbers (C), at a given wavelength, as shown below.

$$k_\lambda = 2\epsilon_\lambda C \tag{2}$$

If several chromophores absorb at the same wavelength then the observed absorption coefficient is the sum of the individual absorptions, as shown below.

$$k_\lambda = 2\Sigma\epsilon_{\lambda i} C_i \tag{3}$$

Theoretically, the Kubelka–Munk remission function is valid at all reflectance levels. Therefore, in pulp samples for which the scattering coefficient remains constant, the k/s ratio becomes a relative measure of the amounts of colored material present in a pulp sample. The change in the k/s values for brightness reversion is generally regarded as a better indicator of color formation in pulp and paper than the change in brightness values. Giertz applied this approach to determine the amount of dye added to pulp samples and the thermal reversion properties of pulps.[23] These changes in color of pulp were defined in terms of the k/s ratio as defined below:

$$Post\ Color\ Number\ (PCN) = ks_{after\ treatment} - ks_{before\ treatment} \tag{4}$$

The post color number has been used by many researchers since Giertz's initially studies, especially in the field of photo- and thermal-reversion of mechanical pulps. It is now well established that all mechanical pulps suffer from relatively rapid photoyellowing effects, and this difficulty has limited their commercial applications. A recent study by Cockram suggested that if the overall rates of brightness reversion of mechanical pulps could be decreased from three months to thirty-six then the market potential for mechanical pulps could increase four-fold.[24]

The overall chemical processes involved in brightness reversion of mechanical pulps has been extensively studied for over fifty years.[25] Based upon these studies it is believed that the photoyellowing process is initiated by the absorption of 300–400 nm light by several lignin chromophores present in mechanical pulp. Secondary post-photolysis reactions then lead to the oxidative formation of colored chromophores. Although to date there are no commercially viable methods of photostabilizing mechanical pulps, the development of radical scavenging additives and UV-active additives for mechanical pulp has shown continuing promise. Ascorbic acid[26] and substituted mercapto compounds[27] are some of the most promising radical scavenging additives developed for mechanical pulps. Likewise, the use of UV-absorbers such as 2,4-dihydroxy-benzophenone[28] or trans,trans-2,4-hexadien-1-ol[29] have also been shown to be effective at retarding the photoyellowing process. Continued development of these types of additives and an improved understanding of their mechanisms of stabilization promises to lead to the development of technologies for photostabilization of mechanical pulps.

The Kubelka–Munk model of absorption assumes that the UV-absorbers are uniformly distributed in the layers; the scattering centers are uniformly distributed throughout the medium; and that the sample is sufficiently thick so that no incident radiation passes completely through the test sheet. For thick, photoaged pulp test sheets, the distribution of chromophores is often no longer uniform and this discrepancy can invalidate the Kubelka–Munk model. Bailey and Lamont[30] proposed an empirical logarithmic relationship to address these concerns, while Schmidt and Heitner[31] have developed a procedure which allows for facile, independent determination of the scattering and absorption parameters.

FLUORESCENCE

The ability of molecules to absorb light of a given frequency and then re-emit light at a lower frequency is commonly known as fluorescence.[32] Most pulp and paper samples have some limited natural fluorescing capabilities and as a result, handsheets irradiated with a near UV/visible light source will appear brighter than those irradiated with a visible light source (see Figure 1).

Fluorescent whitening agents (FWA), optical brightening agents (OBA), and fluorescent brightening agents (FBA) are usually added to the furnish or coating of paper to increase its "bluish tint".[33] In this manner the re-emission of near–UV light (λ_{max} ≈ 350–360 nm) as blue light (λ = ca 440 nm) increases the apparent whiteness of paper. Virtually all optical brighteners developed for pulp and paper are derived from 4,4-diaminostilbene-2,2-disulphonic acid (DASDSA) (see Figure 4). This basic stilbene unit is then derivatized to optimize application properties. The brightness testing procedures discussed above are standardized for naturally-colored pulps, papers, and paperboard. The addition of fluorescence agents can yield higher apparent brightness values due to

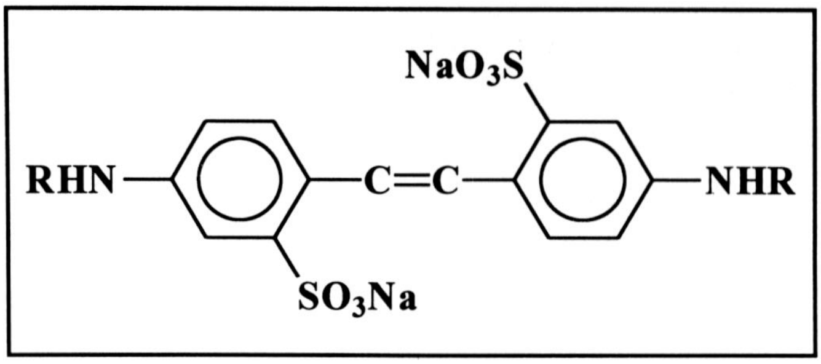

Figure 4. DASDSA

fluorescence. Although no linear relationship exists between the increase in brightness of treated paper products and the addition of fluorescent brighteners, an asymptotic relationship is frequently observed.

Most routine brightness testing procedures can be readily modified to determine the fluorescent component of brightness measurements (see TAPPI Procedure T 452 om–87). The procedure determines the fluorescent contribution to whiteness by differential measurement of reflectance with and without a near–UV cutoff filter. This method provides a reasonably good approximation of the fluorescent contribution to 457 nm reflectance readings, but is not completely accurate since the filters do not completely remove all UV light.

In summary, although current brightness testing procedures were developed approximately sixty years ago and improvements in optically testing procedures have dramatically advanced during this time period, nonetheless these simple measurements are still frequently employed both in research and in the production of pulp and paper.

REFERENCES

1. Smook, G.A. Handbook for Pulp and Paper Technologists. 1982. TAPPI Press, Atlanta, GA.
2. Trotter, P.C. 1990. Biotechnology in the pulp and paper industry: A review. Part 1: Tree improvement, pulping and bleaching, and dissolving pulp. *Tappi J.* 73(4):198.
3. Clark, J.d.A. 1978. Optical characteristics, dirt, and shives. *In*: Pulp Technology and Treatment for Paper, Miller Freeman Publications, Inc., San Francisco, Chapter 29.
4. Davis, M.N. 1935. Instrumentation in brightness grading. *Paper Trade J.* 101(1):36.
5. Akker, V.d., Nolan, P., and Wink, W. 1942. The physical basis of standardization of brightness measurement. *Paper Trade J.* 114(5):46.
6. TAPPI Test Methods. 1991. TAPPI Press, Atlanta, GA.
7. Slinn, R.J. 1992. Bleach Plant Operations Short Course. TAPPI Notes, TAPPI Press, Atlanta, GA, p. 13 .

8. Singh, R.P. 1979. The Bleaching of Pulp. TAPPI Press, Atlanta, GA.
9. Reeve, D.W. 1987. The principles of bleaching. *In*: Bleach Plant Operations, TAPPI Seminar Notes, TAPPI Press, Atlanta, GA, p. 1.
10. Pryke, D.C., Bourree, G.R., Winter, P., and Mickowski, C. 1993. The impact of chlorine dioxide delignification on pulp and manufacturing characteristics at Grande Prairie, Alberta. *In*: Proceedings from the 1993 Non-Chlorine Bleaching Conference, SC. 11, 1.
11. Malinen, R.O., Rasimus, R., andRantanen, T. 1993. ECF bleaching of oxygen-delignified softwood pulp with the minimum charge of ClO_2. *In*: Proceedings of the 1993 TAPPI Pulping Conference, p. 925.
12. Christensen, P.K.and Soteland, N. 1980. Bleiche von sulfitzellstoff mit sauerstoff und ozon. *Das Papier* 34(10A): V23.
13. Byrd, M.Jr., Gratzl, J.S., and Singh, R.P. 1992. Delignification and bleaching of chemical with ozone: A literature review. *Tappi J.* 75(3):207.
14. Libergott, N., Lierop, B.v., and Skothos, A. 1992. A survey of the use of ozone in bleaching pulps, Part 2. *Tappi J.* 75(2):117.
15. Lachenal, D., and Nguyen-Thi, N.B. 1993. TCF bleaching – Which sequence to choose? *In*: Proceedings of the 1993 TAPPI Pulping Conference. Atlanta, GA. p. 799.
16. Gagne, C., Barbe, M.C., and Daneault, C. 1988. Comparison of bleaching processes for mechanical and chemimechanical pulps. *Tappi J.* 71(11):89.
17. Patrick, K.L. (ed.) 1989. Modern Mechanical Pulping in the Pulp and Paper Industry. Miller Freeman Publications, San Francisco.
18. Loras, V. 1980. Pulp and Paper Chemistry and Chemical Technology, Third Ed. Wiley-Interscience Publication, New York. pp. 5, 633.
19. Jones, H.G., and Heitner, C. 1973. Optical measurement of absorption and scattering properties of wood using the Kubelka–Munk equations. *J. Pulp Paper Canada* 74(5):114.
20. Akker, V.d. 1968. Theory of some of the discrepancies observed in application of the Kubelka-Munk equations to particulate systems. *In*: Modern Aspects of Reflectance Spectroscopy. Wendlandt, W.W., ed. Plenum Press, New York.
21. Scallan, A.M. 1985. An alternative approach to the Kubelka–Munk theory. *J. Pulp and Paper Science* 11(3):J80.
22. Akker, V.d. 1949. Scattering and absorption of light in paper and other diffusing media. *Tappi* 32(11):498.
23. Giertz, H.W. 1945. On massans eftergulning. *Svensk Papperstidning* 48(3):317.
24. Cockram, R.A. 1989. CTMP in fine papers. *In*: Proceedings of the 1989 International Mechanical Pulping Conference, Helsinki.

25. Heitner, C. 1993. Light-induced yellowing of wood-containing papers: An evolution of the mechanism. *In*: Photochemistry of Lignocellulosic Materials. Heitner, C. and Scaiano, J.C., eds. ACS Symposium Series 531, ACS, Washington, DC, 2.
26. Fornier d.V., Nourmamode, A., Colombo, N., Zhu, J., and Castellan, A. 1990. Photochemical brightness reversion of peroxide bleached pulps in the presence of various additives. *Cellulose Chemistry and Technology* 24:225.
27. Cole, B.J.W., and Sarkanen, K.V. 1987. Bleaching and brightness stabilization of high-yield pulps by sulfur-containing compounds. *Tappi J.* 72(11):117.
28. Kringstad, K.P. 1969. Degradation of wood and high-yield pulps light: A survey of the present state of knowledge. *Tappi J.* 52(6):1070.
29. Ragauskas, A.J. 1993. Photoyellowing of mechanical pulp, Part 1. *Tappi J.* 76(12): 153.
30. Bailey, A.L., and Lamont, L.J. 1993. A standard procedure for accelerated testing and measurement of yellowing by light in papers containing lignin. *Tappi J.* 76(9):175.
31. Schmidt, J.A., and Heitner, C. 1993. Use of UV–visible diffuse reflectance spectroscopy for chromophore research on wood fibers: A review. *Tappi J.* 76(2):117.
32. Turro, N.J. 1978. Modern Molecular Photochemistry. Benjamin/Cummings Publishing Co., Inc., London.
33. Muller, F., Loewe, D., and Hunke, B. 1993. Fluorescent whiteners – New discoveries regarding their properties and behavior in paper. *Paper Southern Africa,* April 7.

6

FT-IR Spectroscopy

Michael A. Friese and Sujit Banerjee*

Appleton Papers, Appleton, Wisconsin, U.S.A.
**Institute of Paper Science and Technology, Atlanta, Georgia, U.S.A.*

INTRODUCTION

Infrared spectroscopy has a variety of uses in the paper industry, including competitive analysis, quantification of individual components, and identification of contaminants in both the manufacturing process and the product. Successful infrared analysis depends upon discriminating use of sampling techniques and careful interpretation of the data. Detailed descriptions of theory[1] and sampling techniques[2] have been given elsewhere. The primary purpose of this chapter is to review instrumentation and techniques that can be used to study paper or paper-related samples. Data interpretation and some paper-related FT-IR applications are also addressed.

SAMPLING TECHNIQUES

When selecting an instrument for testing or purchase the user should consider function, simplicity, reliability, and availability. There is a wide variety of sampling equipment offered for use in commercial FT-IR spectrometers. When purchasing sampling equipment the user should consider the same traits considered when purchasing a spectrometer: function, simplicity, reliability, and availability. Several vendors offer a large selection.[3–7]

Choosing the proper sampling technique can make the difference between finding a solution to a problem and not finding one; therefore, time taken to define the problem and study the options is time well spent. The first step is to gather as many clues as possible. Find out where the sample came from, and what it is supposed to consist of, or what possible contaminants may be present. For example, a foreign compound taken from a clarification tank or sewer drain is probably not water-soluble, so extracting the sample with water for neat film analysis is not the best approach. Knowing what is supposed to be there or what may be there should help you to select a suitable solvent.

0-8493-8992-5/95/$0.00+$.50

Second, determine whether there are potential hazards and if special handling is required. This will help you assess whether you or your equipment are in danger. For example, highly corrosive samples may require special personal protection equipment and may cause damage to the FT-IR sampling equipment.

Third, determine the goals of the analysis. Decide whether a quick qualitative analysis of the major components will suffice, or whether detailed quantitative analysis is required. Next, determine the best and quickest way to meet the goals by choosing the right sampling technique (be creative!), and implement your plan. If you don't meet the goals of the analysis the first time, review them, select a new sampling technique, and try again!

In the following sections proven sampling techniques are described in greater detail. The advantages and disadvantages of each are discussed. Examples are also given.

NEAT FILM

Equipment used for neat film sampling consists of a transmission window on which the sample is placed and a support system for holding the window in the path of the infrared source. The beam passes through the sample and window and is then directed to the detector. It is critical that the window be transparent to infrared energy in the wavelength region of interest. Most sampling equipment vendors[3–7] sell windows made from numerous materials. Of these, AgCl and AgBr offer versatility, durability, and economy with a wide transparency range.

Neat film sampling is generally a quick and effective way to obtain spectra of liquid samples or extractions of solid samples. If the sample is liquid, neat film preparation is straight forward. For single beam instruments, spectra are collected in three steps: *i*) The clean window is placed in the FT-IR and a spectrum is collected. This establishes a baseline from which the sample absorption is measured; *ii*) The liquid sample is dropped onto the window and the solvent is allowed to evaporate. An infrared heat lamp can be used to speed up the evaporation process, provided that ample time is allowed for cooling before the sample is scanned; *iii*) The window is then placed back in the FT-IR and a spectrum is collected. A diagram of the path taken by the beam is shown in Figure 1, where I_0 is the radiation intensity of the source and I_t is the intensity of the transmitted radiation.

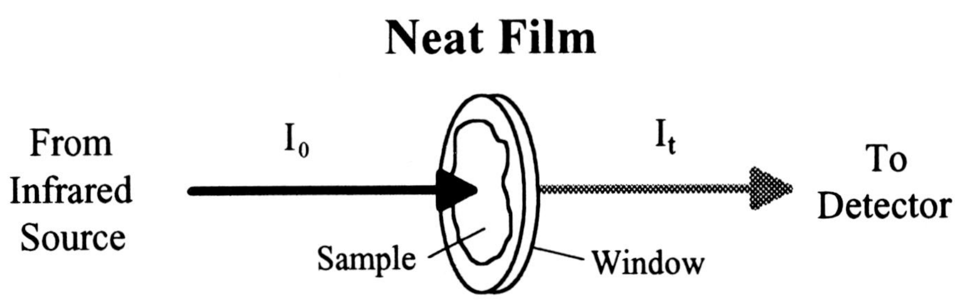

Figure 1. A diagram of the path taken by the beam during analysis via neat film.

There are two sample-related problems which may arise with liquid samples. First, the sample may be too concentrated or too strong an absorber. In this case a few drops will effectively block the source and overload the detector. To solve this problem a portion of the sample can be diluted and reapplied. Second, the sample may be too dilute or too weak an absorber. In this case try to build a thicker sample layer by drying successive drops in the same spot on the window. If this becomes too tedious, the sample may be concentrated. Neat film spectra prepared from polyvinylacetate (PVAc) and hydroxymethylcellulose (HMC) are shown in Figures 2 and 3. Some of the characteristic peaks have been identified.

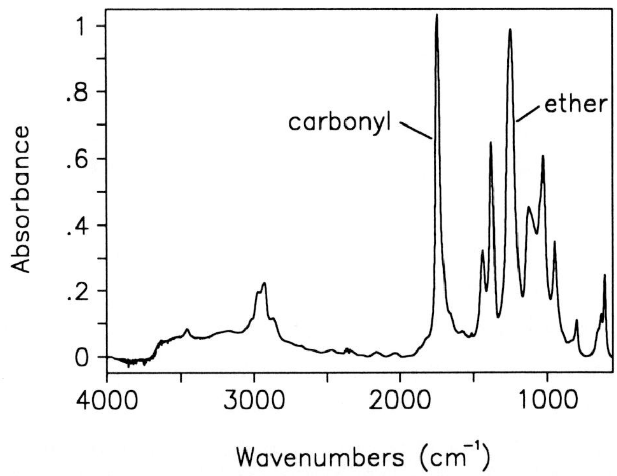

Figure 2. A neat film spectrum of poly(vinyl acetate) with the carbonyl and ether peaks labeled.

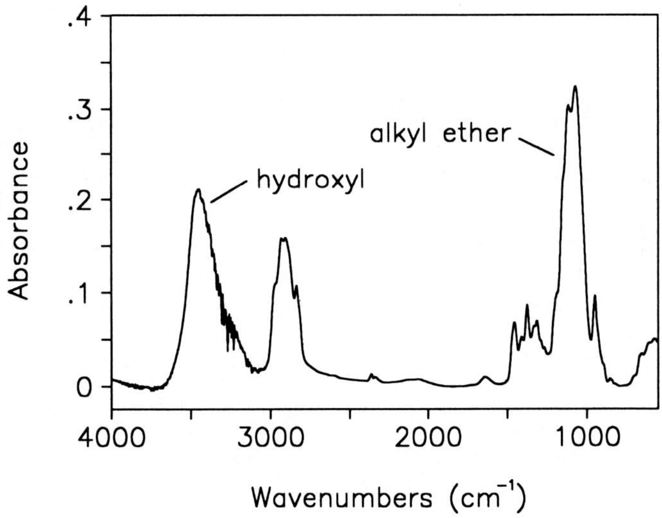

Figure 3. A neat film spectrum of hydroxymethylcellulose with the hydroxyl and alkyl ether peaks labeled.

Solid samples must be extracted for neat film analysis. The success of any extraction will depend on the solvent used. Therefore, it is important to review the goals of the analysis and to choose solvents carefully. Often a series of extractions is performed, beginning with polar solvents (water, hot water) and progressing to less polar (alcohol, acetone) and non-polar solvents (dichloroethane). After extraction, the solution is used to prepare a neat film as described above.

To demonstrate the usefulness of extraction and the associated pitfalls, paper base sheet samples were coated with a mixture of clay, $CaCO_3$, PVAc, and HMC. Samples of the base sheet and the coated sheet were extracted with water, and the solutions were then used to prepare neat film spectra. These spectra are shown in Figures 4 and 5, respectively. Figure 4 shows the water-soluble compounds in the base sheet, while Figure 5 shows the water-soluble components in both the base sheet and the coating. If the spectrum in Figure 4 is subtracted from the spectrum in Figure 5, the water-soluble components of the coating will remain. This difference spectrum is shown in Figure 6. Here the hydroxyl and alkyl ether peaks of HMC are clearly visible. An overlay of the difference spectrum and the HMC spectrum is shown in Figure 7.

The characteristic carbonyl and ether peaks of PVAc are absent in Figure 6. This is because water is not an adequate solvent for PVAc extraction. Subsequent extractions with ether removed only a small amount of the PVAc. Choosing the right solvent was critical when it came to detecting the PVAc in the coating. In fact, it might easily have been missed altogether.

The advantages of using neat film analysis when testing liquids includes minimal sample preparation and well defined spectra. When testing solid samples, it is necessary to perform a solvent extraction to remove analytes from the solid. This step increases sample preparation time and limits detection to soluble components.

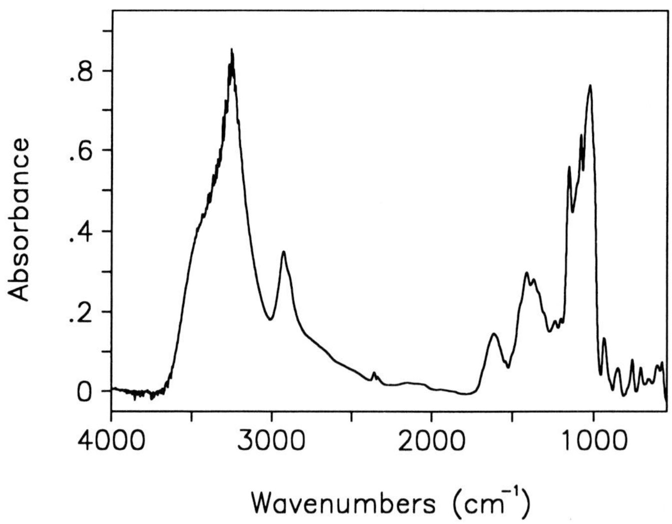

Figure 4. A neat film spectrum prepared from a water extract of the base sheet.

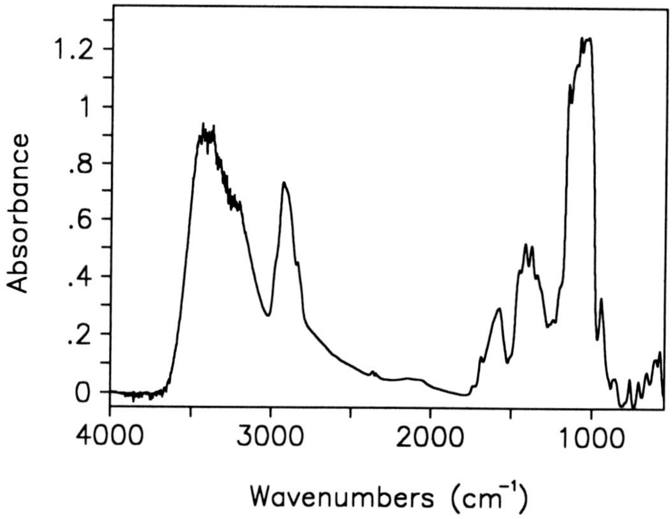

Figure 5. A neat film spectrum prepared from a water extract of the coated sheet.

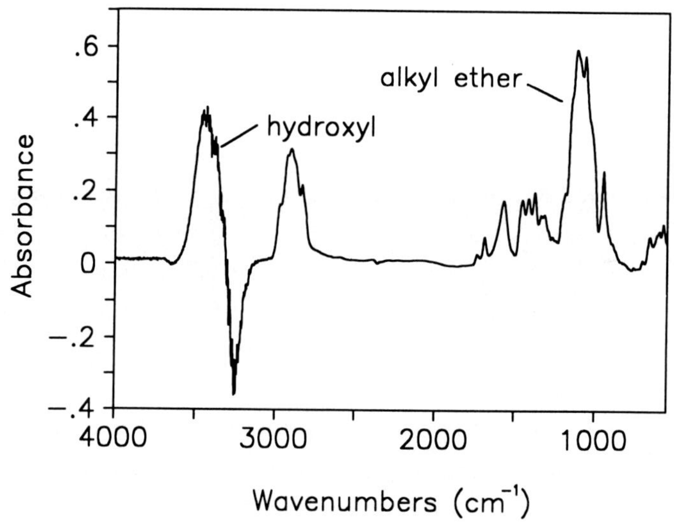

Figure 6. A coated base sheet difference spectrum (Figure 5 - Figure 4) with the hydroxyl and alkyl ether peaks labeled.

Note that it is very difficult to do quantitative work with neat film analysis since it is nearly impossible to reproduce film thickness in subsequent tests. Other sampling techniques must be used to obtain quantitative data.

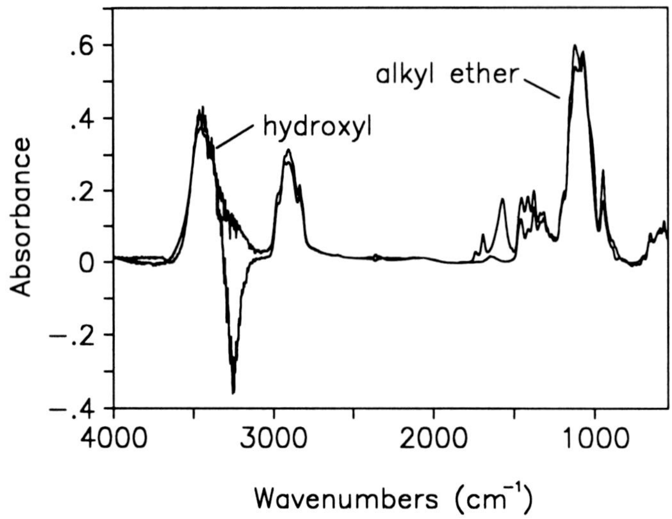

Figure 7. An overlay of the coated base sheet difference spectrum (Figure 6) and the neat film spectrum of HMC (Figure 3).

KBr PELLETS

The preparation of KBr pellets for sample analysis is more time-consuming and requires more equipment than neat film analysis. However, this technique will, in many cases, give the most accurate infrared spectra for solids. The necessary equipment includes dry KBr, a means of grinding the sample, a pellet die and press, and a support for holding the pellet in the path of the infrared source. During analysis, the beam passes through the pellet and then goes to the detector. A diagram of the path is shown in Figure 8, where I_0 is the radiation intensity of the source and I_t is the intensity of the transmitted radiation.

Preparation of the pellet is the most difficult step of the analysis. In essence the sample is mixed with KBr and pressed into a pellet. The pellet then acts as the sample support or window. The basic steps are as follows: *i*) the dry sample is combined with KBr and ground to a fine powder. An automated grinding mill or a mortar and pestle can be used for this purpose; *ii*) the sample is next placed in a pellet die and a vacuum is applied; *iii*) once evacuated, the die is placed in the press and sufficient pressure (normally in excess of 2000 psi) is applied to form a firm pellet; *iv*) the vacuum and the pressure are released and the sample is removed; *v*) finally, the pellet is placed in the FT-IR for analysis.

A pure KBr pellet must be run through the process first to generate a background spectrum. Normally a pure KBr pellet will be transparent. If not, additional grinding, pressure and/or care in removing the vacuum must be used. Pellets should be prepared as quickly as possible in a dry environment to prevent the KBr from absorbing moisture. Once water is absorbed, the intense peaks due to OH stretching and bending vibrations may obscure sample peaks. To avoid these problems, KBr and ground samples must be stored in a desiccator.

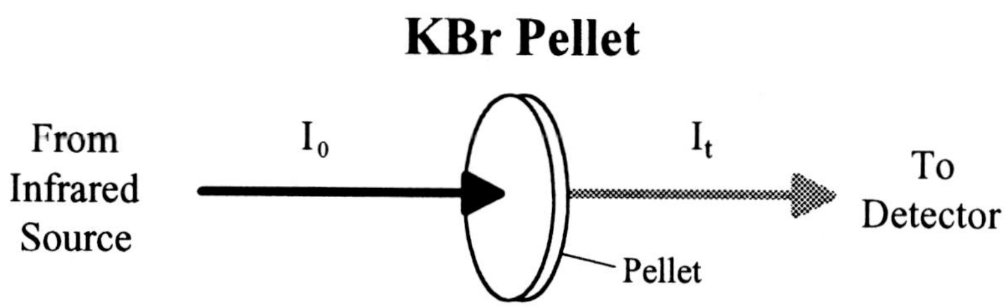

Figure 8. A diagram of the path taken by the beam during analysis via neat film.

The optimum ratio of KBr to sample is usually determined by trial and error. To minimize sample use it is best to begin with a high KBr:sample ratio (*e.g.*, 20:1). If the spectrum is weak, more sample can be added and a new spectrum collected. Frequently, as more and more sample is added, the pellet will begin to cloud or become opaque. In some cases additional grinding, pressure, or care in removing the vacuum may help. In others it may be unavoidable due to the nature of the sample. It is best to find a compromise between peak intensity and transparency of the pellet when these problems arise.

To demonstrate the usefulness of this sampling technique several spectra were collected. First, dry samples of kaolin clay and $CaCO_3$ were ground with KBr to prepare pellets. The resulting spectra are shown in Figures 9 and 10, respectively. Peaks corresponding to the stretching vibrations of the silicates are labeled in Figure 9. Peaks corresponding to carbonate stretching are labeled in Figure 10.

The sheet coated with clay, $CaCO_3$, PVAc, and HMC was also tested. Samples were taken from the base sheet and coated sheet by carefully scraping the surface with a razor blade. The scrapings were then combined with KBr, ground, and pressed into a pellet. The resulting spectra for the base sheet and the coated sheet are shown in Figures 11 and 12, respectively. Figure 13 is the difference spectrum obtained when Figure 11 is subtracted from Figure 12. The carbonate, silicate, and carbonyl peaks of $CaCO_3$, clay, and PVAc are easily detected. There is no evidence of HMC peaks.

The main advantage of KBr pellet sampling is that solid samples can be used directly without significantly altering their state. To obtain coating samples the test sheets had to be scraped. The samples could also have been ground whole and analyzed. However, we were interested in the coating and not the sheet. It is always best to perform as much physical separation as possible prior to analysis, taking care not to alter the sample (*e.g.*, with heat, solvents, etc.). This simplifies the final spectrum.

The main disadvantage of KBr pellets is that more equipment and time is required to prepare the pellets. There are other minor disadvantages. First, care must be taken to prevent the pellet from absorbing too much water. Second, as with neat film analysis, it is virtually impossible to do quantitative analysis because it is difficult to reproduce pellet thickness and the level of sample distribution.

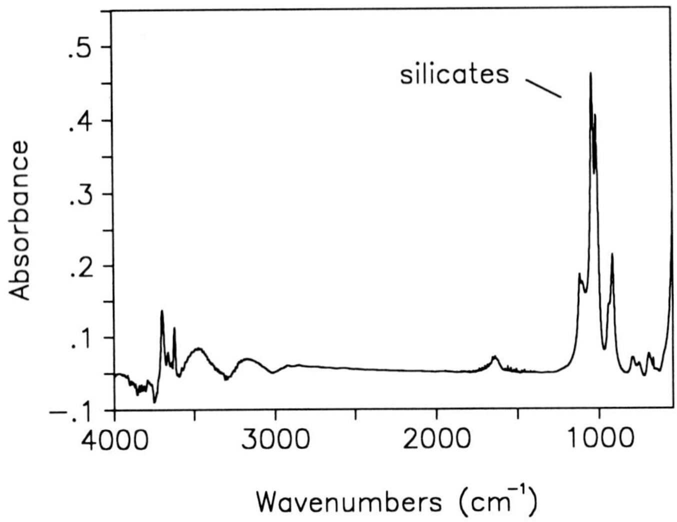

Figure 9. A KBr pellet spectrum of kaolin clay with silicate peaks labeled.

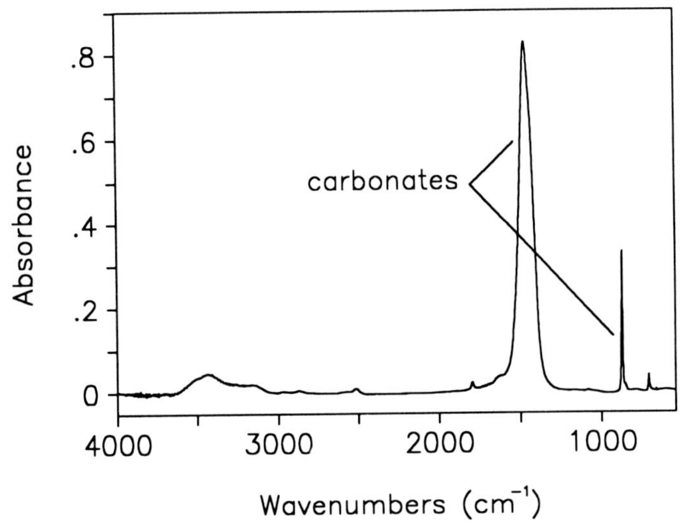

Figure 10. A KBr pellet spectrum of $CaCO_3$ with carbonate peaks labeled.

DIFFUSE REFLECTANCE

Diffuse Reflectance Infrared Fourier Transform Spectroscopy, or DRIFTS, is another sampling technique used for studying solid samples. It is easier to use and requires less equipment than that needed to prepare a KBr pellet. However, it can be more difficult to obtain sharp, clear spectra. Sample preparation is easy compared to the neat film or KBr pellet techniques. The general steps taken during analysis are as follows. In the initial stages, one of two approaches is usually taken. The sample is either ground, mixed with KBr and placed in the sample holder, or it is cut to fit the holder. Once in place, the

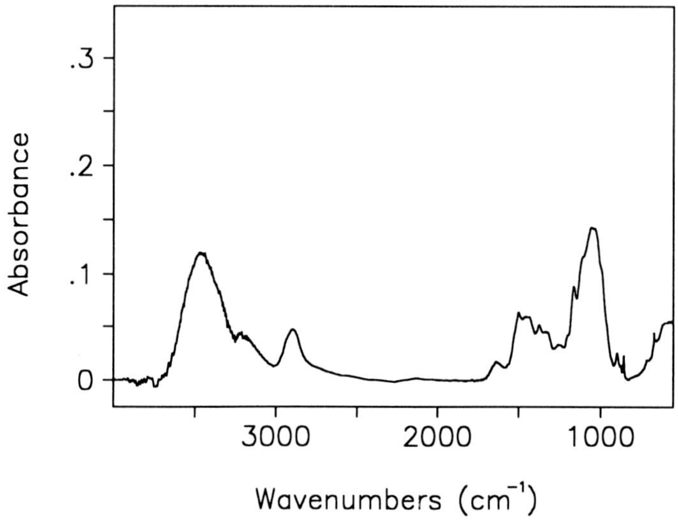

Figure 11. A KBr pellet spectrum of scrapings from the base sheet prior to coating.

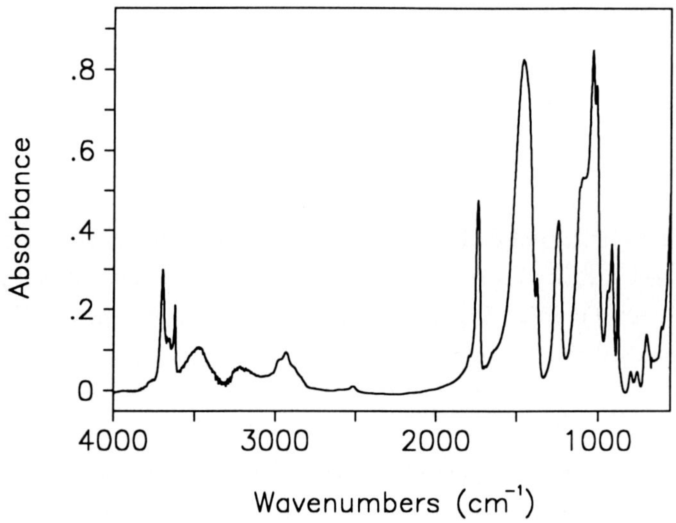

Figure 12. A KBr pellet spectrum of scrapings from the coated sheet.

sample is scanned. Ground KBr, if used, should be used to collect a background spectrum prior to analysis.

Mirrors are used to focus the radiation on the sample; when it strikes the surface, some of the radiation is reflected off the surface of the sample. This is referred to as *specular* or *Fresnel reflectance*. The remainder of the radiation penetrates the sample and reflects back and forth between internal surfaces. A portion of this radiation reflects back out of the sample, and this is referred to as diffuse reflectance. Mirrors around the sample collect the reflected radiation and direct it to the detector. A diagram of the process is shown in

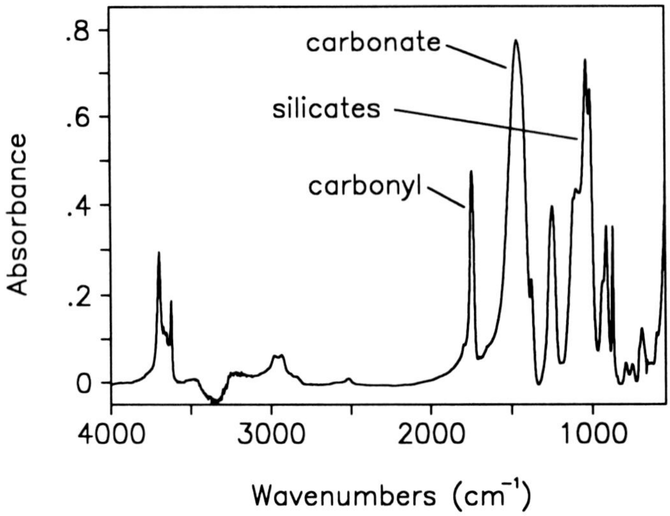

Figure 13. A KBr pellet coated base sheet difference spectrum with the carbonate, silicate and carbonyl peaks labeled.

Figure 14, where I_0 is the radiation intensity of the source, I_s is the radiation intensity due to specular reflectance, and I_d is the radiation intensity due to diffuse reflectance.

DRIFTS sampling works best when the sample is reasonably homogeneous and when it lies flat in the holder. If the spectrum appears weak the sample should be adjusted in the holder. If this does not improve the spectrum the placement of the holder can be adjusted. If all else fails the DRIFTS sampling apparatus may need to be realigned.

The base sheet and coated sheet tested earlier were also tested using DRIFTS. The spectra are shown in Figures 15 and 16, respectively. In each case, the sheet was scraped with a razor blade to obtain sample from the surface. The scrapings were then mixed with KBr, ground, placed in the DRIFTS holder, and scanned. Pure KBr was used to collect a background spectrum. The difference spectrum is shown in Figure 17. The peaks representing all four major components (HMC, $CaCO_3$, clay, and PVAc) are now clearly visible. This was done by simply scanning the background and sample in a nondestructive manner. In this case the DRIFTS technique was extremely useful.

There are many advantages to using DRIFTS. One advantage is that the effective path length increases with increased scattering. This makes DRIFTS particularly good for the study of powder and paper samples. Another advantage is that only minimal sample preparation is necessary. Solid samples can either be analyzed "as is" or ground and mixed with KBr. There is yet another advantage – all compounds present in sufficient quantities should be detected. Their detection is not dependent upon whether they can be extracted or otherwise separated from the rest of the matrix.

The presence of specular reflectance is the main disadvantage of DRIFTS sampling. Specular reflectance occurs when the radiation reflects off the sample surface. In diffuse

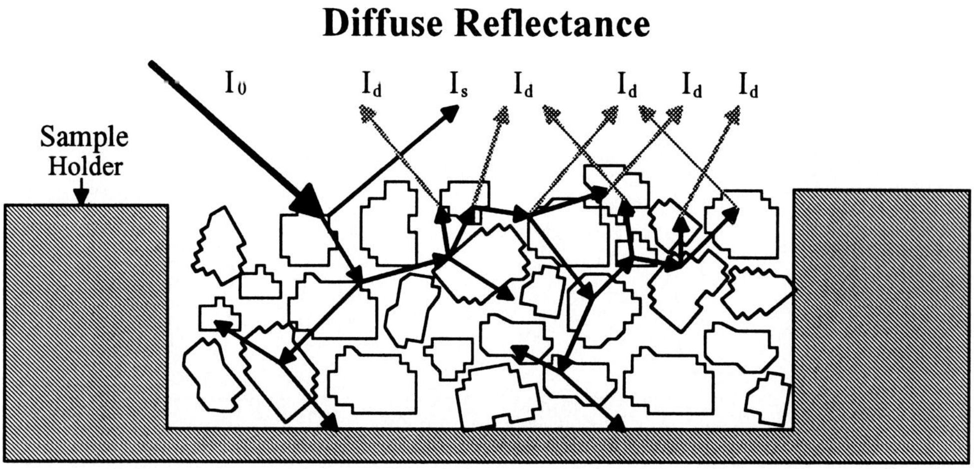

Figure 14. A diagram of the path taken by the radiation during DRIFTS sampling.

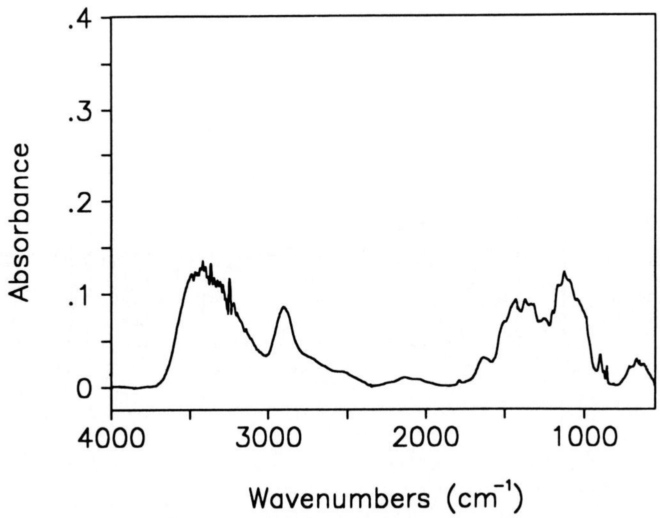

Figure 15. A DRIFTS spectrum of the base sheet.

reflectance the beam penetrates the sample, thereby interacting with all of the sample. If a highly-absorbing species is present at the surface, its absorbance will dominate the spectrum, causing distortion or inversion of peaks. In these instances quantitative work can be very difficult to perform, which is why it is important to have a reasonably homogeneous sample. Grinding can help reduce errors caused by specular reflectance.

In general, the sample must scatter the radiation sufficiently to ensure its interaction with the sample. In addition, it must not be too absorbing, so that escaping radiation can be focused and collected by the detector. For this reason, alignment of the sample holder is critical.

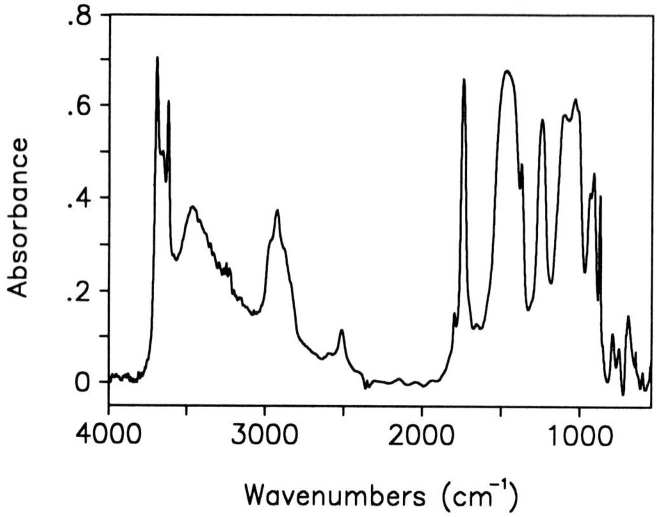

Figure 16. A DRIFTS spectrum of the coated sheet.

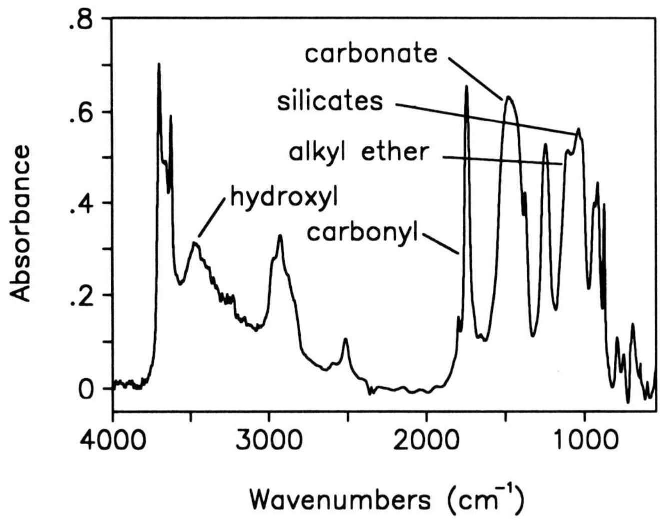

Figure 17. The coated base sheet difference spectrum obtained using DRIFTS.

INTERNAL REFLECTION SPECTROSCOPY

Internal reflection spectroscopy (IRS) is another sampling technique which requires minimal sample preparation. In this technique the sample is placed in direct contact with a prism made of a material (called the Internal Reflection Element or IRE) that is transparent to infrared radiation. Radiation from the source enters the IRE, strikes the IRE/sample interface, and is reflected back through the IRE when the angle of incidence is greater than the critical angle. (The critical angle is measured from the perpendicular axis and is a function of the refractive indices of the IRE and sample.) The radiation is then collected and sent to the detector. A diagram of the process is shown in Figure 18, where I_0 is the radiation intensity of the source and I_i is the internally reflected radiation.

During reflection, the infrared radiation passes through the IRE/sample interface and penetrates the sample a short distance before being reflected back. This portion of the radiation path is called the *evanescent wave*. The intensity of the evanescent wave decays exponentially with distance from the surface giving the analysis a very short effective path length. This property makes IRS particularly effective for analyses of surfaces, thick samples, and strongly absorbing species. Multiple reflections result in an increase the effective path length. This technique is frequently referred to as Attenuated Total Reflection (ATR) spectroscopy. Sample preparation is minimal. When necessary, samples are cut or measured to fit the particular sampling attachment being used. The best spectra will be obtained with smooth, homogenous samples. As always, the appropriate background should be run.

To demonstrate the technique, the base sheet and coated sheet tested earlier were loaded into a continuously variable angle ATR attachment with a KRS-5 element and scanned at 45°. The samples were held in place by rubber platens. The resulting difference spectrum is shown in Figure 19. The carbonyl, carbonate, and silicate peaks of PVAc, $CaCO_3$, and clay are clearly evident. The hydroxyl and alkyl ether peaks characteristic of HMC are absent. The latter may not be present at the surface in sufficient quantity for detection by ATR.

Care must be taken to prevent spectral distortion when using this technique. There are two main causes of distortion which should be considered. The first occurs when the angle of the incident radiation falls below or is near the critical angle. In this case the peaks can shift to lower frequencies or adopt a first derivative shape.[2] The likelihood of this type of distortion can be reduced by selecting an IRE with a high refractive index.

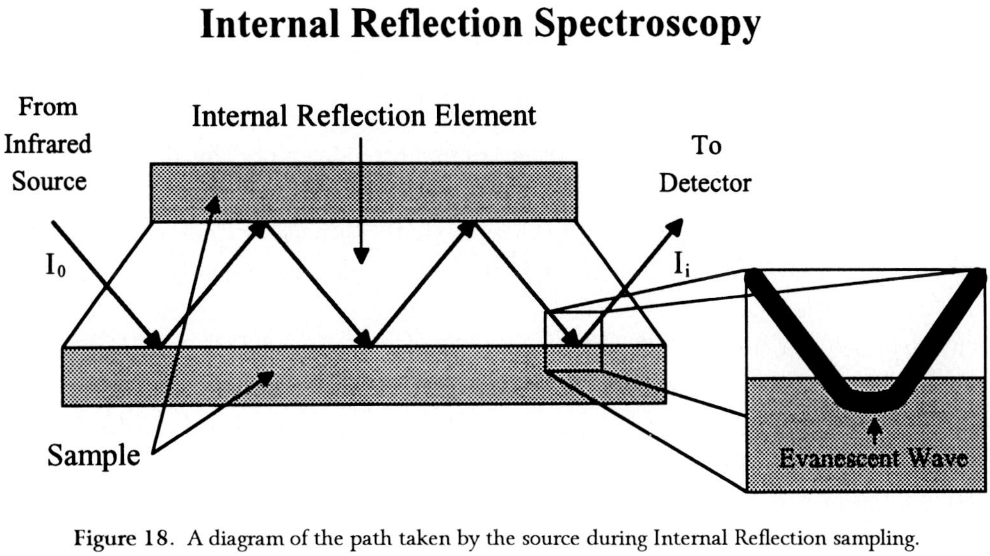

Figure 18. A diagram of the path taken by the source during Internal Reflection sampling.

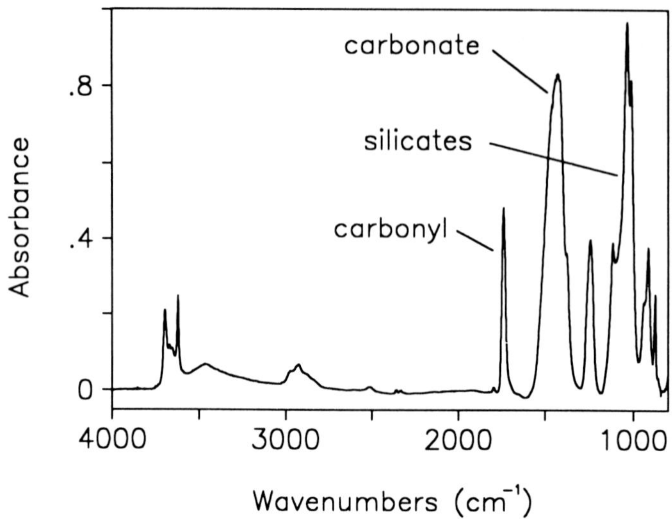

Figure 19. The coated base sheet difference spectrum obtained using ATR sampling.

The second problem arises when studying samples which are not homogeneous. The depth of penetration (d_p) of the radiation is dependent upon the wavelength of the radiation and the refractive indices of the sample and IRE as seen in the equation below:

$$d_p = \frac{\lambda_1}{2\pi(\sin^2\theta - \eta_{21}^2)^{0.5}} \qquad (1)$$

where λ_1 ($=\lambda/\eta_1$) is the wavelength in the denser medium and η_{21} ($= \eta_2/\eta_1$) is the ratio of the refractive index of the rarer medium divided by that of the denser.[8] If the samples change or if they are inhomogeneous in the direction of the radiation's penetration, the spectra could be distorted.

The main advantage of ATR sampling is the ability to study samples containing strongly IR-absorbing species. The effective path length of the ATR sampling is very small when compared with other FT-IR sampling techniques. This provides a means for measuring absorption which would otherwise overload the detector. The ability to analyze surfaces and the ease of sample preparation also sets ATR sampling apart.

Reproducibility depends on the ability to replicate the amount of area scanned and the level of sample/IRE contact. Eliminating bubbles or pockets can reduce error associated with changes in the total amount of area scanned. Adjusting platen pressure with a torque wrench can help provide reproducible sample/IRE contact.

One disadvantage associated with this technique is that ATR attachments can be difficult to clean. More importantly, many of the IREs are fragile and can be damaged quite easily. Finally, ATR attachments and accessories can be expensive.

FT-IR MICROSPECTROSCOPY

FT-IR microspectroscopy combines the usefulness of a microscope for visualizing microstructure with the molecular analysis capabilities of FT-IR spectroscopy. It provides a means to isolate and analyze extremely small samples. This is done in one of two ways, either by collecting reflected radiation or by flattening the sample and collecting a transmission spectrum.

Three different techniques can be used for reflection sampling. They are: diffuse reflectance, specular reflectance and reflection–absorption.[2,9] Diffuse reflectance and specular reflectance have already been described. In reflection–absorption (RA) the sample is placed in direct contact with a reflective backing. Radiation from the source passes through the sample, reflects off the backing, and passes back through the sample. Aligning the sample and microscope for reflection sampling can be difficult. However, the approach is useful when nondestructive testing must be used.

For qualitative work, transmission FT-IR microspectroscopy is the easiest to use. Portions of the sample are usually placed on an infrared-transparent slide. A small utility knife, finely-pointed probe, microforceps, or micropipet (for liquids) can be useful for this purpose. A stainless steel roller can be used to flatten solids for analysis. More resilient samples can be rolled on hard surfaces or flattened in a diamond anvil cell.[2] The source is then focused on the portion to be analyzed, and apertured to isolate it. A spectrum is then collected. Ideally, the apportioned sample should be greater than 20 mm in diameter to minimize diffraction of the source.[2] The thickness can be adjusted to change the effective path length and thereby increase or decrease the intensity of the absorbance peaks.

The coated sheet tested earlier was analyzed to demonstrate transmission FT-IR microspectroscopy. Coating was scraped from the surface coated sheet and flattened with a roller. The resulting spectrum is shown in Figure 20. Again, the carbonyl, carbonate, and

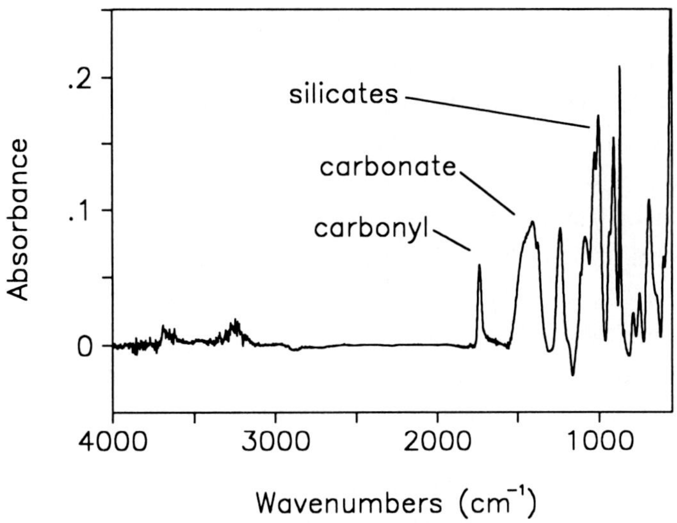

Figure 20. A spectrum of the coating obtained by FT-IR microspectroscopy.

silicate peaks of PVAc, $CaCO_3$, and clay are present and the hydroxyl and alkyl ether peaks characteristic of HMC are absent. In this case, subtraction of the base sheet background was not necessary since the piece taken from the coating contained no fiber.

There are a number of advantages to using FT-IR microspectroscopic sampling. First, the amount of sample preparation needed is minimal. Secondly, only a small amount of sample is necessary. Typically, the source can be focused down on samples as small as 10 mm in diameter. This also allows for the analysis of heterogeneous samples and sample layers. The third and final advantage is that surfaces can be analyzed in a nondestructive manner using reflection sampling.

There are several disadvantages associated with FT-IR microspectroscopy. First, although the light path in many microscope attachments is purged, the sample stage typically is not. The user must therefore account for changes in the surrounding air. Secondly, the sample stage needs to be shielded to prevent stray radiation from entering the detector. Finally, although extremely versatile, FT-IR microscope attachments are expensive when compared to other sampling equipment.

SPECTRAL ANALYSIS

The primary objective of spectral analysis is to isolate and identify components of sample spectra. This can be a difficult task since spectra, and paper spectra in particular, can be complex. In some cases the primary components of a sample will be known. In these instances isolation techniques may be needed to provide quality control or assist in quantification. In other cases the sample may be a complete unknown. Here both identification and isolation techniques may be needed.

ISOLATION TECHNIQUES

Spectral subtraction and deconvolution are common isolation techniques. Two types of subtraction are practiced. The most common of these is interactive subtraction. In this technique the analyst subtracts a reference spectrum in an iterative fashion, thereby removing one of the components. The other approach is to subtract the reference automatically. Subtraction techniques are discussed below.

Traditional deconvolution techniques are used less frequently. Most alter the appearance of the spectra making misinterpretations more likely if they are not used with caution. In general, deconvolution helps isolate or enhance overlapping peaks. Blanket techniques such as Fourier self-deconvolution, maximum entropy, and derivative calculation techniques can be used to check for overlap throughout an entire spectrum. Peak fitting can also be used to estimate areas of partially obscured peaks which are known to be present. Deconvolution techniques have been described elsewhere.[1,2]

Interactive Subtraction: In FT-IR spectroscopy, interactive subtraction is commonly used for removing the peaks of a reference spectrum from a multi-component sample spectrum. The analyst subtracts the reference spectrum in an iterative fashion until a flat baseline is obtained in a region where the only peaks present are due to the reference. The growth of negative peaks during the process is an indication that the subtraction has been carried too far.

Interactive subtraction is suitable for most qualitative work. However, when performing quantitative analyses, individual bias and uncertainties in the technique may result in artificial absorbance peaks and variation in peak absorbance levels. Where reproducibility is important, an objective subtraction technique is needed.

Automatic Subtraction: Mathematical algorithms have been used in the past to study multi-component spectra. Techniques described in the literature include methods based on component ratios,[10] factor analysis,[11] and least square fits.[12–15] Each of these techniques becomes more difficult as the number of components increases, and assumptions made limit their applicability. For example, the component ratio method assumes that band overlap is negligible and that differences in component absorbance levels are small. This can be a problem since large peaks frequently overlap or completely overwhelm small peaks or peak shifts due to component interactions. The problem with least square and factor analysis methods lies in the assumption that some or all of the multi-component spectrum can be described as a linear combination of known component spectra. In many cases, some or all of the components are unknown. In fact, one is rarely sure that all the possible components have been identified. In addition, band shifts due to component interaction can not be measured independently. For these reasons, software supplied by instrument vendors for subtracting spectra usually requires that the user assign the major peaks in a spectrum to a given component. These peaks are then factored out.

In essence, one needs to isolate the spectrum of a component from a background that can be both complex and variable. The problem is one of pattern recognition, with the signature of the pattern being partially or totally obscured by other constituents. If the spectrum can be partially recognized and assigned to a given component, then the entire spectrum of this component can be subtracted out leaving a hypothetical "n–1" difference spectrum.[16] If a second component can now be assigned and subtracted out, an "n–2" difference spectrum results. In principle, each progressive removal of a spectrum reduces band overlap, thereby increasing the chances that a unique spectral feature of some other component will be recognized.

In practice, a distinctive signature can rarely be recognized from a complex background, and the identifications are tentative. Also, subtractions can distort the difference spectrum since the reference spectrum may not exactly match that of the component in the sample, and there will very likely be some degree of over- or under-subtraction. These errors compound, and after only one or two subtractions the difference spectrum may be too distorted to be useful.

We have developed a new technique which allows recognition of a spectrum even in situations where extensive band overlap occurs. The procedure is called *dewiggling,* and has been used in both research and technical service applications at the Institute of Paper

Science and Technology and elsewhere. It has recently been adopted by Galactic Industries as their standard subtraction algorithm in their GRAMS/386 software package.[17]

The procedure is based on the fact that as spectra are overlaid on one another, the complexity of the resulting spectrum increases. In other words, the number of "wiggles" in the spectrum is likely to grow. If the converse operation is carried out (*i.e.,* a spectrum of a component is exactly removed from a multi-component spectrum), then the number of wiggles should decrease. Hence, the wiggles in a spectrum may be used to flag the point at which a component spectrum is exactly removed.

Consider the "spectrum" in Figure 21a which contains two signals and their first derivatives.[18,19] As one signal is progressively removed, the absolute value of the overall intensity (positive and negative) of the first derivative decreases until the signal is exactly stripped out in Figure 21b. Further stripping increases the intensity of the derivative

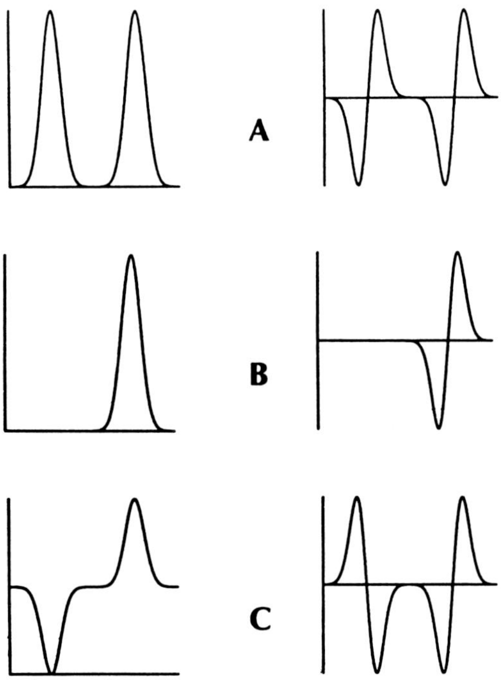

Figure 21. Gaussian signals and their derivatives. Two equivalent Gaussians (A) and the residuals after complete subtraction of one Gaussian (B) and after over-subtraction (C). From Banerjee and Li,[19] used with permission.

(Figure 21c). Thus, the derivative, which relates to spectral complexity, is minimized when a component is exactly removed. This provides the basis of an algorithm for subtracting spectra. All that is required is to subtract out increasing fractions of a suspected component and to take the derivative at each step. If the derivative minimizes, then the component is assumed to be present. The point at which the minimum occurs is a measure of the concentration of the component in the mixture. If no minimum is found, then the component is almost certainly absent.

The technique can be demonstrated using a practical example. Figure 22 shows spectra of softwood kraft pulp (*A*) and softwood lignin (*B*).[20] When the lignin spectrum was subtracted from the pulp spectrum via dewiggling, the curve in Figure 23 was obtained. The minimum occurred at a scaling factor of 0.18. This means that spectrum *A* in Figure 22 includes or contains 18% of spectrum *B*. From another perspective, 18% of spectrum *B* can be subtracted from spectrum *A* without leaving any negative residuals. Note that these subtractions are performed across the board and not at selected wavelengths.

Since pulp is mostly cellulose, one would expect a strong similarity between spectrum *A* and that of cellulose. Indeed, when spectrum *A* was dewiggled with respect to a spectrum of cotton linters (a pure form of cellulose), a minimum occurred at 51%. These minima can be used for quantitative purposes. While the magnitude of the minimum can be affected by factors such as moisture content, the ratio of two minima effectively provides the concentration of one component using the other as a reference. Thus, the ratio of the lignin minimum to that of the cellulose minimum should be an index of kappa number (a measure of the lignin content of pulp). This was shown to be the case, and a strong relationship (r=0.99) between these ratios and kappa number exists for a variety of bleached (oxygen, chlorine) and unbleached hardwood and softwood pulps.[20]

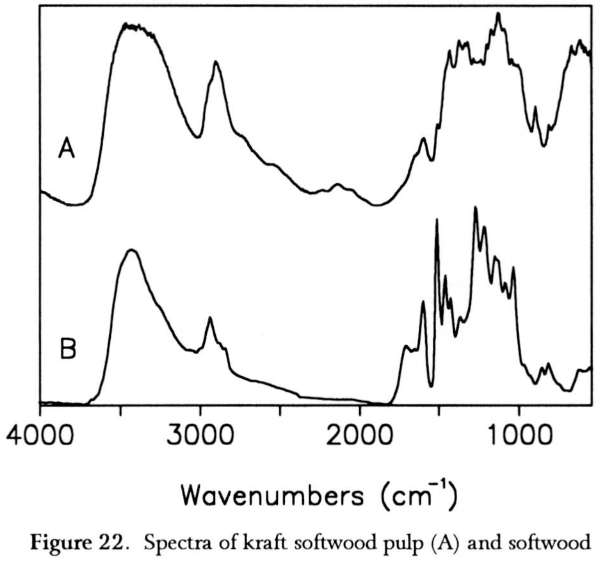

Figure 22. Spectra of kraft softwood pulp (A) and softwood lignin (B).

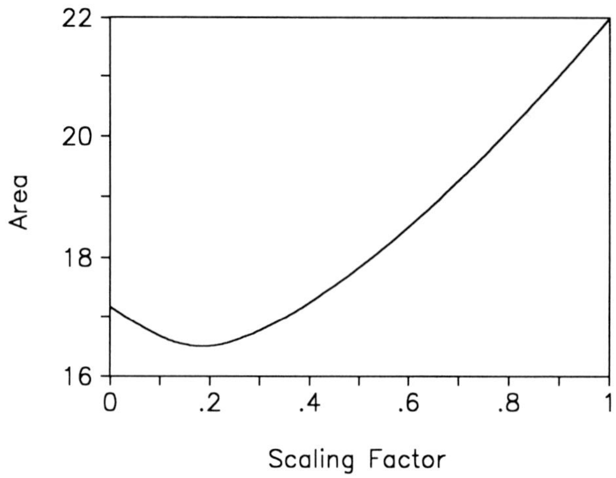

Figure 23. The area of the derivative plotted versus the scaling factor for the subtraction of *A* and *B* in Figure 22 via dewiggling.

Dewiggling owes its success to its simplicity and to the fact that it utilizes all the spectral information, *i.e.*, the intensities at all the wavelengths are utilized. This feature is also its principal drawback, since all the information must be processed in an iterative fashion to find the minimum. This could make a search through a library of several thousand spectra time-consuming to the point of being impractical. A saving grace is that diffuse reflectance (DRIFT) spectra tend to contain broad peaks. In our experience, dewiggling of DRIFT spectra at 20 cm^{-1} resolution works just as well as when higher resolution spectra are used. The lower resolution spectra contain fewer data points making it easier to process. The search time will also decrease as personal computers increase in speed.

Friese[21] recognized that curves like the one in Figure 23 approximated parabolas, and was able to rapidly locate the minimum using a semi-numeric approach. His algorithm begins by calculating the integrated area of the derivative curve at the minimum scaling factor, the maximum scaling factor, and the midway point. The points are fit with a parabola and the resulting function is then used to estimate the minimum in the curve. The process is repeated using the scaling factor at the estimated minimum, and new minimum and maximum scaling factors at ±20% of the previous interval. This is repeated until the net change in the scaling factor is negligible. At this point, the estimated minimum is equal to the optimum scaling factor.

This version of dewiggling was used to calibrate a Spectra Tech Inc., CIRCLE® Cell ATR.[21] The cell was recalibrated after disassembly and cleaning, and the calibration data are shown in Table 1. Aqueous solutions of D_2O at known concentrations were scanned at the beginning of each experiment. The water background was then subtracted away and the area of the D_2O peak was determined. The area and D_2O concentration values were

Table 1
Calibration of an ATR Cell Using Dewiggling[a]

Area	D_2O Conc. (mol/dm^3)	q	N	d_p (mm)
4.310	498.6	51.39	7.076	0.3499
4.338	498.6	51.32	7.094	0.3503
4.249	498.6	51.55	7.036	0.3491
4.319	526.2	51.96	6.932	0.3469

[a] q is the internal reflection incidence angle, N is the number of internal reflections, and d_p is the depth of penetration.

used to calculate the internal reflection incidence angle (q) and the number of internal reflections (N) according to the technique of Sperline, *et al.*[22] The depth of penetration (d_p) at 2508 cm^{-1} was estimated using Equation (1) defined by Harrick.[8] The technique used showed good reproducibility.

Dewiggling is much more tolerant of interferences than conventional techniques that rely on the recognition of a particular spectral feature of a component. The additional signals contribute to the area of the derivative, but the absolute area plays no role in the analysis. Rather, it is the position of the minimum that is important, and this tends to be unaffected by interferences. Since baseline drift is a form of interference, it follows that dewiggling is relatively insensitive to drift, and baseline zeroing is unnecessary.

Another advantage of dewiggling is that it can be performed on the original sample spectrum for each reference tested. In this way distortions due to progressive subtraction are not incurred. In other words, the original spectrum is used for each dewiggling, and in no case does one spectrum need to be subtracted out before another can be recognized.

False minimization is a potential problem when using dewiggling. In our experience, a "match" where the scaling factor is 20% or less is suspect. For example, 11% of the spectrum of lignin is contained in the spectrum of pure cellulose.[20] The probability of an incorrect match decreases as the complexity of a component increases, since a fortuitous overlap of signals then becomes less likely. Conversely, dewiggling is more likely to lead to false positives for spectra containing one or two broad bands since these signals can be contained in several more complex spectra.

In summary, dewiggling allows automated objective subtraction of complex spectra without the necessity of partial identifications and manual spectral subtractions. The technique can be used for quantitative work. Moderate interferences are usually not a problem. The principal drawbacks are the time required to search multiple spectra and the potential for false minimizations.

IDENTIFICATION TECHNIQUES

Earlier it was stated that, in some instances, both identification and isolation techniques may be needed. In any case it is best to first separate the known components from the unknown components. The isolation techniques discussed above may be used for this purpose. Once this is accomplished, identification of the unknown component or components is the main priority.

One of the most common ways to solve this problem is to perform a library search on the unknown spectrum. Most FT-IR software packages come with this feature. The programs search purchased or user-created libraries containing hundreds of spectra, and matches are reported to the user. Library searches can provide quick candidates for comparison. However, the matches are only as good as the program used to obtain them. All candidates identified in this manner should be viewed with a healthy dose of scientific skepticism.

Manual identification is another proven approach to component identification. Individual peaks are identified by position or the presence of companion peaks in particular positions. Ordinary organic or analytical chemistry texts are valuable resources for identification of the more common functional groups. There are two disadvantage to this approach. First the manual nature of the search makes it time-consuming. Secondly, overlapped or obscured peaks may be misinterpreted. All identification techniques can suffer when interactions of sample components cause shifts or peak broadening. More detail on peak identification can be found in the literature.[1,23]

ACKNOWLEDGMENTS

We would like to thank Mr. Mark R. Fisher of Appleton Papers Inc. for his advice on paper coatings and their preparation. We would also like to thank Ms. Sally A. Berben of Integrated Paper Services Inc. for advice on FT-IR sampling techniques and for preparation of the coated and uncoated paper spectra shown in this chapter. This work would not have been possible without their help.

REFERENCES

1. Colthup, N.B., Daley, L.H., and Wiberley, S.E. 1975. Introduction to Infrared and Raman Spectroscopy. Academic Press, New York.
2. Coleman, P.B., ed. 1993. Practical Sampling Techniques for Infrared Analysis. CRC Press, Inc., Boca Raton, FL.
3. Axiom Analytical Inc., 18103-C Sky Park South, Irvine CA 92714. (714) 757-9300.

4. Harrick Scientific Corporation, 88 Broadway, Box 1288, Ossining NY 10562. (914) 762-0020.
5. International Crystal Laboratories, 11 Erie Street, Garfield NJ 07026. (201) 478-8944.
6. Janos Technology Inc., HCR #33, Box 25, Route 35, Townshed VT 05353. (802) 365-7714.
7. Spectra-Tech Inc., 652 Glenbrook Road, P.O. Box 2190-G, Stamford CT 06906. (800) 661-1766
8. Harrick, N.J. 1987. Internal Reflection Spectroscopy. Harrick Scientific Corporation, Ossining, New York.
9. Reffner, J.A. and Wihlborg, W.T. 1990. Microanalysis by reflectance FTIR microscopy. *American Laboratory* (April).
10. Koenig, J.L. and Kormos, D. 1979. Quantitative infrared spectroscopic measurements of mixtures without external calibration. *Appl. Spectrosc.* 33:349–350.
11. Antoon, M.K., D'Esposito, L. and Koenig, J.L. 1979. Factor analysis applied to Fourier transform infrared spectra. *Appl. Spectrosc.* 33:351–357.
12. Haaland, D.M. and Easterling, R.G. 1980. Improved sensitivity of infrared spectroscopy by the application of least squares methods. *Appl. Spectrosc.* 34:539–548.
13. Gillette, P.C. and Koenig, J.L. 1984. Objective criteria for absorbance subtraction. *Appl. Spectrosc.* 38:334–337.
14. Liu, J. and Koenig, J.L. 1987. A new baseline correction algorithm using objective criteria. *Appl. Spectrosc.* 41:447–449.
15. Lacey, R.F. 1989. Elimination of interferences in spectral data. *Appl. Spectrosc.* 43:1135–1139.
16. Gillette, P.C., Lando, J.B., and Koenig, J.L. 1985. *In*: Fourier Transform Infrared Spectroscopy: Application to Chemical Systems. Vol. 3. Freeman, J.R., and Basile, L.J., eds. Academic Press, Orlando, FL.
17. Grams/386 software. Galactic Industries Corporation, 395 Main Street, Salem, NH 03079. (800) 862-6004.
18. Banerjee, S. 1992. Interpreting multicomponent infrared spectra. *Tappi J.* 75(8): 147–149.
19. Banerjee, S. and Li, D. 1991. Interpreting multicomponent infrared spectra by derivative minimization. *Appl. Spectrosc.* 45(6):1047–1049.
20. Friese, M.A. and Banerjee, S. 1992. Lignin determination by FT-IR. *Appl. Spectrosc.* 46(2):246–248.
21. Friese, M.A. 1993. Equilibrium Adsorption of Polyallylamine from Aqueous Media. Ph.D. thesis. Institute of Paper Science and Technology, Atlanta, GA.
22. Sperline, R.P.; Muralidharan, S.; Freiser, H. 1986. *Appl. Spectrosc.* 40:1019–1022.
23. Bellamy, L.J. 1975. The Infra-Red Spectra of Complex Molecules. Third Edition. Chapman and Hall. London.

7

Near-Infrared Spectroscopy of Wood Products

John M. Pope

Byk-Gardner USA, Silver Spring, Maryland, U.S.A.

INTRODUCTION

The near-infrared (NIR) spectral region is usually considered to include light with wavelengths from 800–2400 nm, or 12500–4166 cm^{-1}. In the analysis of surface composition, the technique of diffuse reflectance is generally employed. The intensity of radiation which is reflected from the surface of the sample is analyzed as a function of wavelength. This produces a reflectance spectra characteristic of the material. This is usually presented as an absorbance spectra. Changes in the spectra due to changes in chemical composition can be quantified, and that is the utility of the technique.

Motivations for the use of NIR technology in the analytical laboratory or in a process environment include the speed of the analysis (typically less than 30 seconds), absence of sample preparation, simplicity of the procedure, and improved repeatability in comparison to traditional chemical methods of analysis. Cost savings due to the speed of the analysis alone can often justify the purchase of the equipment. For example, in the determination of cellulose content in wood, the Kurschner method requires about 16 hours. Additionally, minimal sample preparation and simple procedures reduce the need for skilled technicians and analysis variation between technicians. Instrument stability also contributes to improved repeatability and confidence in near-infrared analytical results.

Near-infrared techniques are typically successful for applications in which there are changes in organic chemical composition at concentrations of 0.01% (100 ppm) and higher. One of the inherent features of NIR analysis is that it is a correlative technique, and relies upon a calibration developed using data acquired with some other method, termed the "primary method." The accuracy of the NIR calibration is limited to the precision of the primary method.

The maximum analysis depth of a surface using near-infrared spectroscopy is determined by the penetration depth of the radiation in the sample matrix, typically 1– 3 mm for organic substances. For a surface film, the minimum thickness necessary for analysis

0-8493-8992-5/95/$0.00+$.50

with incident radiation normal to the surface is one quarter of the incident radiation's wavelength. Hence, films thinner than about 300 nm will be transparent to NIR radiation.

COMPARISON WITH OTHER ANALYTICAL TECHNIQUES

Other spectroscopic and spectrometric techniques for the analysis of the chemical composition of surfaces are presented in this volume, including infrared, ultraviolet, Raman, EELS, SIMS, and electron spectroscopy. A detailed comparison with infrared spectroscopy is presented here.

The infrared (IR) region includes radiation with wavelengths of 2500 to 25000 nm, or 4000–400 cm^{-1}. One general advantage of NIR spectroscopy over IR spectroscopy is that the NIR region is associated with higher energy transitions. These have smaller absorptions so that longer pathlengths are possible. This permits little or no sample preparation. Also, quartz fiber optic cables may be used in the NIR region, but begin to absorb strongly at about 2200 nm, so they are not practical for mid-IR measurements. NIR absorption bands are sensitive to their local environment, and distinguish subtle polar, steric, or hydrogen bonding effects.[1] The penetration depth of IR radiation is about 10–100 micrometers, whereas for NIR radiation it is several millimeters. Typical detection limits for IR instruments are as low as 1 ppm, as compared to 100 ppm for NIR instruments.

HISTORICAL DEVELOPMENT

Commercial infrared instruments first became available in 1940s.[2] The work of Karl Norris was central to the development of NIR as a tool for analysis of agricultural products, and later for the analysis of other materials. For example, in 1965 Norris and Hart[3] reported the measurement of moisture in grain and seeds, and in 1968 Ben-Gera and Norris[4] reported the measurement of moisture in soybeans. The significance of Norris' work was due in part to the application of multiple linear regression models to the spectroscopic data. The current emergence of NIR as a tool for the chemical process industries has been due to the recent advances in data analysis (chemometrics), computing power, and the development of fiber optic technology.

PHYSICAL AND CHEMICAL PRINCIPLES

Features of the mid-infrared spectrum are associated with transitions of molecules from their ground states to their first excited vibrational level. The frequency of this transition is known as the fundamental frequency. The major features seen in the NIR region result from overtones of the fundamental vibrations and combinations. These arise because of anharmonicity in the potential energy well. The overtones are due to vibrational transitions to levels above the first excited state. The frequencies of these transitions are approximated by integral multiples of the fundamental frequency. Combinations occur when light is absorbed by more than one bond attached to a common atom. Strong NIR absorbers include the bending and stretching modes of C–H, O–H, N–H, C=O, =C–H, COOH, and aromatic C–H groups.

In many dilute chemical systems, a linear relationship between absorbance and concentration can be established. This is referred to as Beer's Law, and may be expressed as A=abc, where a is termed the absorptivity of the compound (a function of wavelength), b is the pathlength, and c is the concentration. For a transmission measurement, the absorbance is defined as $A = \log_{10}(1/T)$, where T is the relative amount of light transmitted by the sample. Mathematically, $T = I/I_o$, where I_o is the energy of the incident light and I is the energy of the transmitted light. For a reflectance measurement, as in surface analysis, the approximation $A = \log_{10}(1/R)$ is used, where R is the relative amount of light reflected by the sample. Murray and Williams[1] give a concise introduction to the chemical basis of NIR.

INSTRUMENTATION

The spectrometer itself includes a light source, a means of dispersing the light in the near infrared region, optics for focusing the incident light on the sample, optics for focusing the reflected or transmitted light from the sample onto a detector, and the detector itself. The two modes of sample illumination are termed pre- and post-dispersive. In pre-dispersive systems, the incident light is decomposed into its various wavelengths prior to illuminating the sample. In post-dispersive systems, the incident light is decomposed into its various wavelengths after illuminating the sample.

Three distinct optical configurations may be used for a surface reflectance measurement: reflectance, remote reflectance, or fiber optic reflectance. In a reflectance measurement, the sample is placed directly against a reflectance detector attached to the spectrometer. In a remote reflectance measurement, the sample is placed against a reflectance detector which communicates with the spectrometer through a length of fiber optic cable. In a fiber optic reflectance measurement, the tip of a bidirectional fiber optic cable which communicates with spectrometer is placed against the surface of the sample.

Several types of instrumentation are in use, including filter-based instruments (photometers), scanning-grating-based instruments (dispersive spectrophotometers), Fourier transform spectrophotometers, diodes, and acousto-optically tuned filter (AOTF) based instruments. Filter instruments are inexpensive and rugged, but have poor bandwidth, low resolution, and may miss spectral features. Scanning gratings are fast and provide information across the spectral region, but rely on a moving part. Detectors are usually lead sulfide (PbS). Indium gallium arsenide (InGaAs) detectors are also used for wavelengths between 900 and 1700 nm. Fourier transform instruments convert modulated light signals into spectral data. They are fast and provide high resolution, but have a poor signal-to-noise ratio and do not readily provide quantitative information. Systems with diode array detectors are very fast (capable of hundreds of scans per second), and may have no moving optical components. Diode arrays are composed of many individual photodiodes, each of which measures the signal in a different spectral region. The AOTF uses acoustic waves to tune a crystal to a specific wavelength of light. The crystal acts as a diffraction grating. This type of filter is able to change wavelengths as rapidly as 10,000 times per second. Refer to Figures 1 and 2.

Landa[5] gives a good introduction to scanning monochromator systems. McClure[2] describes both filter and monochromator based systems in general terms.

SOURCES OF ERROR

For the purpose of chemical analysis, one desires both accuracy and reproducibility. Sources of error may arise during the sample acquisition, during the analytical laboratory's primary analysis, from the sample presentation, from the NIR instrumentation, from improper assumptions concerning the chemical matrix, or during modeling.

To minimize error during sample acquisition, it is critical that the sample be representative of the process. The sample must be uniform, free of artifacts, and the sample integrity must be maintained prior to analysis. According to Gy,[6] correct probabilistic

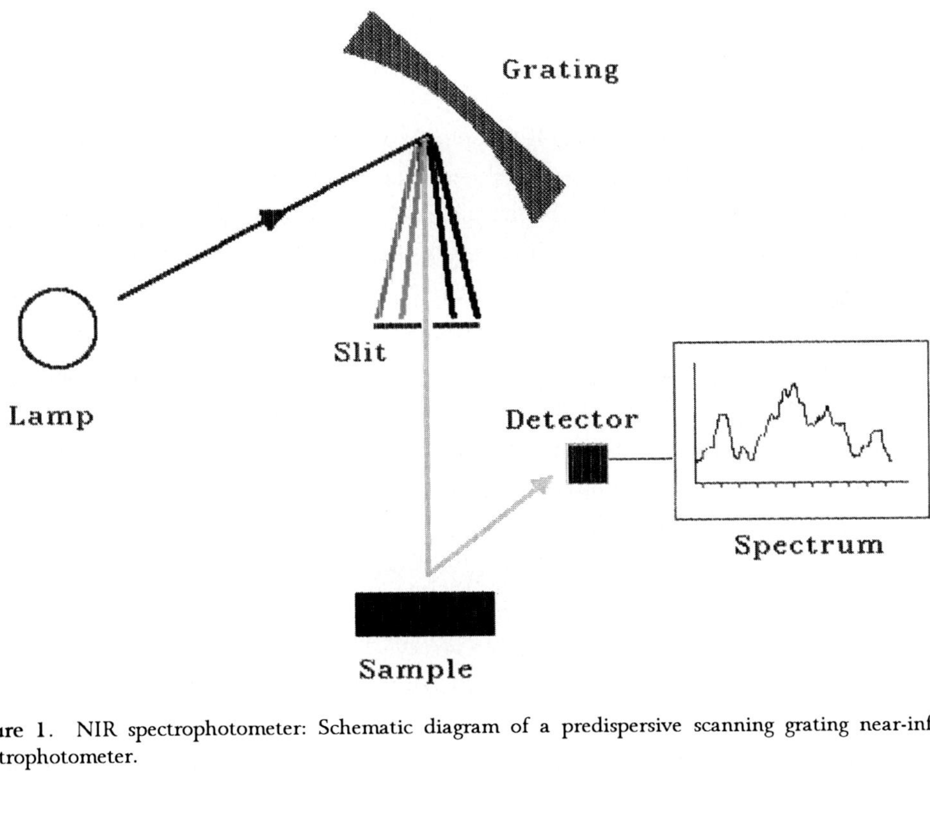

Figure 1. NIR spectrophotometer: Schematic diagram of a predispersive scanning grating near-infrared spectrophotometer.

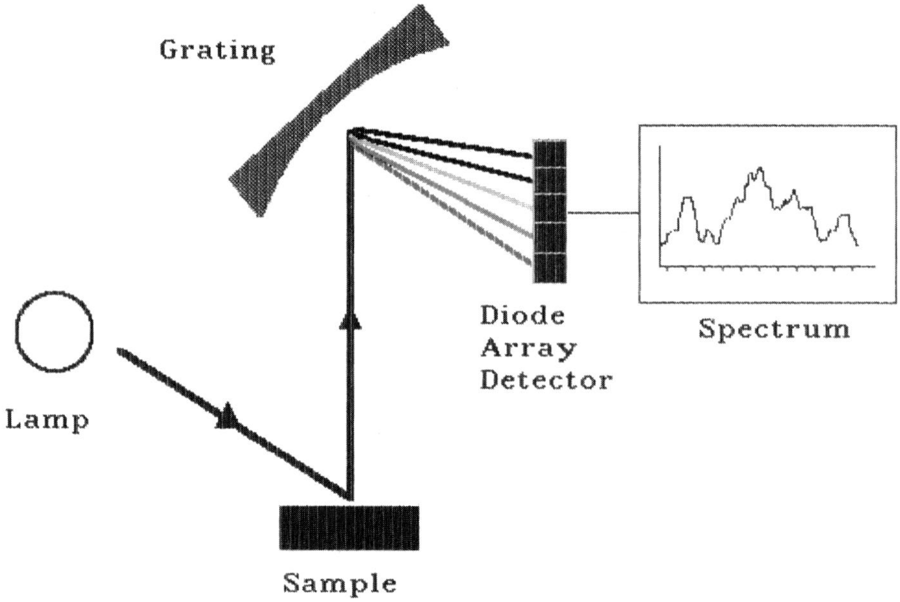

Figure 2. NIR diode array: Schematic diagram of a postdispersive near-infrared spectrophotometer with a photodiode array detector.

sampling requires that all potential sample elements have equal probability of selection, and that the selected elements are unaltered prior to analysis.

The accuracy of the NIR calibration is limited by the precision of the primary analysis method. For example, if the precision of the primary analysis for kappa number is ± 1.0, then the accuracy of the NIR method cannot exceed ± 1.0. Frequently, the precision of the primary analytical test method is the most significant source of error in the NIR analysis. If this is the case, the accuracy of the NIR analysis will approach the precision of the primary test as the size of the calibration set increases to a statistical limit.

As with sample acquisition, sample presentation to the near-infrared spectrometer must be representative and reproducible. This is of particular importance for samples with surface heterogeneities such as wood pulp. For samples of this type, there may be physical surface effects. Pathlength and scattering variations due to differing particle size distributions, orientations, and compacting can affect the spectra. One approach to this problem is to take several spectra of the same sample in different orientations, to remix the sample, or to use multiple samples from the same batch. These spectra can then be averaged, after appropriate normalization, to remove variation due to surface artifacts. Another approach is to use some type of dynamic averaging procedure, such as spinning the sample during the spectroscopic data acquisition.

One common assumption which may lead to analysis errors is that a physical property is linearly related to the spectroscopically measured concentrations of chemical functional groups. An error which may appear during modeling is due to the assumption of linear concentration behavior according to Beer's law. This assumes that concentrations are expressed in moles per unit volume. The use of weight based concentration terms can thus lead to errors. Additionally, nonlinear behavior may be associated with high concentrations of analyte, and/or high absorbance levels. More frequently, prediction errors may arise because of overmodeling. This occurs when adjustable parameters are used to describe the natural variation (noise) of the calibration set. Fortunately, modern statistical packages make this error relatively easy to identify.

GENERAL PROCEDURES

Typical steps in developing a near infrared analysis procedure include optimizing the instrument and sampling system, establishing a background or reference spectra, validating the primary analysis method, collecting spectra from the calibration set, processing the spectral data to improve resolution and remove sampling artifacts, applying a modeling technique to the data (usually a linear model such as multiple linear regression (MLR), partial least squares (PLS), or principal component regression (PCR)), and validating the technique by predicting analytical values for samples independent of the calibration set. Statistical tests may be applied to detect outliers both in the calibration and validation sets.

Optimization of the sampling system and spectrometer for a surface reflectance measurement includes adjusting the pathlength to produce an average absorption level of 0.4 to 0.5. Additionally, a technique is used to fix both the wavelength position and the energy scale of the spectrometer. The wavelength position may be fixed by the use of some standard, such as polystyrene, which exhibits absorbances at specific wavelengths in the NIR region. Establishing a background or reference spectrum is required to adjust for minor optical variations in a single beam spectrometer. The background spectrum might be an empty cell or probe for a transmission measurement, or of a neutral material such as white ceramic for a reflectance measurement. The absorbance due to the background is subtracted from the absorbance due to the sample. The sampling technique may include some method for averaging of surface heterogeneities. The precision of the primary analysis method is determined by multiple analyses of the same sample. The standard error of a valid calibration can never be less than the precision of the primary analysis method. A calibration set is collected which represents the full range of chemical and physical variation expected to be encountered in practice. Spectra for both the calibration and validation sets are obtained and processed in an identical manner. Data are converted to absorbance units by dividing by a reference or background scan and taking the negative logarithm. This is equivalent to subtracting the absorbance of the background scan. First and second derivatives of the absorbance spectra are frequently used to enhance the spectra by removing the effects of baseline offsets caused by small changes in pathlength, light intensity, scattering, etc.

To develop a prediction equation for chemical analysis, single and multiple variable analysis routines are run on a calibration set of spectra with known concentrations. These techniques are usually based on linear regression algorithms. The change in instrument response is correlated with independently determined primary values to develop predictive models. In single linear regression (SLR), changes in absorbance intensity at a single wavelength are correlated with the changing analytical values. In multiple linear regression (MLR), changes in absorbance intensity at several wavelengths are correlated with changing analytical values. In simple chemical systems, it is expected that one wavelength will be required for each independently varying component of the chemical matrix. Other methods of linear analysis, termed chemometrics, may be applied to more complex systems. Among these are included principal component regression (PCR) and partial least squares (PLS). These algorithms calibrate using all available spectral data. Both PLS and PCR represent the original data as linear combinations of "factors" weighted by "scores." The factors are pseudo spectra, and in some cases can be identified as representing specific chemical species. The scores are numeric multipliers. In the notation of matrix algebra, $X = SF$, where X are the original set of spectra, S are the scores, and F are the factors. The number of factors appropriate for a specific system may frequently be determined by minimizing the prediction error with respect to the number of factors. In general, simple chemical systems, such as mixtures, are well represented with simple models, such as MLR. Complex systems, such as biological materials, usually require complex models, such as PLS. Mathematical models are typically evaluated using the correlation coefficient (R), the standard error of calibration (SEC), and the standard error of prediction (SEP). R^2 is interpreted as the fraction of the variation of the calibration set which is explained by the model. The SEC is interpreted as the standard deviation of the true (laboratory analysis) versus the fitted (modeled) values of samples in the calibration set. The SEP is interpreted as the standard deviation of the true versus fitted values for samples independent of the calibration set.

Typical NIR calibration sets range from 20 to 400 samples. The number of samples required for robust calibration will vary with the nature of the calibration set, the number of independent variables, and the precision desired. NIR calibrations frequently have a standard error of prediction approaching that of the primary method.

After confidence in a particular model has been developed through the use of validation samples, the model may be used to predict the properties of unknown samples. In practice, various models might be compared over a period of time to determine which is the most robust with respect to environmental factors such as temperature, humidity, and process variation.

WOOD PRODUCTS: REVIEW OF LITERATURE AND APPLICATIONS

In 1989, Paralusz[7] described the analysis of silicone coating weights on paper using a reflectance detector in the 1200 to 2400 nm range. Coating weights varied from 0.244 to 4.15 g/m^2, with standard errors of about 0.05 g/m2.

Birkett and Gambino[8] used NIR to estimate kappa number for both hardwoods and softwoods. Standard errors of about 1 were achieved over the range of 20 to 90 kappa numbers.

Schultz and Burns[9,10] compared NIR and FTIR techniques for analysis of hardwood and softwood for lignin, hemicellulose, and cellulose content. The NIR technique was judged to be superior for this analysis on the basis of simplicity of sample preparation, shorter scanning time, and smaller standard errors. The best results were for NIR analysis of lignin, with standard errors of calibration less than 1% over the range of 10 to 30% in pine and sweetgum.

Wright *et al.*[11] used NIR to predict pulp yield and cellulose content in powdered wood samples. Using a filter instrument employing five wavelengths, they indicate standard errors of about 2%.

Easty *et al.*[12] used reflectance NIR to quantify hardwood–softwood ratios in paperboard and to estimate lignin content (kappa number) in hardwoods and softwoods. They found that separate calibrations were required for the two types of woods. Using second derivative spectra, they identified wavelengths which correlated with lignin content at 1680 nm and in the 2100 to 2200 nm region.

In 1991, Wallbacks *et al.*[13] compared results for the analysis of birch pulp using NMR, FTIR, and NIR. The pulping process is essentially a delignification reaction in which lignin's phenylpropane units are cleaved at ether linkages with an alkaline reagent. NIR bands from cellulose, hemicellulose, and lignin overlap, but may be resolved by using multivariate mathematical techniques such as partial least squares (PLS) analysis. They developed PLS models for Klason lignin, glucose, and xylose. NIR gave the best prediction results, with standard errors of 1% or less for all three components. The range of the measurements was xylose: 21–26%; glucose: 55–75%; and Klason lignin: 0–25%.

In 1993, Yuzak and Lohrke[14] measured kappa number for kraft batch pulp using surface reflectance techniques. They examined different sample preparation methods, and found the best results with homogenized, dried samples. They concluded that kappa number could be measured consistently with a 95% confidence interval of ± 2 kappa numbers.

Papers with different coatings have been easily distinguished in laboratory studies.[15] In related work, Backa and Brolin[16] have used DRIFT (diffuse reflectance infrared Fourier transform spectroscopy) for the determination of lignin and carbohydrates in pulp. They used a PLS analysis, and found a standard error of prediction for kappa number of 2.75% over the range of 20 to 90 kappa numbers (kappa number goes to zero as lignin content goes to zero). They believed this was consistent with the precision of the primary test method (the SCAN standard). They also found that homogenization of the sample improved the precision of the measurement.

Figure 3 illustrates absorbances of several paper specimens in the near-infrared region.

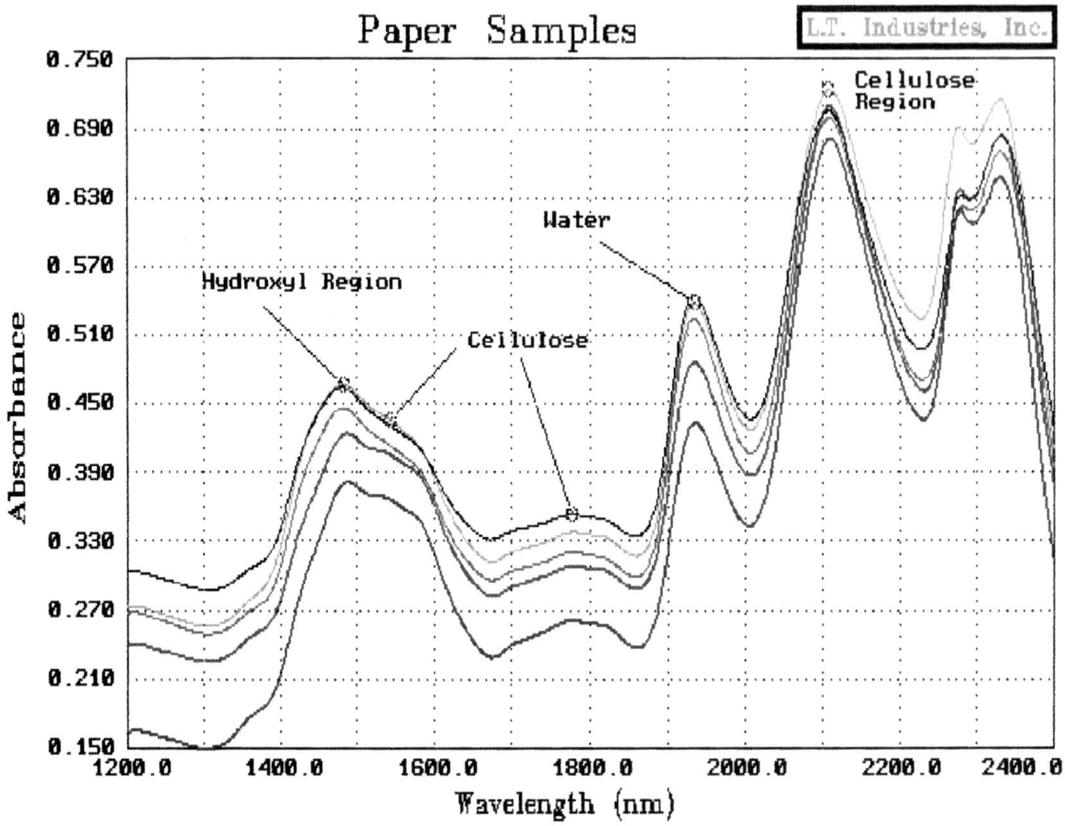

Figure 3. Typical absorbance spectra of paper, showing absorbance regions attributed to cellulose and water.

ACKNOWLEDGEMENTS

Portions of this chapter were written while the author was employed as a Senior Scientist at LT Industries, Rockville, MD. The author would like to thank his colleagues at LT Industries, especially James P. Malone, Dr. Glenn Merberg, and Dr. George Asimopoulos for their help in preparing this article.

REFERENCES

1. Murray, I., and Williams, P.C. 1987. Chemical principles of near-infrared technology. *In*: Near-Infrared Technology in the Agricultural and Food Industries. Williams, P. and Norris, K. eds. American Association of Cereal Chemists, Inc., St. Paul, MN. Chapter 2.
2. McClure, W.F. 1987. Near-infrared instrumentation. *In*: Near-Infrared Technology in the Agricultural and Food Industries. Williams, P. and Norris, K. eds. American Association of Cereal Chemists, Inc., St. Paul, MN. Chapter 3.
3. Norris, K.H., and Hart, J.R. 1965. Direct spectrophotometric determination of moisture content of grain and seeds. *In*: Principles and Methods of Measuring Moisture in Liquids and Solids. Vol. 4. A. Wexlar, ed., Reinhold, NY. pp. 19–25.
4. Ben-Gera, I., and Norris, K. 1968. Determination of moisture content in soybeans by direct spectrophotometry. *Isr. J. Agric. Res.* (18):125–132.
5. Landa, I. 1979. High-energy spectrophotometer for rapid constituent analysis in the range of 0.25–2.4 μm. *Rev. Sci. Instrum.* 50(1):34–40.
6. Gy, Pierre M. 1990. Do not forget the sampling theory! *Process Control and Quality* (1): 15–22.
7. Paralusz, C.M. 1989. Near-infrared reflectance analysis for silicon coating weights. *Applied Spectroscopy* 43(7):1273–1279.
8. Birkett, M.D. and Gambino , M.J.T. 1989. Estimation of pulp kappa number with near-infrared spectroscopy. *Tappi J.* (9):193–197.
9. Schultz, T.P., and Burns, D.A. 1990. Rapid secondary analysis of lignocellulose: comparison of near infrared (NIR) and Fourier transform infrared (FTIR). *Tappi J.* (5):209–212.
10. Burns, D.A. and Schultz, T.P. 1992. FT/IR vs. NIR: A study with lignocellulose. *In*: Handbook of Near-Infrared Analysis. Burns, D.A. and Ciurczak, E.W., eds. Practical Spectroscopy Series, Vol. 13, Marcel Dekker, Inc. pp. 663–671.
11. Wright, J.A., Birkett, M.D., and Gambino, M.J.T., Prediction of pulp yield and cellulose content from wood samples using near infrared reflectance spectroscopy. *Tappi J.* (8):164–166.
12. Easty D.B., Berben, S.A., DeThomas, F.A. and Brimmer, P.J. 1990. Near-infrared spectroscopy for the analysis of wood pulp: Quantifying hardwood–softwood mixtures and estimating lignin content. *Tappi J.* (10):257–261.
13. Wallbacks, L., Edlund, U., Norden, B., and Berglund, I. 1991. Multivariate characterization of pulp using solid state ^{13}C NMR, FTIR, and NIR. *Tappi J.* (10):201–206.
14. Yuzak, E., and Lohrke, C. 1994. Submitted for publication. *Tappi J.*
15. Pope, J.M. 1993. Unpublished results.
16. Backa, S. and Brolin, A. 1991. Determination of pulp characteristics by diffuse reflectance FTIR. *Tappi J.* (5):218–226.

8

Raman Spectroscopy

Umesh P. Agarwal and Rajai H. Atalla

USDA Forest Products Laboratory, Madison, Wisconsin, U.S.A.

INTRODUCTION

Since the discovery of the Raman effect in 1928, Raman spectroscopy has been a field of major advances. In the early years, the technique was primarily used to study vibrational states of simple molecules. This was due to the unavailability of intense monochromatic light sources and the lack of suitable optical components and detectors. However, with the discovery of lasers in 1960s and the development of high-throughput monochromators and sensitive detectors, Raman spectroscopy has been increasingly applied to solve problems of both fundamental and technological interest. In the area of wood, pulp, and paper research, however, Raman spectroscopy has not been used very much. Primarily, this seems to be due to the fact that in conventional Raman spectroscopy sample-generated laser-induced fluorescence (LIF) overwhelms the Raman signal. It was only when means to suppress the LIF contribution (to a Raman spectrum) were found, that useful spectra were obtained. Additional factors responsible for limited Raman use in pulp and paper research were instrumentation costs, spectrum acquisition time, the lack of trained personnel, and a perception that Raman and infrared (IR) spectroscopies provide the same information. In reality, the two techniques are complementary.

With the development of near-IR FT Raman spectroscopy, LIF is no longer a problem for most paper and lignocellulosic samples. This is evident from the fact that even unbleached kraft pulps (strongly fluorescent in conventional Raman) have been successfully analyzed using the near-IR FT Raman approach.[1,2] Another advantage with the FT technique is that the time needed to acquire a spectrum has been drastically reduced. With these developments behind us, no barriers exist to using Raman spectroscopy and its application in pulp and paper research is expected to increase. Indeed, FT Raman systems are now available in industrial laboratories affiliated with the pulp and paper industry.

None of the experimental difficulties encountered in complex lignocellulosic systems are present when cellulose and lignin models are studied using conventional Raman

0-8493-8992-5/95/$0.00+$.50

spectroscopy.[3–10] Normal coordinate analyses of such models have been carried out to provide a conceptual framework for interpretation of spectra of cellulose and lignin. Although such studies contributed significantly to the Raman studies of wood, pulp, and paper, additional studies of models and other simpler systems will be needed in the future to solve problems using Raman spectroscopy.

PRINCIPLES

The phenomena underlying Raman spectroscopy can be described with reference to infrared spectroscopy as shown schematically in Figure 1. The primary event in infrared absorption is the transition of a molecule from a ground state (M) to a vibrationally excited state (M*) by absorption of an infrared photon with energy equal to the difference between the energies of the ground and the excited states. The reverse process, infrared emission, occurs when a molecule in the excited state (M*) emits a photon during the transition to a ground state (M). In infrared spectroscopy, one derives information by measuring the frequencies of infrared photons a molecule absorbs and interpreting these frequencies in terms of the characteristic vibrational motions of the molecule. In complex molecules, some of the frequencies are associated with functional groups that have characteristic localized modes of vibration.

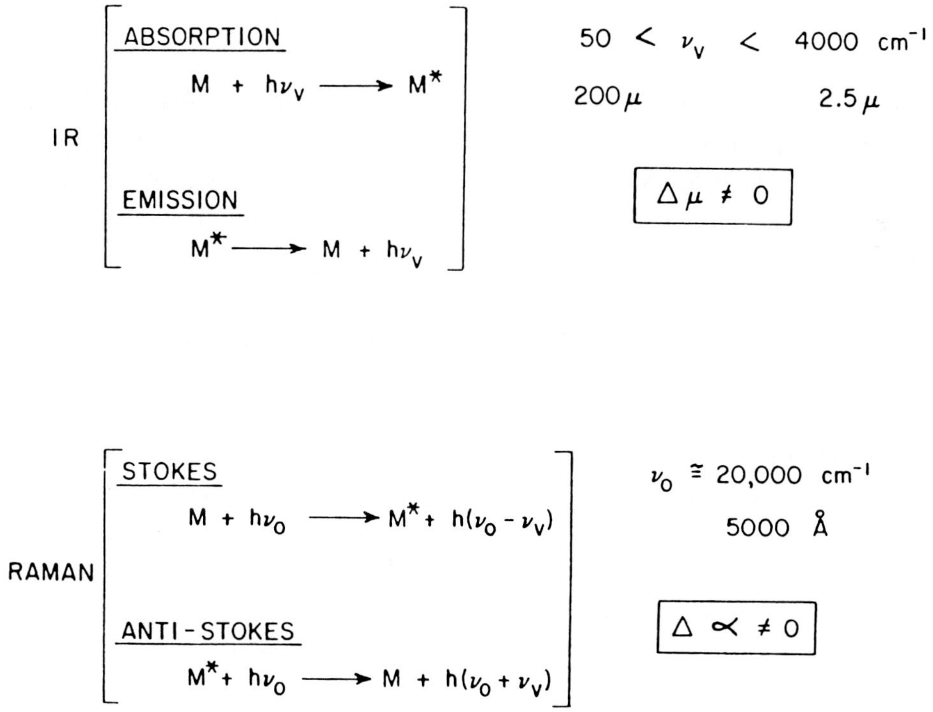

Figure 1. A comparison of infrared and Raman phenomena; μ = dipole moment, α = polarizability, ν_v = vibrational frequency, ν_0 = exciting frequency.

As shown in Figure 1, the same transitions between molecular vibrational states (M) and (M*) can result in Raman scattering. A key difference between the Raman and infrared processes is that the photons involved are not absorbed or emitted but rather shifted in frequency by an amount corresponding to the energy of the particular vibrational transition. In the *Stokes process*, which is the parallel of absorption, the scattered photons are shifted to lower frequencies as the molecules abstract energy from the exciting photons; in the *anti-Stokes process*, which is parallel to emission, the scattered photons are shifted to higher frequencies as they pick up the energy released by the molecules in the course of transitions to the ground state. In addition, a substantial number of the scattered photons are not shifted in frequency. The process which gives rise to these photons is known as *Rayleigh scattering*. This scattering arises from density variations and optical heterogeneities and is many orders of magnitude more intense than Raman scattering.

Figure 1 depicts another major difference between Raman scattering and infrared processes. To be active in the infrared spectra, transitions must have a change in the molecular dipole associated with them. For Raman activity, in contrast, the change has to be in the polarizability of the molecule. These two molecular characteristics are qualitatively inversely related.

A Raman spectrum is obtained by exposure of a sample to a monochromatic source of exciting photons, and measurement of the light intensity at the frequencies of the scattered light. Because the intensity of the Raman scattered component is much lower than that of the Rayleigh scattered component, filters and diffraction gratings are used to suppress the latter component. A highly sensitive detector is required to detect weakly scattered Raman photons.

In conventional Raman spectroscopy where visible laser radiation is used, the exciting photons are typically of much higher energies than those of the fundamental vibrations of most chemical bonds or systems of bonds, usually by a factor ranging from about 6 for O–H and C–H bonds to about 200 for bonds between very heavy atoms, as for example in I_2. The 514.5 and 488 nm lines from an argon ion laser are often used as exciting frequencies. In contrast, in the case of FT Raman spectroscopy, excitation with a laser is usually carried out in the near-IR region (1064 nm). The energies of the exciting photons are higher by factors that are one-half or less of the above-mentioned values. Another point to consider in near-IR excited Raman is that, for some samples, absorption of Raman scattered photons due to overtone and combination modes of molecular vibrations can occur.

An important difference between the conventional and near-IR FT approaches is the difference in the scattering efficiencies at the absolute Raman frequencies. Because Raman scattering is directly proportional to the fourth power of the absolute frequency of a Raman line, in the FT approach, Raman scattering is lower by factors anywhere between 18 to 53 when compared to the 514.5 nm excited spectra. However, this is not a problem because of the "multiplex" and throughput advantages of the interferometer in the FT approach.[11]

INSTRUMENTATION

CONVENTIONAL RAMAN

A number of stages are involved in the acquisition of a Raman spectrum using a conventional approach. A sample is mounted in the sample chamber and laser light is focused on it with the help of a lens. Generally, liquids and solids are sampled in a Pyrex capillary tube. Solids are sampled either as pellets or are examined directly without any sample preparation. The scattered light is collected using another lens and is focused at the entrance slit of the monochromator. Monochromator slit widths are set for the desired spectral resolution. The monochromator effectively rejects Raleigh scattering and stray light and serves as a dispersing element for the incoming radiation; sometimes more than one monochromator is used to obtain high resolution and/or better suppression of the Rayleigh line. The light leaving the exit slit of the monochromator is collected and focused on the surface of a detector. This optical signal is converted to an electrical signal within the detector and further manipulated using detector electronics. In a conventional Raman system using a photomultiplier tube (PMT) detector, the light intensity at various frequencies is measured by scanning the monochromator. In contrast, when a multichannel detector is used a spectral range is simultaneously recorded.

A typical conventional Raman system consists of the following basic components: *i*) an excitation source, usually a visible-light laser; *ii*) optics for sample illumination and collection of sample-scattered light; *iii*) a monochromator; and *iv*) a signal processing system consisting of a detector and a data processing unit. A diagram showing the various components of a Raman spectrometer is shown in Figure 2.

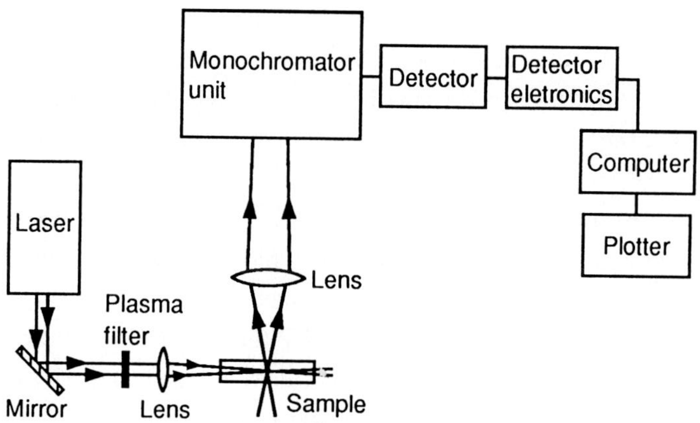

Figure 2. Schematic of a conventional laser Raman system.

Although argon and krypton ion lasers are routinely used for sample excitation, for special purposes UV lasers have also been used.[12] The choice of a suitable monochromator is dictated both by the spectral resolution desired and by the spectral range in which the spectrometer is to be used. A number of detector types exist in conventional Raman spectroscopy. For the scanning instruments, the most frequently used detector is the PMT. More recently, however, multichannel detectors mounted in the plane of the exit slit have been increasingly used. These consist of charge coupled devices and silicon photodiode arrays. These are especially desirable if high resolution is not required and rapid spectral acquisition is needed.

FT RAMAN

The instrumentation needs of conventional and FT Raman techniques are different. Although the former technique has been available for quite some time and has benefited from developments, FT Raman instruments have only recently become available. In fact, the feasibility of FT Raman was demonstrated for the first time in 1986.[13] It is likely, therefore, that further developments will improve the available FT instrumentation.

Conventional Raman systems are based upon the use of dispersive monochromators of various kinds. FT Raman instruments, on the other hand, are built around an interferometer. Interferometer-based instruments have several advantages over dispersive spectrometers using gratings. Two of the most important advantages associated with the FT approach are the Jacquinot and Felgett advantages.[11] The latter is also known as the "multiplex" advantage and has to do with the fact that all wavelengths of light are detected simultaneously. This is the main reason why an FT Raman spectrometer records a spectrum in a shorter time than a grating instrument. The high throughput advantage of the interferometer is known as Jacquinot advantage. An additional advantage for FT instruments is one of accuracy of the wavenumbers in a spectrum. This is important when Raman spectra are averaged to increase the signal-to-noise ratio. The wavenumber values in sequentially acquired spectra should match exactly to avoid problems of band broadening and/or peak shifting during signal averaging. This requirement is not precisely met in scanning instruments because of limits on the precision of mechanical scanning devices.

In addition to the interferometer, an FT Raman spectrometer requires a laser for sample excitation, one or more filters to effectively block the Rayleigh scattering, a highly sensitive detector, and the capability to do a fast Fourier transform on an acquired interferogram. During the development of the FT technique, the suitability of all these components (with the exception of the FT capability) needed to be demonstrated. Early in the development effort (1986–88), a continuous wave Nd:YAG laser (1064 nm) was found to be the most suitable for sample excitation. When such a laser was coupled with the existent IR and near-IR optical and electronic components, it was found that even highly fluorescent samples produced good Raman spectra.[14] That was, and still is, the single most important reason why the FT technique was developed. Since 1986 better Rayleigh-line filters and signal detectors have been developed. It was also found that gold-coated optical components provided higher signal throughput. A layout of an FT Raman system is shown in Figure 3.

METHODS

SCATTERING GEOMETRIES

Generally speaking, there are two geometries in which a sample is studied in Raman spectroscopy (Figure 4). In the 90 degree geometry, the laser beam direction and the axis of the collection lens are at 90 degrees to each other. On the other hand, in the 180 degree scattering geometry (also called back scattering mode) these two axes are coincident. The 90 degree scattering geometry is frequently used in the conventional approach, whereas both 90 and 180 degree modes are used in the FT Raman spectroscopy. However, when a microprobe is used to study small regions of a sample the geometry is invariably set at 180 degrees.

MACRO- AND MICRO-SAMPLING

Until recently, samples in Raman spectroscopy were almost always studied at the macro level. With the development of a Raman microprobe[15] however, sample regions as small as one micron can now be examined. When a sample is analyzed using the macro-sampling mode, the obtained spectrum provides average information over a large sample

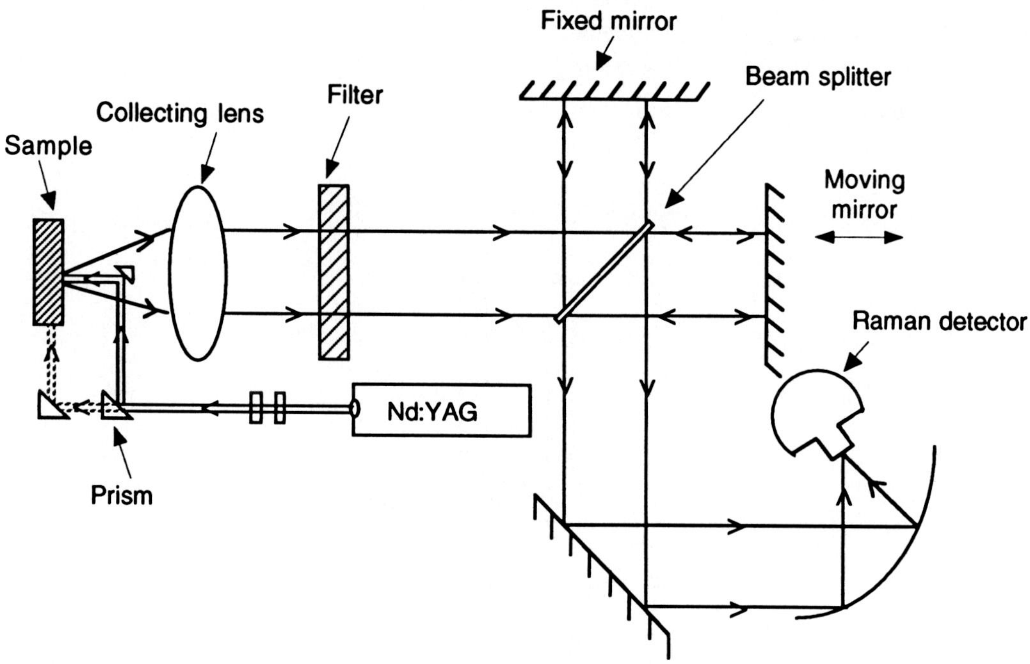

Figure 3. Schematic of an FT Raman instrument.

area. This is so even in cases where a sample is heterogeneous at the micro level. For example, a woody-tissue Raman spectrum acquired in the macro mode provides an average spectrum that contains varying and simultaneous contributions from morphologically distinct regions. On the other hand, spectra that are characteristic solely of the middle lamella, the secondary wall, or any other morphologically distinct area in the woody tissue can only be obtained using a Raman microprobe.[16] A microprobe consists of an optical microscope coupled to a Raman spectrometer. The microscope objective not only focuses the incident laser beam but also collects the scattered light. Depending upon the spatial resolution desired, different magnification objectives have been used in Raman microprobe studies. In our laboratory both the 40 and 100X objectives have been used.

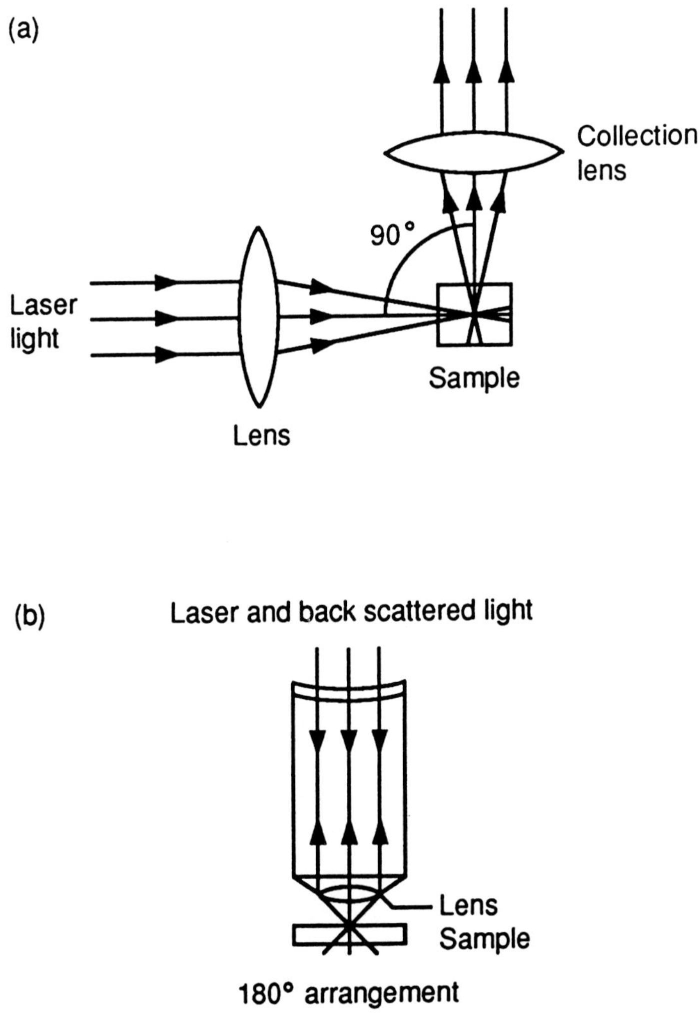

Figure 4. (a) 90 degree and (b) 180 degree (back scattering) geometries in Raman spectroscopy.

LASER-INDUCED FLUORESCENCE

In conventional Raman spectroscopy, a large number of pulp and paper samples exhibit laser-induced fluorescence (LIF). It was found that for lignin-rich mechanical- and chemi-mechanical-pulps (and paper grades containing such pulps), lignin was responsible for the LIF (unpublished results). Furthermore, in the case of unbleached kraft pulps, residual lignin structures produced most of the fluorescence. In contrast, for fully-bleached chemical pulps where lignin was nearly completely removed, LIF was attributed to the presence of low levels of fluorescent impurities.[17] However, the lignin-based and impurities-caused LIF contributions to Raman spectra were significantly reduced by using molecular oxygen[18] and by using a "drench-quenching" technique, respectively.[19] The latter is based on the fact that for some samples irradiation with focused laser light causes LIF to decay with time. The decay in LIF is believed to be caused by the degradation of fluorescing impurities in a sample. This leads to better Raman signal-to- background ratio in the spectrum. The "drench-quenching" effect has been used to obtain better quality Raman spectra from samples that do not have intrinsic fluorescence. Once an acceptable level of fluorescence suppression is achieved, Raman spectra can be obtained. Figure 5 shows TMP (thermomechanical pulp) spectra that were obtained with and without the use of molecular oxygen as a fluorescence quencher. The spectrum in Figure 5a was divided by 50 to reduce signal intensity so that it could be plotted along with the spectrum in Figure 5b. As seen in the spectrum in Figure 5a, when no oxygen is used, the Raman peaks are barely discernible. Sampling under oxygen dramatically reduces the background signal and spectral features of cellulose and lignin can be clearly seen. Likewise, the effect of "drench quenching" is shown in Figure 6. Although the bleached pulp spectrum in Figure 6a showed Raman features, significant fluorescence remained in the spectrum. After exposing the sample to a laser beam for several hours, a much improved spectrum was obtained (Figure 6b). Such fluorescence suppression methods, however, did not work well with samples containing residual lignin (unbleached or partially bleached kraft pulps). Nevertheless, with the availability of the FT Raman technique, where the 1064 nm laser light does not excite many structures in residual lignin, fluorescence is no longer a problem. Bleached chemical pulps and lignin-containing samples required no fluorescence suppression methods when studied using FT instruments .

SPECTRAL INFORMATION ON PAPER COMPONENTS

Cellulose and hemicelluloses are the basic constituents of fully bleached chemical pulps. Raman features due to hemicelluloses are not readily identifiable because these carbohydrate polymers are not crystalline and their primary structure is similar to cellulose (*i.e.*, composed of C–O and O–H bonds). Therefore, Raman contributions of hemicelluloses are expected to be broad and to occur in spectral regions where cellulose Raman bands

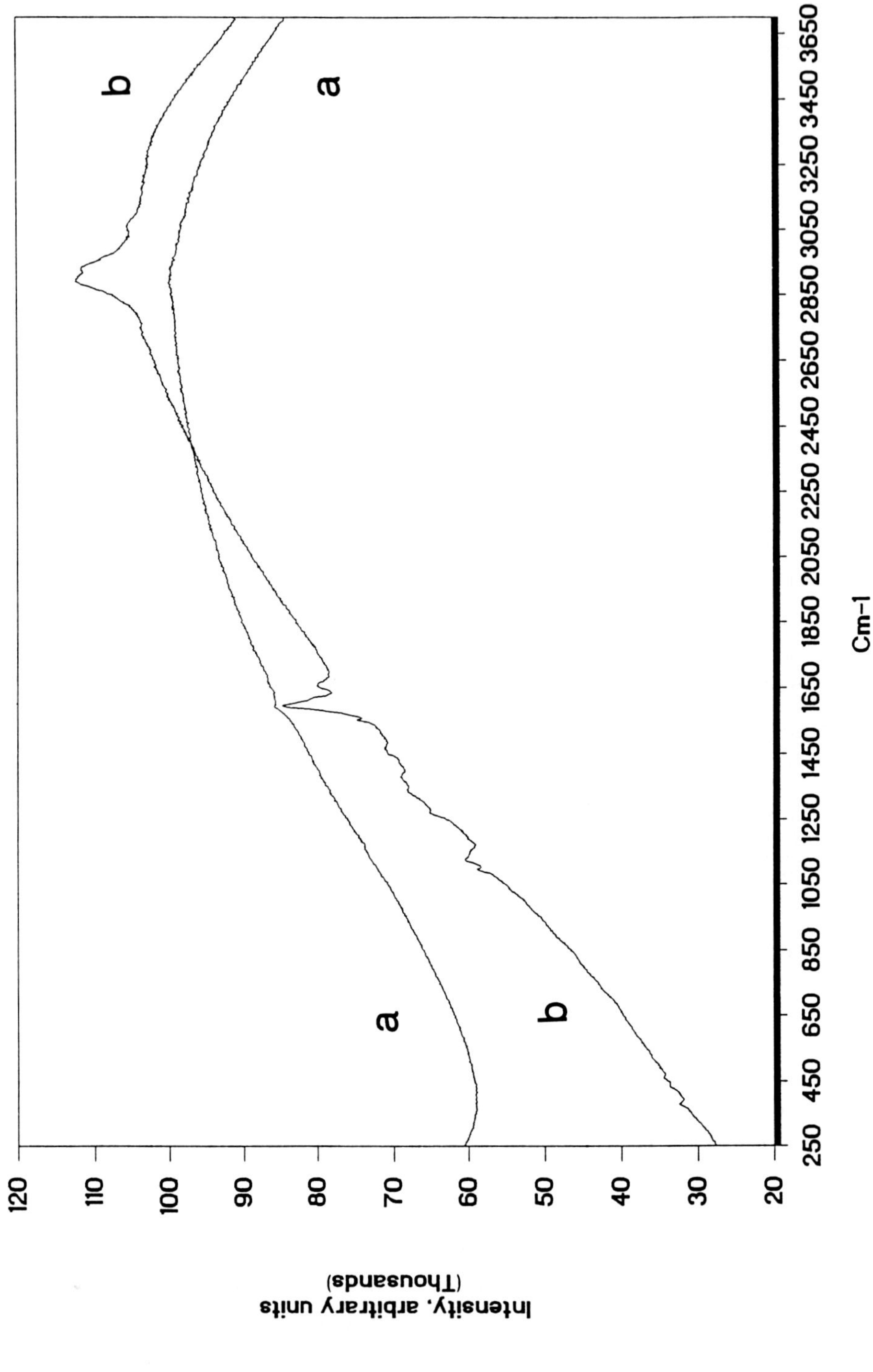

Figure 5. Effect of molecular oxygen on background fluorescence: (a) thermomechanical pulp (TMP) spectrum acquired under ambient conditions and (b) under 50 psi oxygen.

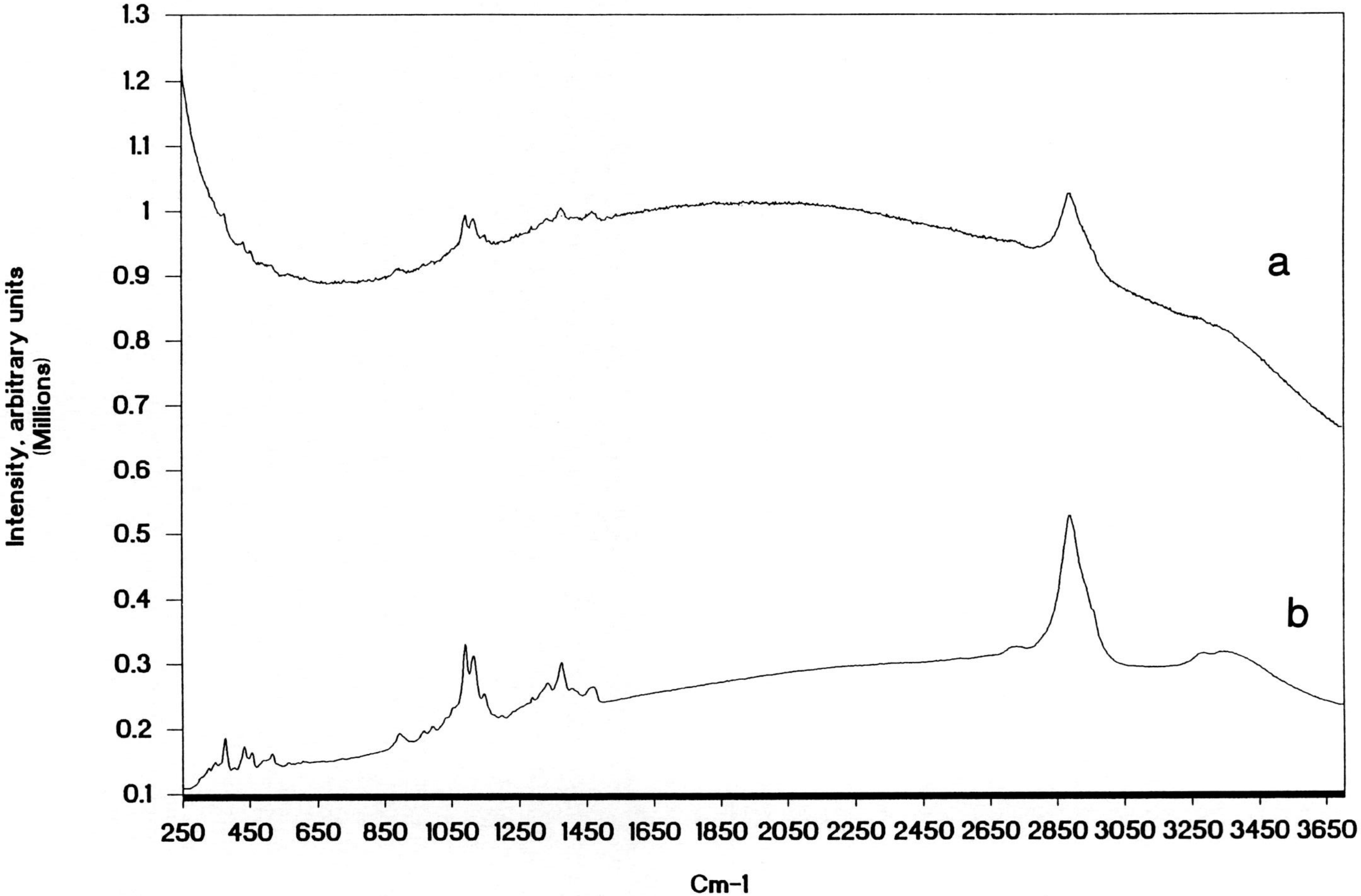

Figure 6. Effect of "drench quenching" on Raman signal-to-background ratio: (a) bleached kraft pulp spectrum recorded without sample exposure to laser beam and (b) after a 5 hour laser beam exposure.

are detected. In addition to these carbohydrate components, lignin is present in mechanical and chemimechanical pulp-based paper products. Furthermore, in unbleached kraft pulps, a modified form of lignin called residual lignin is detected. Lastly, depending upon their intended use, finished paper products may contain a number of paper additives. Some of these (binders, pigments, and fillers, etc.) can be detected at applied concentrations.

CELLULOSE

Raman spectra of native celluloses and cellulose polymorphs have been reported in the literature and their spectral features have been assigned.[20,21] In addition, molecular structures of celluloses have been studied using Raman spectroscopy.[22,23] In Figure 7a, we show an FT Raman spectrum of a cellulose filter paper (Whatman 1). The band positions are listed in column 1 in Table 1. The FT spectrum is almost identical to that obtained in the conventional Raman (Figure 7b). Two subtle differences are: (*i*) the conventional Raman spectrum contains a fluorescence background, and (*ii*) the relative intensities of cellulose bands in FT Raman spectrum may not be identical to that observed in a conventional spectrum due to the dependency of scattering on the fourth power of the absolute Raman frequency.[24] This effect is more pronounced at higher wavenumbers in near-IR FT Raman spectroscopy because the Raman shifts are spread over a larger wavelength region as compared to a visible Raman spectrum.

LIGNIN

In addition to cellulose and hemicelluloses, mechanical and chemimechanical pulps contain lignin. Consequently, lignin spectral features need to be distinguished from those of the other main components. There are a number of complementary ways to derive this information. First, there are regions of the spectrum of wood where cellulose and the hemicelluloses do not contribute and where only features due to lignin are observed. This is the case, for example, in the 1600 cm^{-1} region where only aromatic ring stretching vibrations and lignin's C=C and C=O stretching modes are observed. However, there are other regions of the Raman spectrum where all the components are represented and the interpretation is somewhat complex. To identify lignin bands in these regions, the Raman spectrum of a completely delignified wood sample is subtracted from the native wood spectrum. The difference spectrum is considered to represent the native lignin (Figure 8a). This information, along with that obtained directly from native wood and milled-wood lignin (MWL) spectra (Figure 8b) is then interpreted using Raman studies of lignin models.[10,25–28] The MWL molecular structure is considered to be very similar to that of native lignin. However, the nature of small structural differences and the origin of MWL (middle lamella *vs* secondary wall) is controversial. Raman spectral peak positions of lignin are listed in column 2 of Table 1. Bands in the spectra of wood and pulp samples can be assigned to lignin on the basis of this table.

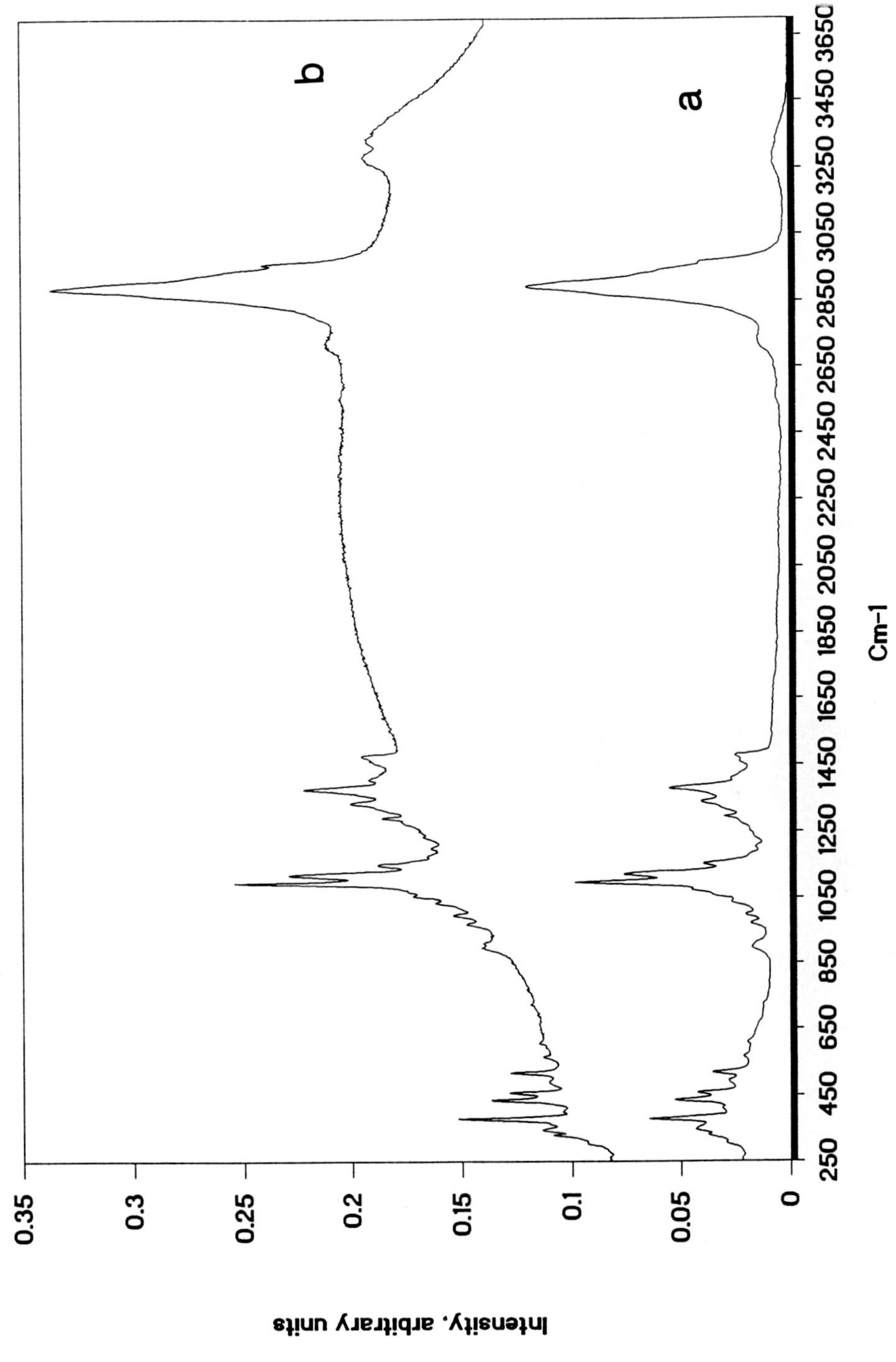

Figure 7. (a) FT and (b) conventional Raman spectra of Whatman No. 1 cellulose filter paper.

Table 1
Band Maxima (cm^{-1}) for Cellulose, Lignin, Softwood Bleached Kraft Pulp (SWBKP) and Thermomechanical Pulp (TMP)

Cellulose	Lignin	SWBKP	TMP
331		331	
348	357	350	
381	384	380	379
		405	403
437		436	434
459	463	458	457
	492	495	492
			499
520		519	519
	555	567	556
			589
	596		599
	638		635
	731		730
	787		809
898		898	897
971	926	972	970
997		997	997
1033, 1037	1037	1036	
1073	1073	1073	
1096	1096	1095	
1121	1120	1122	
1134, 1152	1150	1150	
	1191		1191
	1271		1271
1294	1294	1294	
1333, 1339	1338	1337	
1380	1380	1377	
1409	1409	1409	
1454, 1456		1456	
1478, 1508	1470	1503	
	1602		1601
	1620		1620
	1658		1657
			1766
2739	2734	2738	
2843, 2895	2895	2895	
2938, 2966		2932	
	3007		
	3065		3064
3264	3272		
		3337	

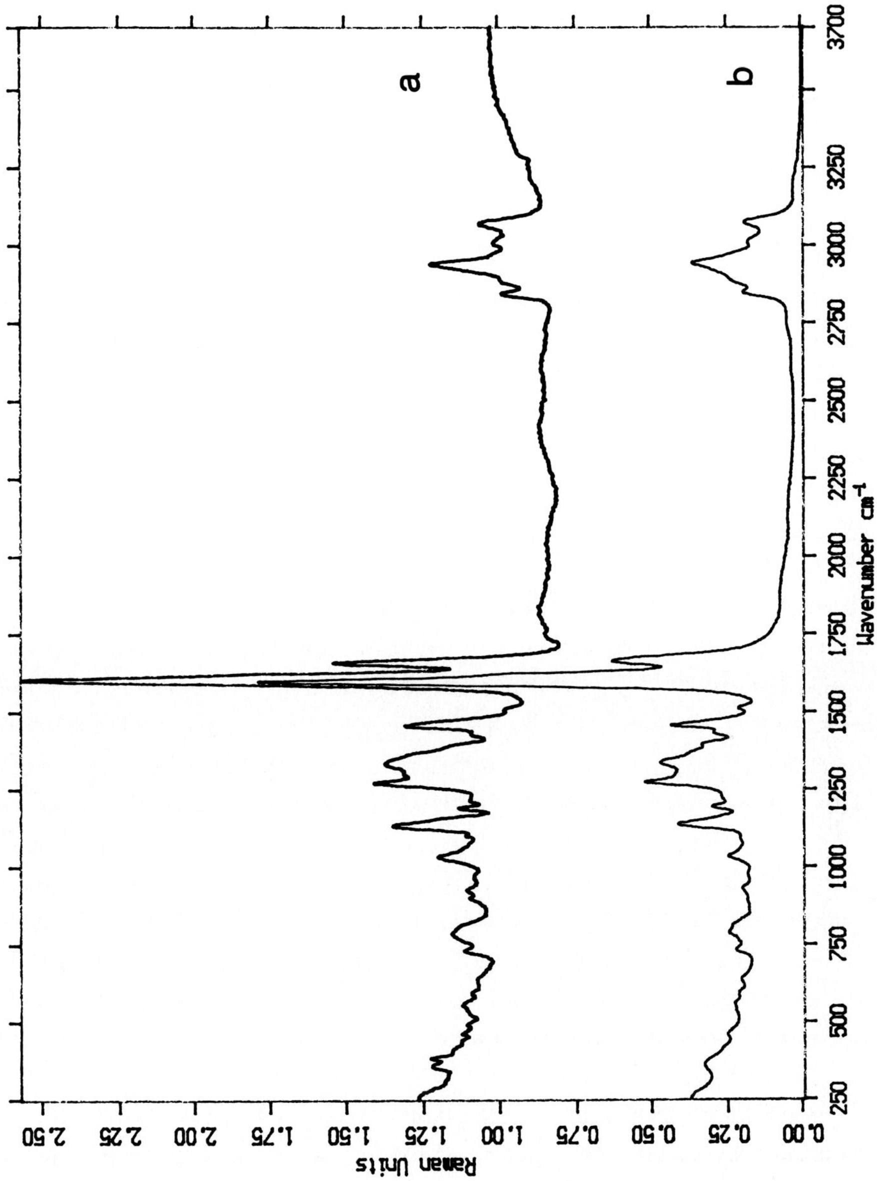

Figure 8. FT Raman spectra of (a) native spruce wood lignin and (b) milled wood lignin (MWL).

BLEACHED AND UNBLEACHED KRAFT PULPS

FT Raman spectra of fully bleached kraft pulps and cellulose filter paper are shown in Figure 9. Except for subtle differences in the low frequency region, the Raman spectra of the hardwood and softwood pulps (Figure 9a and 9b) are very similar.[29] The peak positions of Raman bands for bleached pulps are given in Table 1 (column 3). From a comparison of peak position data and spectral features between cellulose and bleached pulps it becomes obvious that most of the cellulose bands are present in the pulp spectra. This supports the view that the Raman features of hemicelluloses (present in pulps) are broad and are hidden under the spectral contributions of pulp cellulose.

When unbleached kraft pulps are studied using FT Raman spectroscopy, spectra similar to the one shown in Figure 10 are obtained. In addition to features of carbohydrate components (mainly cellulose), a spectral feature is detected at 1600 cm^{-1} due to residual lignin. Recently, this band was used to quantify residual lignin amount in a series of unbleached and partially bleached pulps.[2] When the relative band area of the 1600 cm^{-1} band (representing lignin amount) was correlated to kappa numbers a good linear correlation was obtained as shown in Figure 11. The correlation coefficient of the linear regression was 0.983. Therefore, FT Raman is capable of quantifying residual lignin in kraft pulps.

MECHANICAL AND CHEMIMECHANICAL PULPS

Papers that do not require permanence, like newsprint and magazine papers, are usually produced using lignin-rich high-yield pulps; mechanical and chemimechanical pulps are two representative classes of such high-yield pulps. The spectrum of lignin differs somewhat depending upon whether or not a pulp has been bleached (or chemically treated). This difference can be seen in Figure 12 where TMP spectra are shown before and after bleaching (Figures 12a and 12b respectively) along with the spectrum of a chemimechanical pulp (CMP, Figure 12c). The differences in spectra were evaluated by further manipulations of data (*e.g.*, spectral subtraction, abscissa and ordinate scale expansion). Such analyses clearly indicated that changes occurred only in the lignin spectral features. A bar graph in Figure 13 represents some of these results. For TMP, band maxima are reported in Table 1 (column 4).

COATING, BINDERS, PIGMENTS, FILLERS, AND OTHER PAPER ADDITIVES

Characteristic Raman spectra of a coating mixture, latex, calcium carbonate, TiO_2, starch, and clay are shown in Figure 14. The results are summarized in Table 2. The coating formulation examined consisted of a latex and one or more paper additives. As can be seen from the data in Table 2 and the spectra in Figure 14, the coating spectrum has contributions from latex, $CaCO_3$, and TiO_2. The only detected coating bands that could not be assigned to any of the Table 2 listed coating components were at 3614,

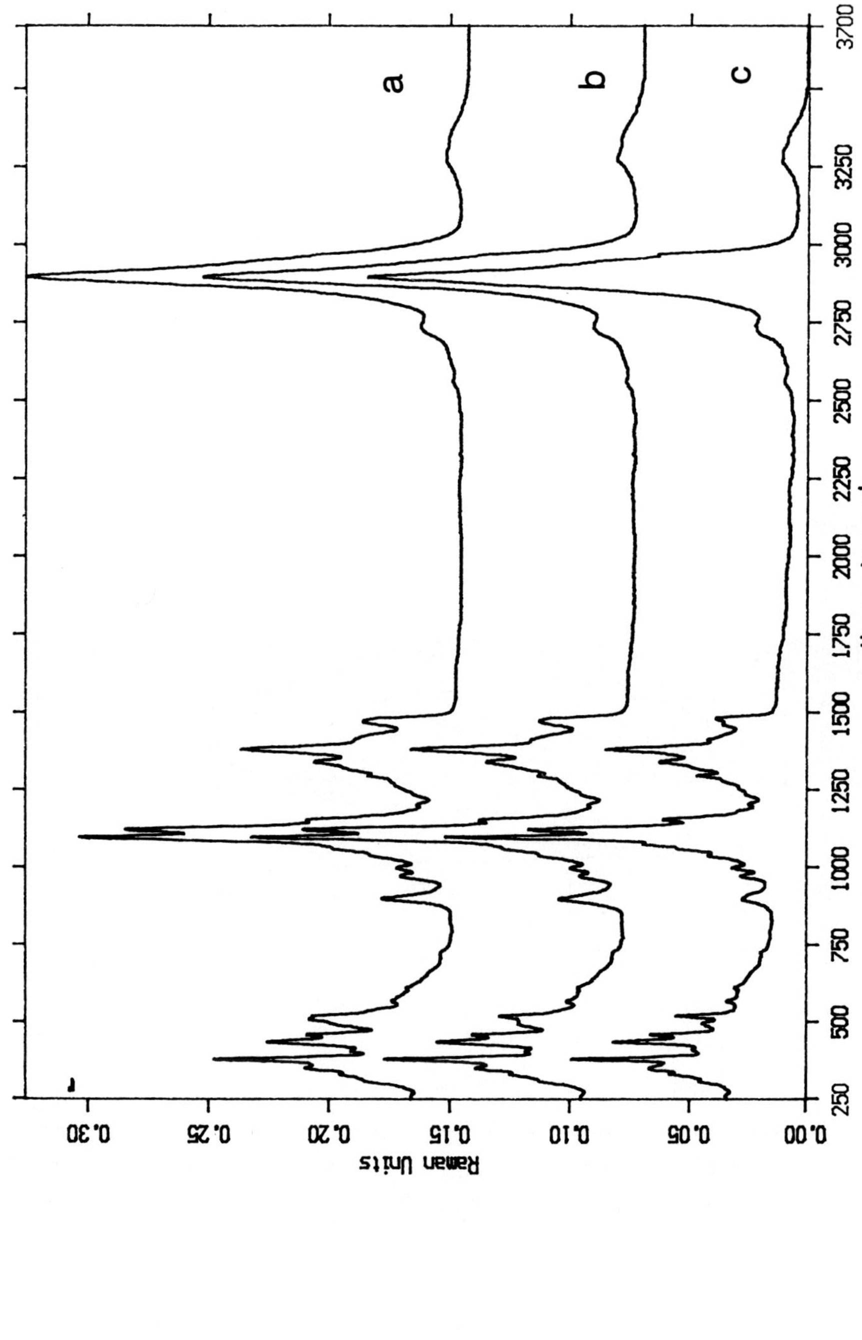

Figure 9. FT Raman spectra of (a) bleached hardwood kraft pulp, (b) bleached softwood kraft pulp, and (c) cellulose filter paper.

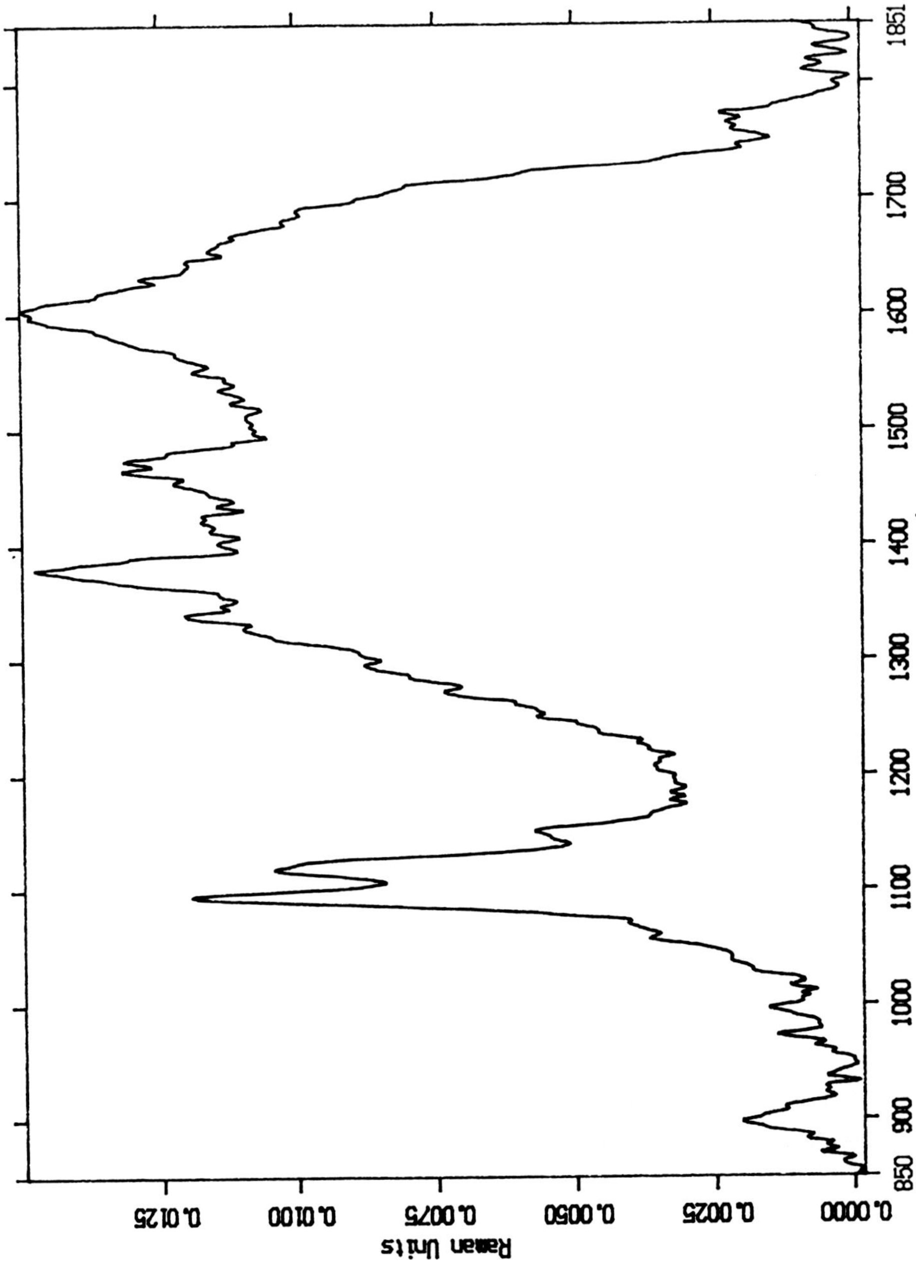

Figure 10. FT Raman spectrum of unbleached kraft pulp in the spectral region 850–1850 cm^{-1}.

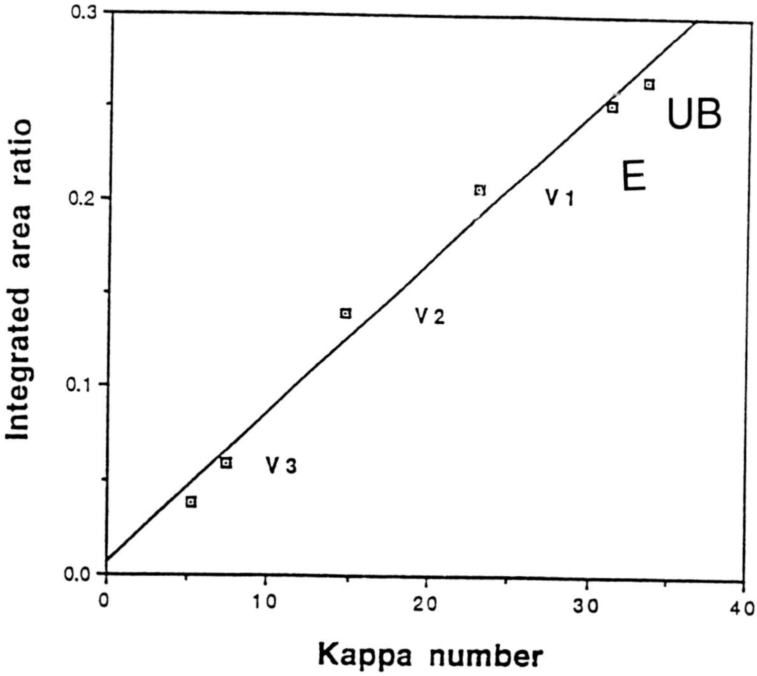

Figure 11. Linear regression between 1600 cm^{-1} Raman band intensities and microkappa numbers. The 1600 cm^{-1} band is due to residual lignin in kraft pulps. V1, V2, V3 data points correspond to the number of times the pulp was treated with a vanadium catalyst.[2]

3645, and 3689 cm^{-1}. These were found to be present in the Raman spectrum of another sample of clay (spectrum not shown). This indicated that this type of clay was also part of the coating formulation.

RAMAN SPECTROSCOPY AS A SURFACE ANALYSIS TECHNIQUE

A considerable body of information has been obtained on surfaces using Raman spectroscopy in areas such as catalysis,[30] semiconductors,[31] and electrochemical processes.[32] In contrast, only a few reports exist in the literature of the application of Raman spectroscopy to paper analysis.[33–37] In our laboratory, Raman spectroscopy has been used to study photoyellowing of lignin-rich pulps,[34,38] the surfaces of coated papers, and knuckled and nonknuckled regions on paper sheets[35] (microscopic domains produced on paper surface during press drying). Conventional Raman microscopy has also been used to study coated paper mottling[36] and to identify pigment grains in printed medieval manuscripts.[37]

Although Raman spectroscopy provides information from the sample surface, the sampling depth depends upon both the wavelength of excitation and the nature of the sample. It is estimated that in conventional Raman the spectra represent only a few micrometers of the specimen thickness. Additionally, in a Raman microprobe the

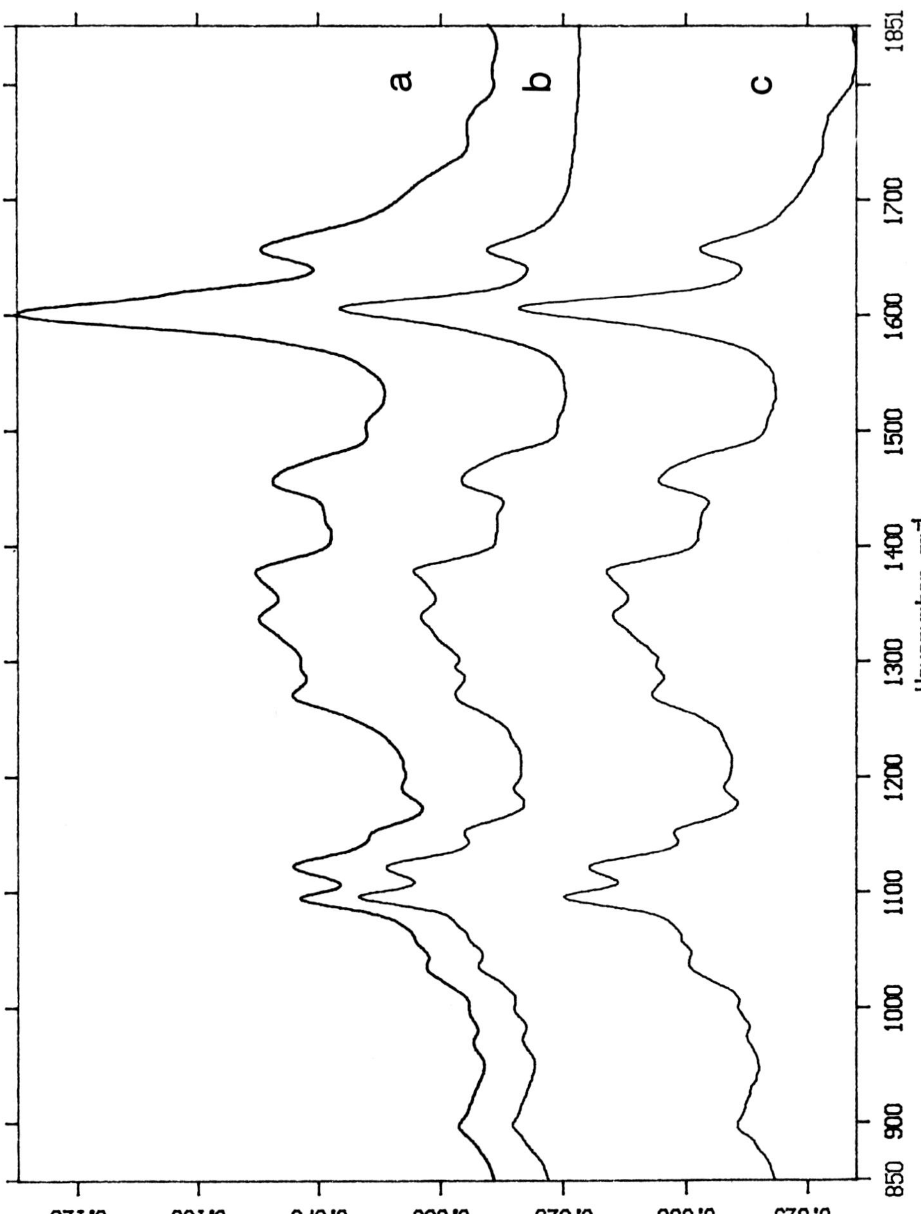

Figure 12. FT Raman spectra in the region 850–1850 cm^{-1}; (a) thermomechanical pulp (TMP), (b) hydrogen peroxide bleached TMP, (c) chemimechanical pulp.

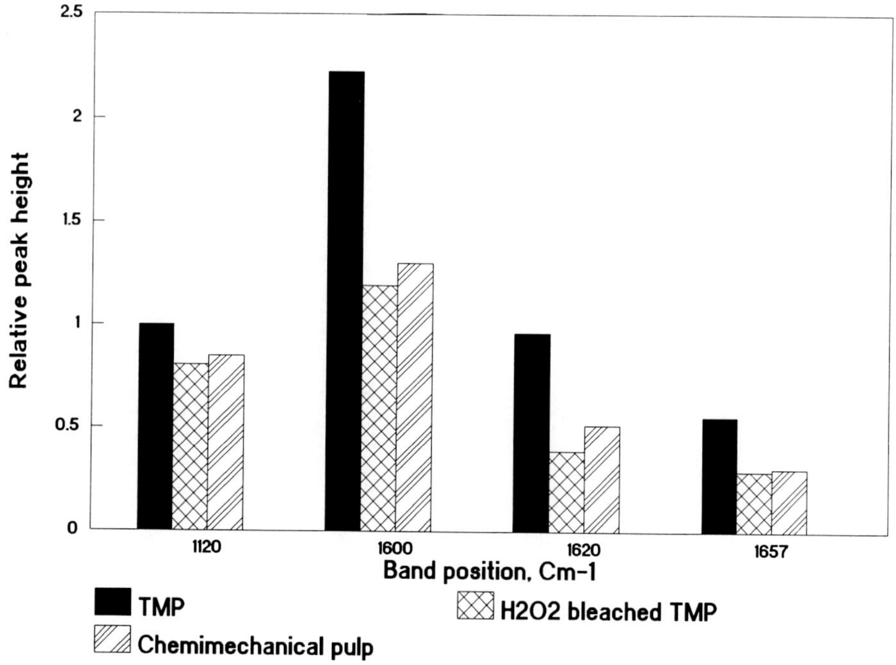

Figure 13. Differences among spectra, shown in Figure 12, represented as bar graphs. The FT Raman band positions 1600 and 1657 cm^{-1} are slightly higher than those obtained in conventional Raman.

characteristics of a microscope objective are important. Using a 160X objective, the axial resolution was reported to be approximately two micrometers.[39] Similarly, when a confocal technique was used in the studies of transparent polymer samples, not only the lateral resolution was enhanced but depth resolutions between 0.7 and 3.0 micrometers were obtained.[40]

APPLICATIONS IN PAPER ANALYSIS

COMMERCIAL PAPERS

Figure 15 shows Raman spectra of coated and uncoated sheets of paper that were produced from fully-bleached kraft pulps (brightness in the 90s). When the spectrum of coated paper (Figure 15a) is compared with spectra of the uncoated paper (Figure 15c) and bleached kraft pulp (Figure 9a and 9b), several coating features in the spectrum can be easily identified. Moreover, these features can be assigned to various components in the coating formulation. From Table 3, which lists positions and assignments of bands, it can be noted that latex and calcium carbonate contribute strongly to the Raman spectrum of the coated paper. The detectability of other components at the applied concentrations depends upon their Raman scattering coefficients.

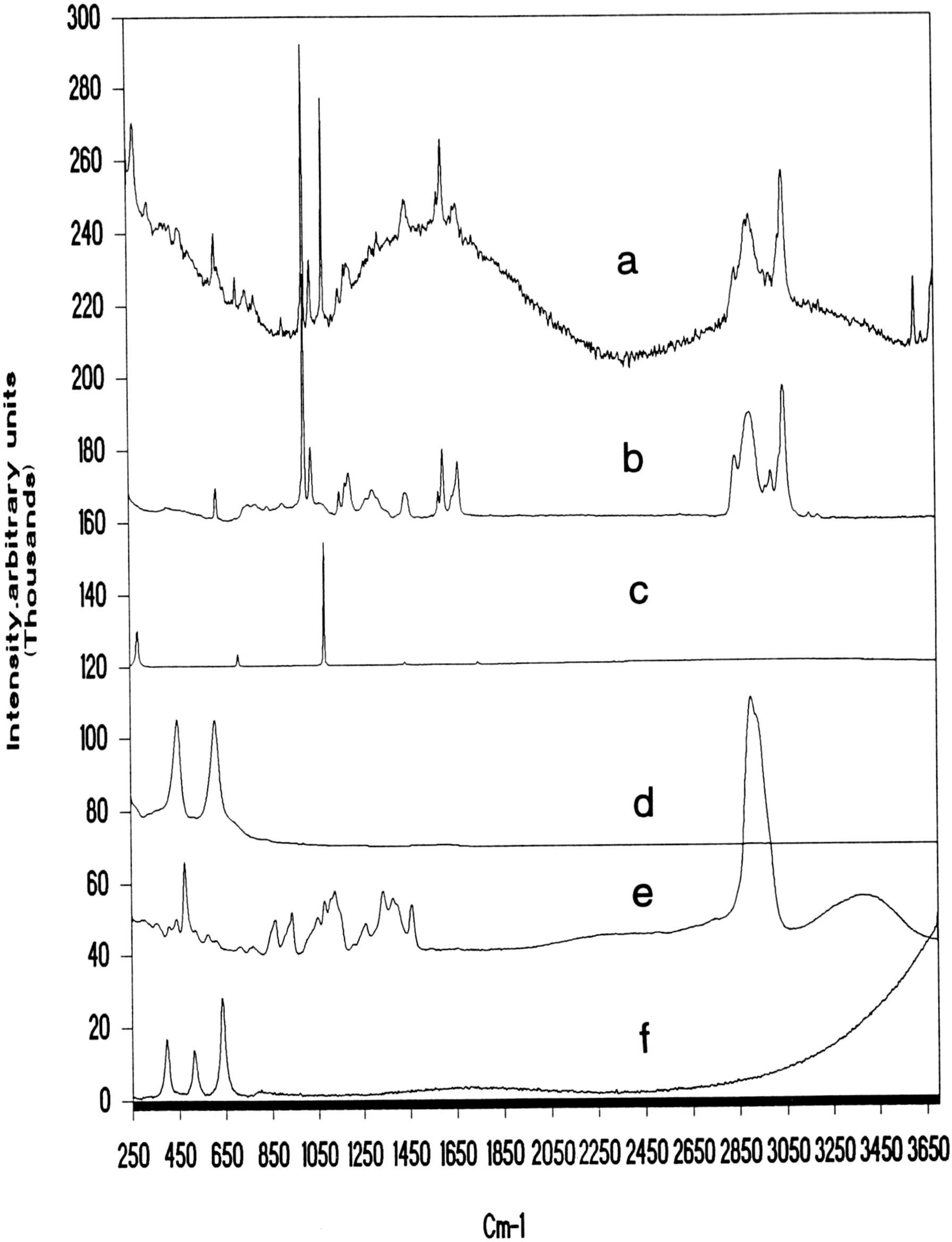

Figure 14. Conventional Raman spectra of (a) coating mixture, (b) latex, (c) $CaCO_3$, (d) TiO_2, (e) starch, and (f) clay. Spectra were shifted with respect to one another to avoid overlapping of features.

Table 2
Band Maxima (cm^{-1}) of Coating and Paper Additives

Coating	Latex	$CaCO_3$	TiO_2*	Starch	Clay
280		282			
336				357	
				408	394
474			446	440	
				476	
				522	514
				576	636
620	620		608	612	
710		710		714	
752	752			766	
788	790				
	838				
				866	
908	906				
				938	
1000	1000				
1030	1030			1048	
1084		1088		1078	
1152	1154			1123	
1180	1181				
1202	1197				
	1270			1261	
	1301				
	1324			1355	
				1379	
1442	1441	1435		1457	
1581	1581				
1598	1601				
1665	1667				
		1747			
2845	2846				
2913	2910			2903	
				2931	
	2996				
3050	3050				
	3158				
	3198				
				3387	
3614					
3645					
3689					

* These were the most prominent features.

An interesting application of Raman microscopy was recently reported in the area of paper surface analysis. Researchers used a Raman microscope to study the causes of the appearance of alternating light and dark printed areas (called mottling) on the paper surface.[36] The influences of latex concentration (one of the binders in the coating formulation) and distribution were studied and their role in the phenomenon of mottling was evaluated. It was found that mottling was related to the heterogeneity of the coating

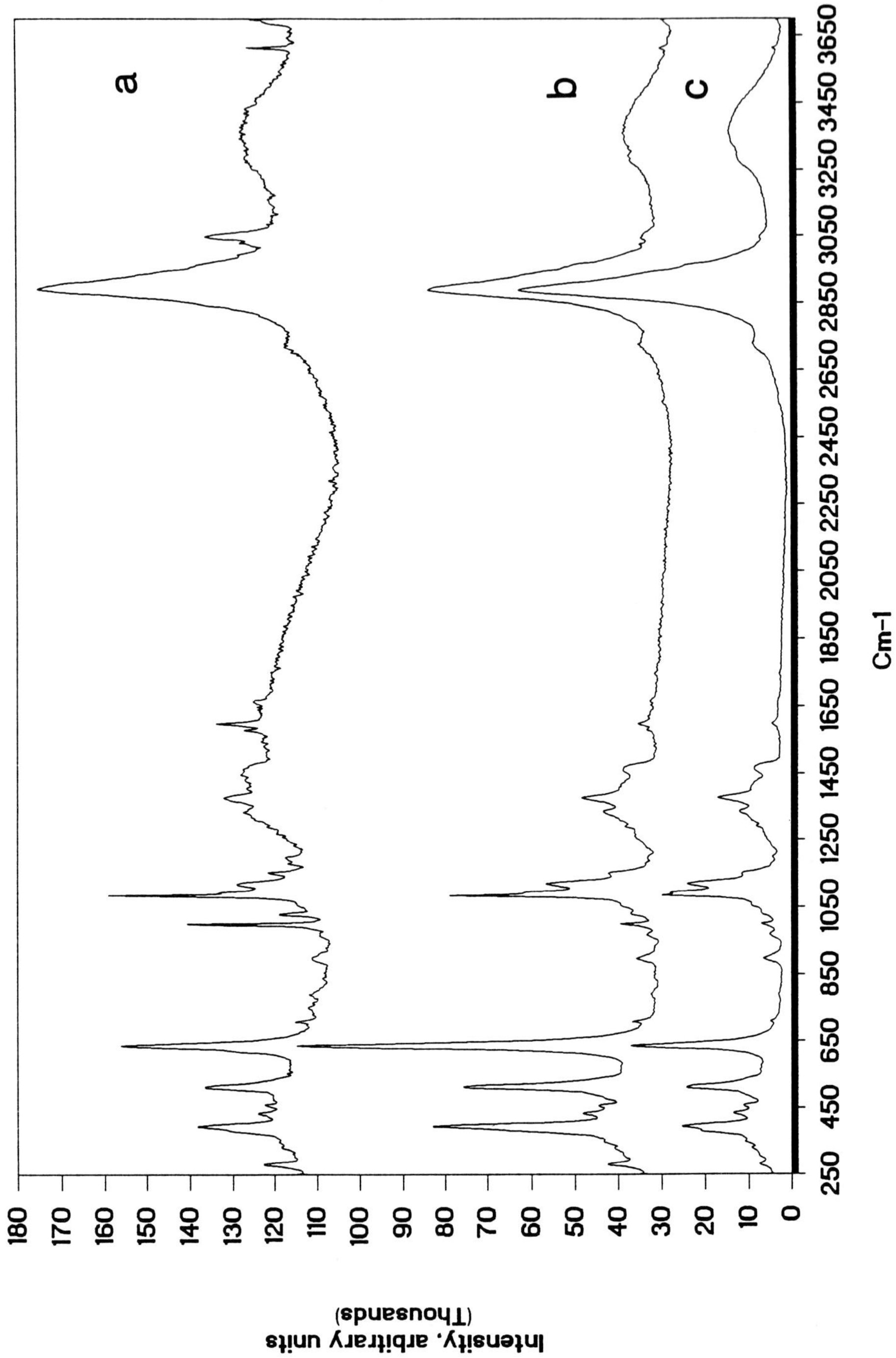

Figure 15. Conventional Raman spectra of (a) coated paper surface, (b) non-coated side of sheet in (a), and (c) base paper. Laser-induced fluorescence contributions were removed from some of the spectra.

thickness and that it can not be attributed to the heterogeneity of the latex distribution at the surface of the paper.

Table 3
Band Maxima (cm^{-1}) for Coated Paper and Spectral Assignments

Coated Paper	Assignment
280	Coating ($CaCO_3$)
380	Cellulose
394	Clay
434	Cellulose
458	Cellulose
512	Clay
636	Clay
710	Coating ($CaCO_3$)
898	Cellulose
1000	Coating (latex)
1030	Coating (latex)
1084	Coating ($CaCO_3$)
1098	Cellulose
1121	Cellulose
1151	Cellulose + Coating (latex)
1196	Coating (latex)
1377	Cellulose
1581	Coating (latex)
1601	Coating (latex)
1667	Coating (latex)
2723	Cellulose
2890	Cellulose
3050	Coating (latex)
3614	High Brightness Clay
3688	High Brightness Clay

PHOTOYELLOWING OF MECHANICAL AND CHEMI-MECHANICAL PULP-CONTAINING PAPERS

When lignin-containing pulps (and papers made from them) are photoexposed they undergo yellowing (also known as brightness reversion). Because yellowed papers are less bright, and therefore undesirable, mechanical and chemimechanical pulp-containing papers are generally used in applications not requiring permanence. Nevertheless, the production of these pulps is attractive because of their low cost and comparatively small impact on the environment.

To find a solution to the problem of photoyellowing the nature of the photochemical processes at the paper surface needs to be understood. We have applied conventional and FT Raman spectroscopy to develop further understanding of this phenomenon. Some of the findings, to be reported elsewhere,[38] are shown in Figure 16. It was found that light-induced changes involve the photodestruction of aromatic ring-conjugated ethylenic double bonds in lignin (*e.g.*, bonds present in coniferaldehyde and coniferyl alcohol units). Moreover, it was found that the heat-induced changes also involved these same structures (see Figure 16 for heat-aged and heat-plus-light-aged pulp data). These results were made possible due to the sensitivity of Raman spectroscopy to ring-conjugated structures in lignin.[27,28]

ADVANTAGES OF RAMAN SPECTROSCOPY

It is clear that Raman spectroscopy can be used to provide information about surfaces. Of course, like many other techniques, it is equally capable of providing data from the interior of the samples. Considering that Raman spectroscopy deals with molecular vibrations, it is appropriate to compare it with IR spectroscopy. The nature of the information that can be obtained from Raman experiments, vibrational frequencies and band intensities, is similar but not identical to that of IR spectroscopy. To better appreciate the differences and similarities between these two techniques, the physical phenomena involved will be briefly considered. As mentioned previously, for a molecular vibration to be IR-active a change in dipole moment accompanying the vibrational transition is needed. On the other hand, a vibration is Raman-active whenever a change in molecular polarizability occurs during the transition. In light of this, polar bond systems with a high dipole show up strongly in IR, whereas bond systems with highly covalent character are quite easily seen in Raman. Water, as a consequence, is a weak scatterer in Raman spectroscopy but absorbs strongly in IR. Indeed, we use water as a part of our sampling procedure for some of the experiments. Because of the difference in selection rules between these two types of spectroscopic transitions, the information obtained from the two types of spectra is complementary.

Ordinary Raman scattering is an inefficient process, and in general it is less sensitive than IR absorption. However, in certain samples, conjugation, resonance or preresonance

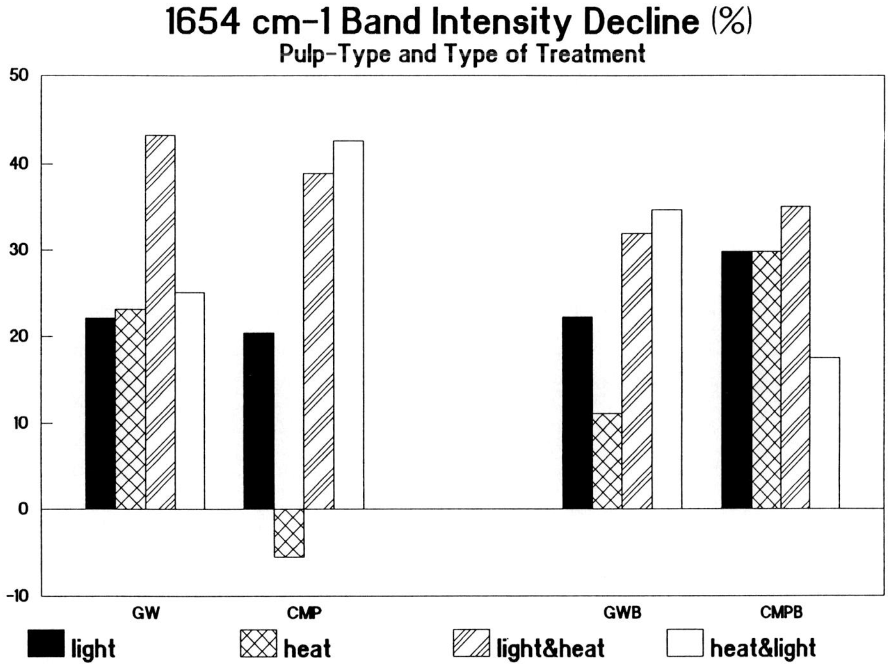

Figure 16. Effect of light, heat, and combinations of light and heat treatments on the intensity of the 1654 cm^{-1} Raman band. GW = ground wood pulp, CMP = chemimechanical pulp, GWB = GW bleached pulp, and CMPB = CMP bleached pulp.

Raman effects can arise and for particular vibrational modes a higher level of sensitivity can be achieved.[41,42] Sometimes, the latter effect can be induced by proper laser frequency selection.

Another Raman advantage is the ease of working with heterogeneous samples. In IR, this is difficult because of Rayleigh scattering of infrared photons. The degree of Rayleigh scattering depends upon differences in the refractive indices at optical heterogeneities. Because the refractive index varies with wavelength in regions of strong IR absorption, it is difficult to separate the extinction due to molecular absorption from the extinction due to Rayleigh scattering. In contrast, discrimination against Rayleigh scattering is relatively simple for Raman measurements.

There are additional advantages in using Raman spectroscopy to study pulp and paper samples. Since it is possible to use a number of excitation wavelengths in conventional Raman, the choice of a proper frequency allows the selective excitation of sample components. Time-resolved Raman studies of an excited component can be carried out using pulsed laser-based Raman spectrometers. This capability opens up a new area of research in pulp and paper science. IR spectroscopy does not have this capability as the frequencies involved are far removed from those which cause electronic excitation.

Some samples are difficult to study using conventional instrumentation. Samples may give rise to high levels of fluorescence or may be thermally degraded by laser excitation. With the availability of FT Raman instrumentation, however, this problem no longer exists. In addition to having solved the most pressing problem in Raman spectroscopy, namely that of fluorescence, the FT approach has made rapid spectra acquisition a reality.

ACKNOWLEDGEMENTS

The authors wish to acknowledge help provided by Dr. Jim Bond (Consolidated Papers, Wisconsin Rapids, Wisconsin) during the writing of this chapter. They also would like to thank Sally Ralph for recording some of the Raman spectra.

REFERENCES

1. Weinstock, I., Atalla, R. H., Agarwal, U. P., Minor, J., and Petty, C. 1993. FT Raman spectroscopic studies of a novel wood pulp bleaching system. *Spectrochimica Acta* 49A:819–829.
2. Weinstock, I., Agarwal, U. P., Minor, J., Atalla, R. H., and Reiner, R. 1993. FT Raman and UV-visible spectroscopic studies of a highly selective polyoxometalate bleaching system. *In*: Proceedings, TAPPI Pulping Conference, Tappi Press, Atlanta, GA. pp. 519–532.
3. Pitzner, L. J. 1973. An Investigation of Vibrational Spectra of the 1,5-anhydropentitols. Ph.D. thesis. The Institute of Paper Chemistry, Appleton WI (now Institute of Paper Science and Technology, Atlanta, GA), 402 pp.
4. Pitzner, L. J. and Atalla, R. H. 1975. An investigation of vibrational spectra of the 1,5-anhydropentitols. *Spectrochimica Acta* Part A:911–929.
5. Watson, G. M. 1974. An Investigation of Vibrational Spectra of the Pentitols and Erythritol. Ph.D. thesis. The Institute of Paper Chemistry, Appleton WI (now Institute of Paper Science and Technology, Atlanta, GA), 178 pp.
6. Edwards, S. L. 1976. An Investigation of Vibrational Spectra of the Pentose sugars. Ph.D. thesis. The Institute of Paper Chemistry, Appleton WI (now Institute of Paper Science and Technology, Atlanta, GA), 245 pp.
7. Williams, R. M. 1977. An Investigation of Vibrational Spectra of the Inositols. Ph.D. thesis. The Institute of Paper Chemistry, Appleton WI (now Institute of Paper Science and Technology, Atlanta, GA), 377 pp.
8. Williams, R. M. and Atalla, R.H. 1984. Vibrational spectra of the inositols. *J. Phys. Chem.* 88:508–519.

9. Wells, H. A. 1977. An Investigation of the Vibrational Spectra of Glucose, Galactose, and Mannose. Ph.D. thesis. The Institute of Paper Chemistry, Appleton WI (now Institute of Paper Science and Technology, Atlanta, GA), 431 pp.
10. Ehrhardt, S.M. 1984. An Investigation of Vibrational Spectra of Lignin Model compounds. Ph.D. thesis. The Institute of Paper Chemistry, Appleton WI (now Institute of Paper Science and Technology, Atlanta, GA), 332 pp.
11. Hendra, P. J., Jones, C., and Warnes, G. 1991. Fourier Transform Raman Spectroscopy. Ellis Horwood, Chichester, England. Chapter 4.
12. Killough, P. M., DeVito, V. L., and Asher, S. A. 1991. Applications of ultraviolet resonance raman spectroscopy: residual olefins in polypropylene. *Appl. Spectrosc.* 45(7):1067–1069.
13. Hirschfield, T. and Chase, D. B. 1986. FT-Raman spectroscopy: Development and justification. *Appl. Spectrosc.* 40(2):133–137.
14. Chase, D. B. 1986. Fourier transform Raman spectroscopy. *J. Am. Chem. Soc.* 108(24):7485–7488.
15. Rosasco, G. J. 1980. Raman microprobe spectroscopy. *In*: Advances in Infrared and Raman Spectroscopy. Clark, R. J. H. and Hester, R. E., eds. Heyden, London. Chapter 4.
16. Agarwal, U. P. and Atalla, R. H. 1986. In-situ Raman microprobe studies of plant cell walls: Macromolecular organization and compositional variability in the secondary wall of *Picea mariana* (Mill.) B.S.P. *Planta* 169:325–332.
17. Atalla, R. H. and Nagel, S. C. 1972. Laser-induced fluorescence in cellulose. *J. Chem. Soc. Chem. Comm.* pp. 1049–1050.
18. Agarwal, U. P. and Atalla, R. H. 1986. Oxygen sensitive background in the Raman spectra of woody tissue. *In:* Proceedings Xth International Conf. Raman Spectroscopy, University Printing Dept., Univ. Oregon, Eugene, OR. Paper 14–46.
19. Freeman, S. K. 1974. Applications of Laser Raman Spectroscopy. John Wiley & Sons, New York, NY. pp. 45.
20. Wiley, J. H. and Atalla, R. H. 1987. Band assignments in the Raman spectra of celluloses. *Carbohydrate Research* 160:113–129.
21. Wiley, J. H. and Atalla, R. H. 1987. Raman spectra of celluloses. *In*: The Structures of Cellulose. Atalla, R. H., Ed. ACS Symposium Series 340, American Chemical Society, Washington, DC. Chapter 8.
22. Atalla, R. H. 1976. Raman spectral studies of polymorphy in cellulose. Part I: Celluloses I and II. *Applied Polymer Symp.* 28:659–669.
23. Atalla, R. H. 1990. Patterns of aggregation in native celluloses: Implication of recent spectroscopic studies. *In:* Cellulose: Structural and Functional Aspects. Kennedy, J. F., Phillips, G. O., and Williams, P. A., Eds., Ellis Horwood, Chichester, England. Chapter 4.

24. Schrader, B. and Hoffman, A. 1991. Can a Raman renaissance be expected via the near-infrared Fourier transform technique? *Vibrational Spectroscopy* 1:239–250.
25. Agarwal, U. P. and Atalla, R. H. 1993. Raman spectroscopic evidence for coniferyl alcohol structures in bleached and sulfonated mechanical pulps. *In*: Photochemistry of Lignocellulosic Materials. Heitner, C. and Scaiano, J. C., Eds. ACS Symposium Series 531, American Chemical Society, Washington, D.C. Chapter 2.
26. Agarwal, U. P. and Atalla, R. H. 1992. Raman spectroscopy of native lignin: raman bands associated with C=O, C=C and phenyl groups. *In*: Proceedings XIIIth International Conf. Raman Spectroscopy, John Wiley & Sons, Chichester, England. pp. 456–457.
27. Agarwal, U. P. and Atalla, R. H. 1993. Sensitivity of Raman spectroscopy to aromatic-ring conjugated structures in lignins and model compounds. Presented at Cellulose, Paper, and Textile Division Symposia, ACS National Meeting, Denver, CO, March 28–April 2.
28. Bond, J. S., Agarwal, U. P., and Atalla, R. H. 1994. Contributions of preresonance Raman and conjugation effects to the Raman spectrum of native lignin (submitted).
29. Agarwal, U.P. (unpublished results).
30. Proceedings XIIIth International Conf. Raman Spectroscopy, Section 10 – Surface and Interfacial Phenomena, and SERS. 1992. John Wiley & Sons, Chichester, England.
31. Proceedings XIIIth International Conf. Raman Spectroscopy, Section 13 – Papers on Semiconductors and Semiconductor Microstructures. 1992. John Wiley & Sons, Chichester, England.
32. Ganter, B. B. E., Muller, H. G., Steinert, D. and Ache, H. J. 1992. Process analysis of electroless nickel deposition baths using Raman spectroscopy. *In*: Proceedings XIIIth International Conf. Raman Spectroscopy, John Wiley & Sons, Chichester, England. pp. 1018–1019.
33. Kenton, R. C. and Rubinovitz, R. L. 1990. FT-Raman investigations of forest products. *Applied Spectroscopy* 44(8):1377–1380.
34. Agarwal, U. P. and Atalla, R. H. 1990. Formation and identification of cis/trans ferulic acid in photoyellowed white spruce mechanical pulp. *J. Wood Chem. Tech.* 10(2):169–190.
35. Atalla, R.H, Woitkovich, C.P., and Setterholm, V. C. 1985. Raman microprobe study of fiber transformations during press drying. *Tappi J.* 68(11):116–119.
36. Guyot, C. and Amram, B. 1993. Application of Raman microscopy to the study of coated paper mottling. Presented at 16th PTM Coating Symp., Aubervillers, France, Sept. 14–17, 1993.
37. Best, S. P., Clark, R. J. H., and Withnall, R. 1992. Identification of pigments on medieval manuscripts by Raman microscopy. *In*: Proceedings XIIIth International

Conf. Raman Spectroscopy, John Wiley & Sons, Chichester, England. pp. 1042–1043.

38. Agarwal, U. P., Atalla, R. H., and Forsskahl, I. 1994 Sequential treatment of mechanical and chemimechanical pulps with light and heat: A Raman spectroscopic study. *Holzforschung* (submitted).
39. Dhamelincourt, P., Delhaye, M, Truchet, M., and da Silva, E. 1991. Etude en microspectrometrie Raman de l'efficacite de detection en fonction de l'epaisseur des echantillons: Determination de la profondeur de champ et consequences pour le couplage avec un microscope electronique en transmission. *J. Raman Spectrosc.* 22:61–64.
40. Tabaksblat, R., Meier, R. J., and Kip, B. J. 1992. Confocal Raman spectroscopy: Theory and applications to thin polymer samples. *Appl. Spectrosc.* 46(1):60–68.
41. Long, D. A. 1977. Raman Spectroscopy. McGraw-Hill, New York.
42. Schmid, E. D., Brosa, B. 1971. Raman-intensitat und Konjugation I. Substituentenabhangigkeit der Raman-intensitaten der 1600 cm^{-1} ringschwingungen Monosubstituierter Benzolderivate. *Ber. Bunsenges Phys. Chem.* 75:1334–1343.

9

Energy Dispersive X-Ray Spectroscopic Analysis

Glynis de Silveira and Terrance E. Conners*

University of Cambridge, Cambridge, U.K.
**Mississippi State University, Mississippi State, Mississippi, U.S.A.*

INTRODUCTION

The impingement of an electron beam onto a specimen causes the emission of various types of signals from which topographic and chemical information may be obtained, as previously discussed in the chapter on scanning electron microscopy. Of particular interest for the purposes of this chapter is the fact that each element in a specimen produces x-rays with characteristic energies when excited by an electron beam. When the specimen is in a good vacuum (as in an electron microscope), there is little to attenuate the x-rays given off; detectors capable of acquiring these x-rays can be added to any electron microscope, expanding their usefulness from microscopes to spectroscopes. Spectroscopic analysis of the intensity and distribution of the x-ray signal generated within a sample provides elemental and chemical information of a qualitative nature which may be quantifiable under strictly defined conditions. What makes the combination of SEMs or STEMs (scanning transmission electron microscopes) with x-ray spectroscopic analysis so useful is the fact that the analyst can direct the electron beam to specific points, lines, or regions, and x-ray data are collected from those specified locations. Ideally one would like to use a single detector to collect the x-ray signal emitted by a sample, resolve it at an energy of a few electron volts and precisely analyze the chemical composition of a specimen. Two types of instruments are available: the WDS (wavelength dispersive spectrometer) and the EDS (energy dispersive spectrometer).

The invention of the lithium-drifted silicon Si(Li) solid state x-ray detector and its installation on an electron probe microanalyzer as part of an energy dispersive spectrometer[1] revolutionized the chemical analysis of specimens in the SEM. Although the WDS had been in use for a number of years its slowness in collecting data (tens of minutes)

0-8493-8992-5/95/$0.00+$.50

compared to EDS (minutes) has relegated it to a secondary position. WDS does, however, have a number of advantages (such as its superior geometry, overall quantum efficiency, its higher resolution and its relative freedom from spectral artifacts) over the EDS. The speed with which data can be collected and interpreted in EDS (as a result of the continuous acceptance of a large range of spectral energies by the detector) has distinct advantages for qualitative analysis. Quantitative analysis can also be conducted, but ideally requires a polished sample and comparison to element standards; this type of analysis is at best a semi-quantitative evaluation when used on rough, fibrous and coated materials (*e.g.*, paper products). Thus, it is currently recognized that a combination of EDS and WDS is necessary if a system with maximum capabilities for optimum spectrometry analysis is desired. If, on the other hand, one needs to carry out quick, qualitative analyses of specimens whose overall chemical composition is more or less known, then the EDS alone can fulfil the need; this method is the one most commonly used in the pulp and paper industry. This chapter will deal superficially with the EDS technique and will dwell on the specifics of x-ray spectroscopic analysis of paper.

FUNDAMENTAL PRINCIPLES

When an element loses an inner-shell electron through interaction with a high-energy electron beam, an electron from one of the outer shells gives up energy and fills the vacant position, thereby producing an x-ray. In order of increasing energy (distance from the nucleus), the shells are designated by the letters K, L, M, and N; when an x-ray is produced, it is named according to the shell with the vacant position and the shell from which the vacancy is filled. For example, if an electron vacancy in the K shell is filled by an electron from the L shell, the x-ray is designated as a $K\alpha$ x-ray; if it is filled by an electron from the M shell it is designated as a $K\beta$ x-ray, and so on. Most EDS analysis deals with KLM lines; x-rays are produced by each element at several different characteristic energies, and it is by comparing the characteristic energies produced by the specimen with a computer library "snapshot" of each element that element identification can occur. X-ray spectra would ideally be collected as a group of lines with varying amplitudes within the energy range sampled, but the detector electronics that convert monochromatic x-ray pulses to analyzable digital signals produce artifacts that tend to broaden or distort the observed energy distribution;[1] non-characteristic radiation, called *bremsstrahlung* or the *continuum*, is also present as background noise (Figure 1). Without getting into the details of how x-rays are converted into digital signals, it is sufficient to note that some detectors have better resolution than others. The window covering the x-ray detector can also have an important effect on the detector's sensitivity and ability to detect light elements. Many windows are constructed of beryllium and the detectors are consequently insensitive to elements lighter than sodium (Z=11), but alternative

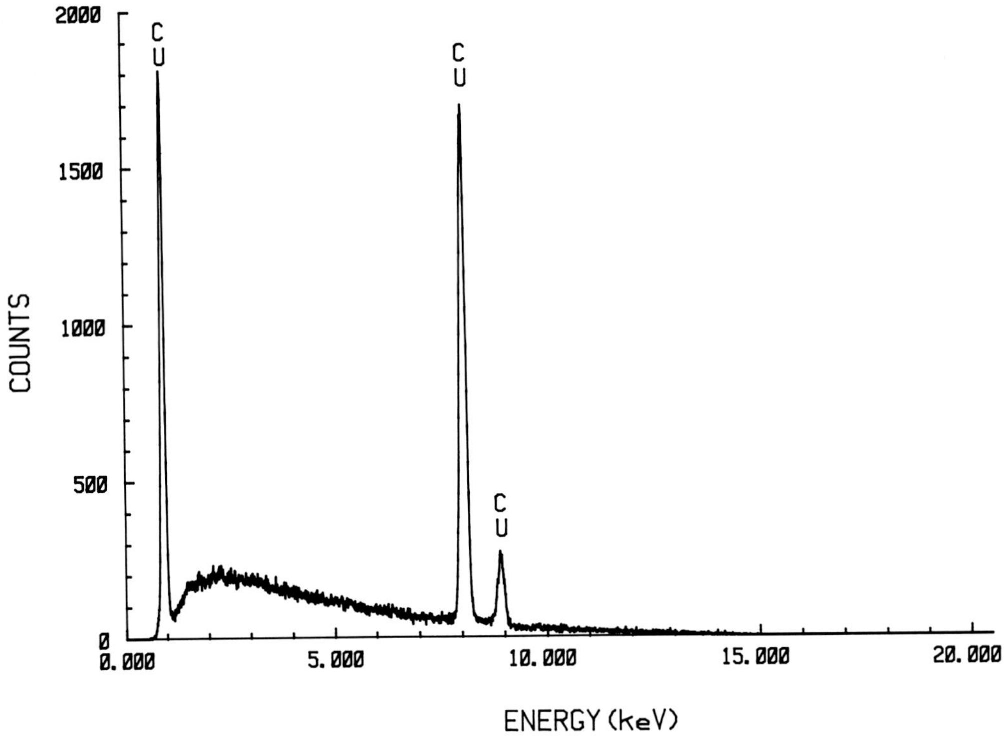

Figure 1. EDS spectrum of copper specimen grid for a transmission electron microscope. Copper L and Kα, Kβ lines are represented. The unlabeled "hump" is the bremsstrahlung. Total acquisition time was 100 seconds live time.

windows made from ultra-thin polymers and even windowless detectors are available for light elements (down to and including beryllium).

Details about the design and operational characteristics of the EDS detector have been described by a number of authors,[2,3] and will not be dealt with here. However, certain aspects specific to the nature of fiber and paper samples and the energies at which these samples are generally analyzed must be considered in detail. One such detail is the low x-ray yield from biological samples which necessitates an EDS detector that has a high geometrical collection capability and overall quantum efficiency.

ELEMENTS DETECTED AND INTERACTION VOLUME

Uncoated paper *per se* is a composite material constituted of pulped wood fibers, air that is trapped within the uncollapsed fiber lumens and the fiber network and contaminants originating from the tree itself or the pulping process. The main constituent elements of wood fibers are carbon, oxygen and hydrogen, low atomic weight elements constituting compounds (cellulose, hemicellulose and lignin) with average atomic weights that are themselves low. These specimens have a very low x-ray yield and, as discussed in the chapter on scanning electron microscopy, the depth of penetration of the incident beam can be quite significant. Depending on the energy of the incident electrons many of the x-rays produced will also originate from deep within the sample, thus providing chemical information about particles lying deep within the paper sheet as well as others on the surface (Figure 2). The interaction volume producing x-rays is usually much greater than that which produces secondary electrons, and the lateral resolution is therefore not as good.

With the advent of highly sensitive (ultra-thin polymer window or windowless) EDS detectors most of the low atomic number elements ($5<Z<11$) (including some of the main constituents of wood fibers, carbon and oxygen) can be detected. It is very difficult to quantify these due to problems of sample surface contamination and absorption of weak x-rays. However, pulp additives such as organic or inorganic polymers, mineral pigments and contaminants can be monitored using EDS analysis if they are present in detectable quantities (about 0.1 weight percent in bulk materials). Crystals of silica and/or calcium occur naturally within the lumen of fibers while they are part of the tree. If not dislodged by the pulping process these particles can be detected in paper during EDS analysis, leading the analyst to believe that they are contaminants introduced at some stage of the papermaking process.

As a rule, the mechanical pulping of wood chips does not add any chemical compounds to the pulped fibers. Occasionally one may find mineral and/or metal particles stemming from the refiners in paper made from groundwood and thermo-mechanical pulps. Chemical pulping, on the other hand, leaves a significant amount of chemical compounds in the pulp, even in well-washed samples. The alkaline kraft process ($NaOH + Na_2S$) based on the use of sodium sulfate deposits sodium and sulfur in the

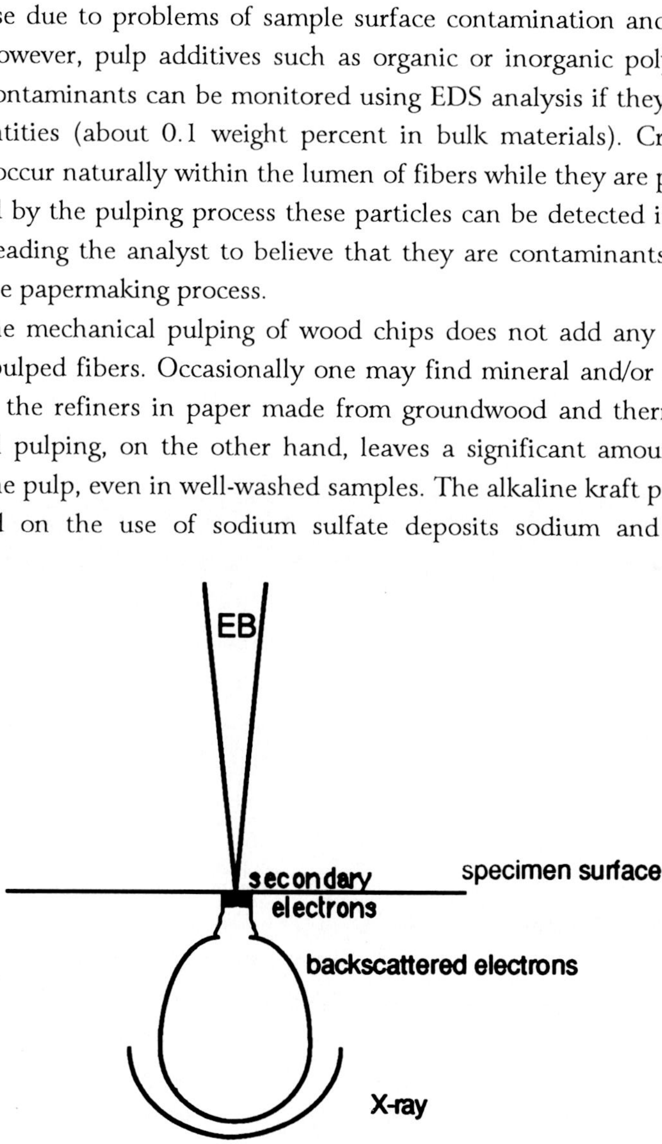

Figure 2. Depth of penetration of the incident beam and the resulting interaction volume.

fiber wall interstices, while the acidic sulfite process (H_2SO_3 + HSO_3) may be detected by the residual sulfur left in the fibers. However, it is difficult to use EDS analysis to distinguish between the two types of chemical pulp as the pulping liquors contain compounds that are constituted by the same elements (*i.e.*, it is difficult to differentiate between sulfate and sulfite as both are made up of sulfur, hydrogen and oxygen but in different proportions). The only element that, if present in the pulp in sufficient amount, gives a clue as to which pulping process has been used is the sodium (Na) in alkaline krafts. One comes across the same problem when trying to determine whether elemental chlorine or an organic chlorine compound is present as a result of bleaching.

POINT ANALYSES, LINE SCANS AND DOT MAPS

As mentioned previously, SEMs and STEMs can direct the electron beam to specific locations on the specimen. This capability significantly enhances the usefulness of EDS analyses as either points, lines or regions can be scanned and spectra collected. Point analyses are fairly straight-forward (pick a point, direct the beam to that point, collect a spectrum, identify the peaks for the elements present) and are very useful for contaminant identification, but scanning along lines or regions needs more explanation to show how these are performed and why the data can be so valuable.

When a spectrum is acquired, the x-ray analyzer displays the spectrum of the entire range of x-ray energies detected, as in Figure 1. (The spectrum will look similar to Figure 1 when more than one element is present, but each element will exhibit one or more peaks). Once this initial spectrum is collected, the analyst can specify one of more *regions of interest* (ROI) on the spectrum, *i.e.*, specific energy ranges corresponding to one or more peaks or background areas in the spectrum. Setting up ROIs essentially tells the x-ray analyzer to only display x-ray counts if they fall within the specified energy ranges(s). This provides a means to qualitatively measure the relative concentration of specified elements at prescribed locations on a specimen surface exposed to the electron beam.

Consider how a line scan might be used: perhaps one wished to know the distribution of coating pigment in a cross-section of paper, and an initial analysis confirmed that the pigment (*e.g.*, the titanium or calcium peak) was present. The analyst would first specify the ROI for the pigment x-ray peak to be observed, and then the endpoint coordinates for the line to be scanned would be entered into the analyzer (perhaps between the paper surface and some interior location). The x-ray analyzer would assume control of the electron microscope beam and direct it to scan along the prescribed coordinates. (The scan would be made point-by-point with a specified dwell time at each location; longer dwell times increase the number of x-rays collected, but the electron beam may damage sensitive specimens). A graph of the number of ROI x-ray counts recorded at each point along the line specified would then be displayed or printed out by the x-ray analyzer. (See Figure 3).

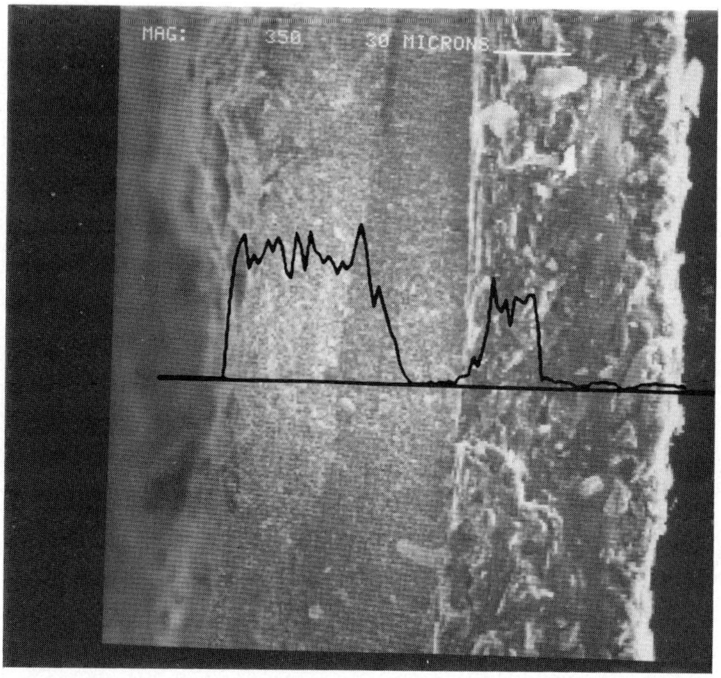

Figure 3. Linescan for lead (in black) overlaid on an SEM micrograph of a multi-layer paint chip from the USDA Forest Products Laboratory in Madison, Wisconsin.

An x-ray dot map would be collected using a procedure similar to that used for line scans, but the x-ray analyzer would collect data for an area, not just a line. Several maps can be acquired simultaneously (the exact number depends on the x-ray analyzer). Often a background region of the spectrum adjacent to an ROI will be collected and a corresponding number of counts subtracted from each elemental map to minimize non-characteristic x-ray counts due to bremsstrahlung. The number of counts at each point can sometimes be displayed in false color, assisting in interpreting the image. Figure 4 is an example of an x-ray map for iron and the corresponding SEM secondary electron image of the same area. This particular image is of the eye at the top of the pyramid on the back of a United States one-dollar bill (The Great Seal). The ink pattern (iron pattern) is clearly visible in the x-ray image, although it is practically invisible in the secondary electron image.

SAMPLE ANALYSIS

One faces several decisions when doing EDS analysis of paper resulting from the need to irradiate the sample with sufficient energy to cause the emission of x-rays by all the elements present and from concomitant complications created by the structure of paper.

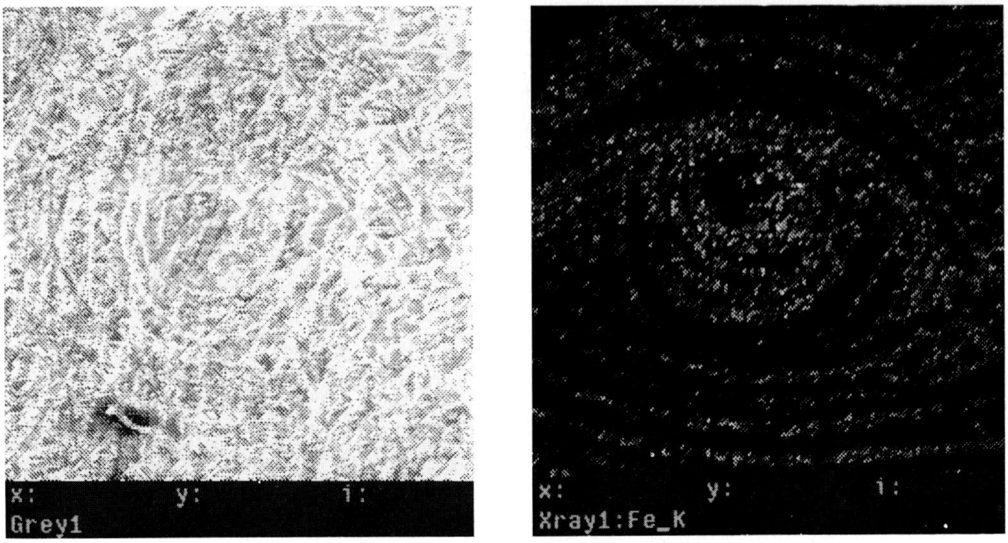

Figure 4. Secondary electron image (left) and x-ray map (Fe K peak, right) of the eye above the pyramid on the reverse of a United States one-dollar bill (The Great Seal).

The first requirement is met by obtaining a number of spectra taken at different incident energies, thus creating the ideal conditions for the analysis of each element. An understanding of the three-dimensional structure of paper and the distribution of additives and/or contaminants is essential in the latter case. The analyst is urged to study Chapter 6 in Goldstein *et al.*,[2] "Qualitative X-Ray Analysis," because of the ease with which the elemental constituents of a sample may be misidentified, rendering the final qualitative analysis meaningless. When used circumspectly, however, the automated peak identification routines provided by EDS manufacturers can be quite useful.

ENERGY OF THE INCIDENT BEAM

The most significant problem one faces when using EDS to analyze paper is how to choose the energy of the incident beam. While the total x-ray spectrum of interest (0.1 to 20 keV) can be acquired in a short period (200 seconds) with an incident beam energy of 20 keV, the considerable size of the interaction volume makes it impossible to determine the depth and consequently the particle(s) from which the x-rays originate. To overcome this problem one should first use a high energy incident beam (20 keV or more) so that all possible lines of an element (normally 0.1–10 keV range) are excited. This allows a high degree of confidence in the identification of each element, because all the members of a family of lines of any particular element are present in the spectrum. Once the elements present are known, the incident beam energy may be tailored to make it possible to analyze each element within the structure of the paper by locating all the lines of each element. The following rule of thumb is generally used: the energy of the incident beam should be equal to the characteristic energy of the K lines of the element (*e.g.*, for Ti Kα = 4.51 keV and Kβ = 4.98 keV) plus 3 to 5 keV (3 keV in the case of elements with $Z<10$ and 5 keV for all the other elements). Although modern computer programs do much of the work, all the lines of an element (*i.e.*, all the peaks) should be identified, verified and labelled, as this is the only way to insure that misidentifications have not occurred. Artifacts such as escape and sum peaks should be checked also. Fiori and Newbury[4] published excellent tables of all the x-ray lines observed in EDS spectra which are useful in evaluating possible interferences.

PREPARATION OF SPECIMENS FOR EDS ANALYSIS

SEM topographical images of paper are obtained from specimens that are generally stuck onto aluminum stubs with conducting tape and coated with a thin layer of a heavy metal. These samples are inappropriate for EDS analysis because x-rays produced by the specimen have to overcome the barrier created by the metal film and x-rays characteristic of the coating metal are also produced. To eliminate this problem samples are usually coated with a thin conductive layer of evaporated carbon, if carbon is not one of the elements sought in the analysis. If carbon is sought the sample may be analyzed without a coating layer, but this makes it difficult to image the sample and induces charging unless a very low incident energy (<5 keV) is used.

The size of the interaction volume may be such that the x-rays generated are produced by the aluminum stub or brass holder rather than the paper (Figure 5). This problem may be overcome in two ways: *i*) a stack of paper specimens from the same sample (glued together with carbon paint) may be used so that the interaction volume is generated within the paper stack; *ii*) alternatively, the paper specimen may be glued with

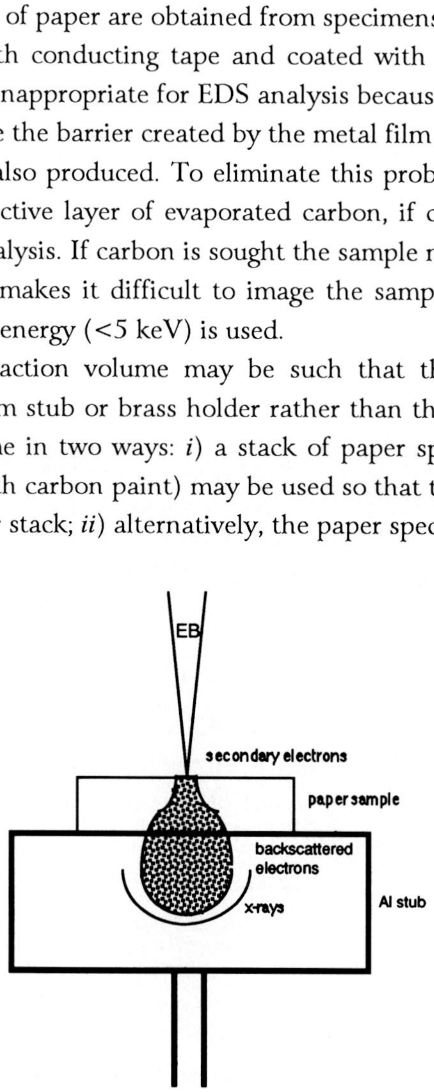

Figure 5. Diagram of a sample of linerboard (1.0 mm thick) stuck with carbon paint onto an aluminum stub. When examined with a high-energy incident beam (*e.g.*, 20 keV), the size of the interaction volume is such that x-rays are emitted by the underlying stub.

carbon paint onto a carbon substrate, usually a disc of pure carbon several millimeters thick which may in turn be glued onto the aluminum stub. Carbon stubs similar to aluminum stubs may also be used. In the latter case the x-rays within the carbon planchette will not interfere with the x-rays generated from the heavier elements within the paper sample. The preferred configuration will vary among microscopes, as different manufacturers use different types of stubs or specimen holders.

Sample preparation procedures for TEM (embedding) are useful for SEM–EDS analysis as well; specimens are sometimes embedded in order to accurately map the cross-sectional distribution of coating or filler components, for example. Sample preparation is somewhat more involved than it is for SEM–EDS examination. Fixation with glutaraldehyde or osmium tetroxide (OsO_4) is often used for TEM examination of soft tissue samples to toughen cell walls, etc., but for wood fibers this step can often be omitted (and may not be desirable for EDS anyway). For most embedding resins, specimens are successively immersed in a graded series of hygroscopic solvents (*e.g.*, ethanol) followed by a dehydration agent (*e.g.*, acetone), then embedded in the chosen resin. Many resins work well, but the ultra-low viscosity variations of Spurr's resin may be a good initial choice. Some resins (such as the Spurr's formulations and other epoxies) are very sensitive to small amounts of water and may become brittle if dehydration is not carefully carried out.

To embed specimens, place the dehydrated specimen in a capsule and fill it with the embedding mixture. A 50% resin/50% dehydration solvent mixture may be used at first, especially if the specimen is difficult to infiltrate with the resin by itself. Label the capsule with a small piece of paper on which you have written the specimen number (use pencil, as many resin formulations contain solvents that will cause ink to bleed and become illegible) and place the paper in the capsule with the label facing outward. Place the capsules in a desiccator. If a 50:50 mixture is used, the resin mixture should cover the specimen for several hours or more, and after this initial infiltration the resin should be changed. Exchange the 50:50 mixture for fresh resin, return the capsules to the desiccator, and place under enough vacuum to draw entrapped air from the surface of the specimens. At this point, leave the specimen for twelve hours to two days (depending on the permeability of the specimen and the resin viscosity) and cure the resin according to the manufacturer's directions. Some materials (such as latex pigments) are soluble in certain embedding materials, and the solvents used may cause dislocation of material. Experimentation is a usual requirement for each new type of specimen and each type of embedding material.

Following embedding the embedded specimens are removed from the capsules and trimmed with a glass knife. They may be left like this if the analyst is satisfied with the cut, but the surface is often recut with a diamond knife for a cleaner face. The surface is then analyzed using x-ray line scans or digital maps.

EDS analyses in a STEM are usually conducted more for research purposes than for product or contaminant examination. We have used this type of analysis to determine the distribution of sulfur across a tracheid cell wall, for example. STEM specimens are dehydrated and embedded as above and thin sections are prepared from the encapsulated specimens with a diamond knife; TEM specimens for imaging are often about 100 nm thick, and for STEM–EDS they are sometimes cut a little thicker. After the sections are microtomed, they must be mounted on grids; these come in various configurations (100 mesh, 200 mesh, hole grids, etc.) and several different metals (*e.g.*, nickel and copper). Carbon and nylon grids are also available. The analyst must choose a grid that will support the specimen without unduly interfering with the image, and the grid material must not have KLM lines that conflict with the element of interest in the specimen. Casting a formvar coating on the grid prior to mounting the specimen will help to support and stabilize the specimen in the microscope, but carbon coating and other preparative procedures are not usually required. Depending on the detector position and angle in the microscope, the specimen may need to be tilted in the STEM to facilitate x-ray collection.

Storage and Safe Use of Embedding Materials: Embedding resins have a definite shelf life, and it is better to make several small orders as work progresses than to order larger quantities to take advantage of economies of scale and end up disposing of unused resins and catalysts. The resin kits should be refrigerated until the embedding material is mixed, and we *highly* recommend that all relevant safety precautions be observed in measuring and mixing these materials; some components are carcinogenic or cause skin irritation. The curing oven required for many resins should be vented to the outside air and all waste should be hardened before disposal. Wear gloves when handling resin components and do not use alcohol to remove resins from skin, as this only increases the resin penetration; use soap and water.

LOCATION AND ANALYSIS OF CONTAMINANT PARTICLES

The location of contaminant particles, usually present as agglomerates within a paper specimen, can present a special problem to the spectroscopist. The energy of the incident beam should be tailored to detect the elements present and the size of the interaction volume should be such that particles can be detected. X-rays and backscattered electrons produced deep within a paper sample (especially those produced by higher atomic number elements with correspondingly high energy x-rays) will interact with particles on or close to the paper surface, and will in turn cause x-rays characteristic of these particles to be emitted. However, x-rays produced by fibers and particles within the range of the interaction volume are not likely to excite particles deeper within the paper cross-section. It is therefore essential to obtain multiple spectra of the same sample at different energies to insure that all particles present within the paper are analyzed, but this method will not localize the position of the contaminant within the paper structure.

The precise location of contaminant particles and/or fillers can be determined by producing rough cross-sections of a paper sample and then analyzing the surfaces obtained. The fastest method consists of freezing a paper specimen (~ 10 x 2 mm) in liquid nitrogen, placing it on a chilled metal plate (sitting on dry ice) and making parallel cuts (3 or 4) with a frozen blade. For best result the cutting motion should be quick and perpendicular to the surface to avoid plastic flow of the cut fiber surfaces and subsequent dislocation of the particles. The cut specimens are then mounted on edge on carbon planchettes using carbon paint (perhaps between carbon rods for support) and coated with evaporated carbon. EDS analysis of the newly-cut faces is carried out to determine the elemental composition of the various particles, and their location within the paper structure can then be mapped. Micrographs are usually acquired for comparison purposes.

The advantages of this sample preparation method are: *i*) multiple specimens from various locations of a paper roll can be produced for EDS analysis within a short time period; *ii*) the specimen to be analyzed is not subjected to solvents or embedding media which could dissolve or translocate particles or introduce foreign elements; and *iii*) (most importantly) the paper structure is preserved with its original arrangement of contaminant and/or filler particles so that their position can be accurately mapped.

ANALYSIS OF POWDERS

The analysis of mineral pigments and/or contaminant particles in the form of powders may be carried out by analyzing a sample of the powder on a carbon planchette onto which a thin, continuous film of carbon has been evaporated (Figure 6). Sample preparation consists of dusting a representative sample of the powder (ground with a pestle and mortar made of agate to avoid contaminating the sample) onto wet carbon paint. After the initial spectrum taken at high voltage (20 keV) and low magnification (covering most of the sample), it is recommended that a minimum of ten spectra be taken at voltages tailored to the elements present in different areas of the specimen. If a particle presents areas of differing topography, point spectra of the different areas should be obtained, thus insuring that all the particles present have been analyzed.

ANALYSIS OF ADDITIVES

The large variety of organic and inorganic additives as well as mineral and plastic pigments currently being used by the paper industry can for the most part be detected and analyzed by EDS. The elements and the proportions in which these are present can be rapidly analyzed qualitatively, but the compounds they form can not always be identified.

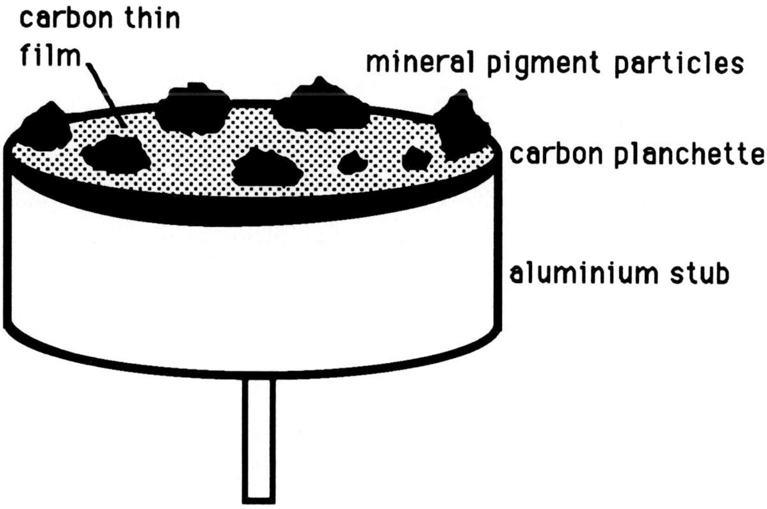

Figure 6. Diagram of a continuous thin film of carbon covering powdered contaminant particles which have been stuck onto a carbon planchette with wet carbon paint for EDS analysis. The carbon film is continuous to prevent the particles from charging.

Mineral pigments commonly used in North America (clay, calcium carbonate and titanium dioxide) can be identified and mapped because one or more specific elements in each of these compounds is not shared, *i.e.*, aluminum and silicon are only present in clay, calcium is exclusive to calcium carbonate and titanium is restricted to titanium dioxide. Thus, differences in concentration in any one of these compounds in a particular area of a sheet of paper can be detected by comparing spectra . Their distribution can also be analyzed by creating elemental maps. However, the spatial resolution of the paper additives is somewhat diffuse due to the size of the interaction volume; as mentioned previously, the x-rays of a particular element are not generated at the surface of the specimen but originate at various depths. Secondary x-rays generated by backscattered electrons interacting with particles closer to the surface will also contribute to the mapping of these elements, giving them a diffuse outline, further decreasing the resolution.

Polymer additives are generally more difficult to detect due to the fact that they share the same elements (carbon, oxygen and hydrogen) as the fibers and therefore are indistinguishable from one another. Their presence can only be analyzed by comparing spectra acquired with windowless detectors (see Goldstein *et al.*[2]) which are highly sensitive to the presence of low atomic weight elements. By acquiring point spectra taken under the same microscope conditions on fiber cross-sections (>5 μm thick) near the center of the fiber wall and at the periphery, it is possible to determine if an additive polymer is present on the surface of the fibers by noting differences in the size of the peaks for carbon and oxygen. It should be noted that this type of analysis is strictly qualitative and that it is only possible if there is enough material (polymer additive) present to generate a sufficient number of x-rays.

Due to poor energy resolution and the inability of the EDS detector to separate the various components (L-, M- and N- lines of relatively heavier elements and K-lines of lighter elements) of x-ray families which occur at low energies (<3 keV) there are frequent spectral interference problems when dealing with spectra of carbon, nitrogen and oxygen compounds. Other artifacts of the EDS detection process also contribute to difficulties in resolving the spectra of low atomic number elements. Refer to Goldstein *et al.*,[1] Chapter 5 for a detailed description of these artifacts.

CONTAMINANT IDENTIFICATION

EDS is an essential tool for the rapid analysis of paper contaminants. Metal particles can be readily identified and by comparing the spectra obtained with the spectra of standard materials used in the industry (such as stainless steel alloys used in a specific section of a mill) it is possible to determine the origin of the particle. This method makes it possible to determine where excess wear is taking place within the pulping and/or papermaking systems. Often these particles will contain metals of high atomic number (Z>30) which have to be identified by the presence on the spectrum of L- or M- lines, some of which are superimposed on the K-lines of lighter elements. Examples are molybdenum (present in some alloys) L-lines which coincide with the K-lines of sulfur (present in kraft pulps). Especial care should be taken to correctly deconvolute the spectrum and identify all the lines of each element.

Contaminant mineral particles which may be introduced in the paper by the water system or which originate in recycled pulp can also be identified and traced by comparing spectra of the contaminant and a white water filtrate.

Organic contaminants present in paper are generally more difficult to analyze. However, small quantities of mineral compounds are usually associated with the contaminant and by comparing spectra of the elements present in the contaminant with spectra taken from deposits in various locations of the system it is possible to pinpoint the origin of the contaminant. Special care should be exercised during this type of analysis because trace quantities of elements can lead to misidentification due to problems of spectral interference, artifacts and insufficient material to generate x-rays (as previously discussed).

EXAMINATION OF WOOD FIBERS

Seldom is one required to analyze wood fibers before they have undergone some type of pulping process. However, a study of wood chip impregnation under different conditions[5] presented an opportunity to verify how accurately it is possible to determine variations in concentration of a particular element by acquiring a series of spectra using the line scan mode from the periphery to the center of each chip. For this experiment chips of a predetermined size were cut and impregnated by chemical compounds under specific conditions. Subsequently, the chips were sectioned and EDS analyses were conducted to determine the concentration of a particular element known to be present in the liquor.

The main problem encountered during the testing was the size of the interaction volume which produced x-rays originating deep within the sample and not exclusively from the cut surface. To overcome this problem a low atomic number element was added to the pulping liquor so that by using a relatively low energy incident beam (with a relatively small interaction volume) it was possible to track the concentration of this element on the cut surface. Thus, it was possible to determine the optimum impregnation parameters (time, temperature and pressure) for chips of particular size and wood species.

EXAMPLES OF EDS APPLICATIONS

X-ray analysis in the SEM can be very useful in solving contaminant identification problems, or for determining the depth of surface treatments (on cross-sections), etc. The ability of the SEM to direct an electron beam at a point, along a line or across a defined area makes it possible to acquire a spectrum and to map the spatial distribution of selected elements within the rastered area. The presentation of the spectra and the maps will depend on the x-ray analyzer used to compile the data.

Figure 7 shows a secondary electron image of a softwood kraft paper sample with a contaminant on the surface. EDS analysis can be quickly carried out to obtain the spectrum of the contaminated area (Figure 8). This spectrum indicates: *i*) the presence of carbon and oxygen from the fibers; *ii*) a small sulfur peak, perhaps a residue from the pulping process; *iii*) a large titanium peak, indicating the presence of a titanium filler in this uncoated sheet; and *iv*) aluminum and silicon peaks. Aluminum and silicon together probably indicate that clay is present, but the unexpected height of this aluminum peak perhaps indicates that alum $[Al_2(SO_4)_3]$ may have been added to the pulp. This was subsequently confirmed by performing spot analyses on different areas of the sample away from the contaminant which indicated that aluminum was present over the entire paper surface. Figure 9, a composite EDS map of the contaminated area, shows the distribution of titanium, aluminum and silicon in the region. Aluminum and titanium were more or less equally distributed over the whole area with a very high concentration of titanium in particle (1), indicating a filler agglomerate, and a predominance of silicon in particle (2), perhaps a particle of sand or clay. A higher concentration of titanium on the right side of the area indicates an uneven distribution of the filler pigment within this particular region of the sample. This may or may not be a valid conclusion to draw for the entire sample – analysts all too often are asked to remove a tiny snippet of material

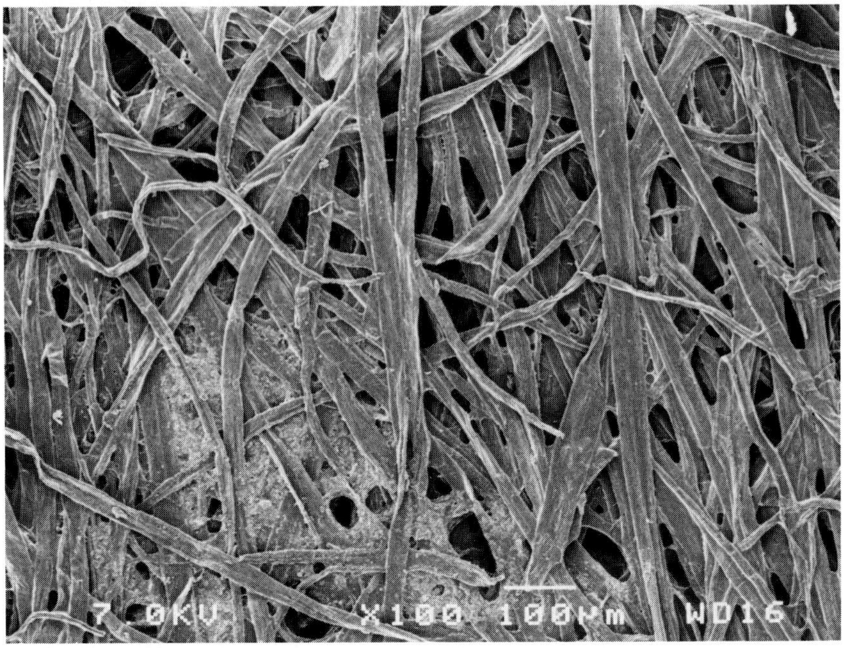

Figure 7. SEM micrograph of a softwood kraft paper sample with a surface contaminant. The sheet contains titanium dioxide as a filler.

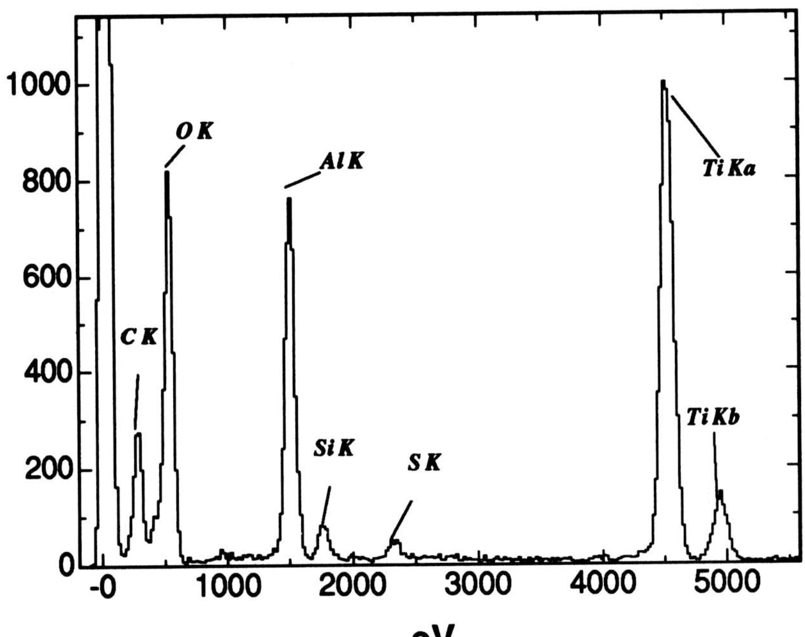

Figure 8. EDS spectrum of the contaminated surface in Figure 7.

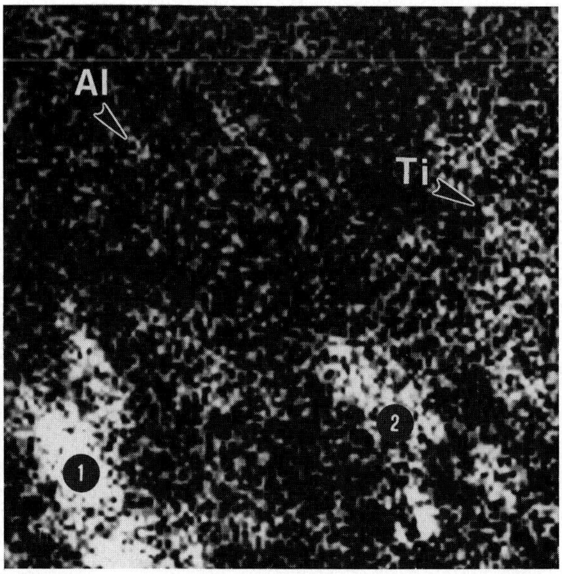

Figure 9. EDS composite map showing the distribution of titanium (Ti), aluminum (Al), and silicon in the central portion of Figure 7.

from square meters of a paper sample and make definitive problem assessments based on a single examination, but this may not be a valid technique for every problem.

The next set of micrographs is presented to demonstrate how comparison spectra and maps can be used. Figure 10 shows a secondary electron image of a control area of a paper sheet. No contaminant spots are visible, and an EDS spectrum taken over this area (Figure 11) indicates that the mineral pigments are in proportion. This spectrum was observed with a beryllium window detector, so no light elements are present in the spectrum. The proportions of aluminum and silicon indicate the presence of clay, titanium indicates titanium dioxide, and calcium carbonate is also present. Figure 12 is an SEM micrograph taken at a contaminant spot. The micrograph shows a large agglomerate of mineral particles and ray cells, indicating the possible presence of resinous material holding the cluster together. The corresponding spectrum (Figure 13) indicates that the agglomerate is formed mainly by calcium and that the other mineral pigments are present in nearly the same proportions as in the control area.

Figures 14a and 14b are x-ray maps of the area surrounding the contaminant in Figure 12. Figure 14a is a map of aluminum and Figure 14b is a map for silicon. Note the diffuse outline of the particles and how the relative image intensities reflect the peak heights observed in Figures 11 and 13.

Table 1 may be useful as a starting point to identify possible sources of x-ray peaks in pulp, paper (including recycled paper), or equipment (*e.g.*, fabric or felt) samples submitted for analysis. We have only included a sampling of ink pigments and driers, but

Figure 10. SEM micrograph of the control area of the filled paper sample.

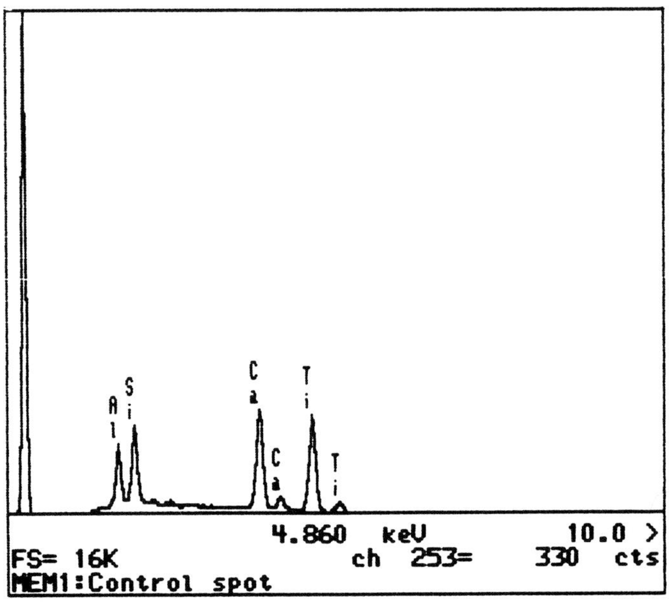

Figure 11. EDS spectrum for the control spot seen in Figure 10. Total acquisition time was 200 seconds live time.

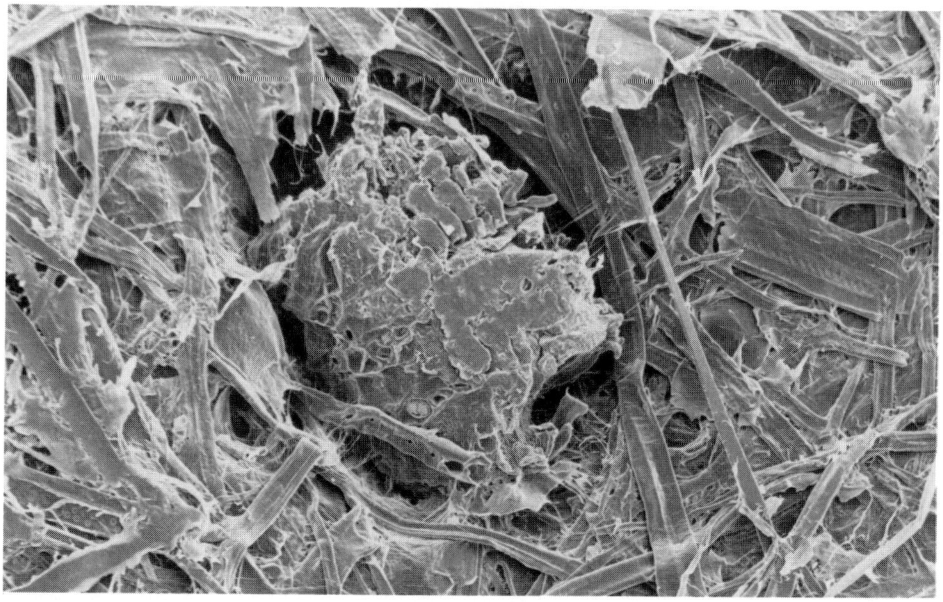

Figure 12. SEM micrograph of a contaminant spot showing a large agglomerate of what appears to be mineral particles and ray cells, presumably held together by some resinous material.

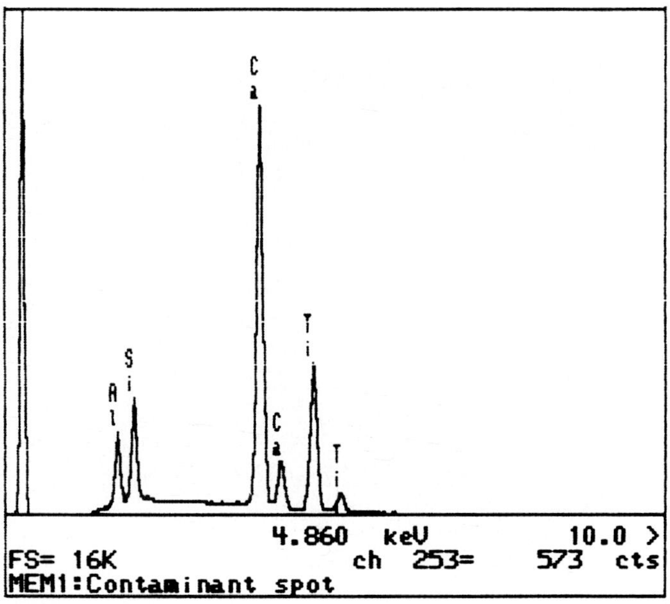

Figure 13. EDS spectrum of the contaminant in Figure 12. This spectrum indicates a much higher concentration of calcium than in the control area; refer to Figure 11. Total acquisition time was 200 seconds live time.

(a)

(b)

Figure 14. (a) EDS map of aluminum in the contaminant spot in Figure 12; (b) EDS map of silicon in the contaminant spot in Figure 12. Note the diffuse outline of the particles and the relative intensities of the two maps, corresponding to the proportions of these two elements in clay (*i.e.*, there is more silicon than aluminum in clay).

many modern pigments are organic molecules with no metallic components that would be analyzable with EDS. Interested readers are referred to a book published by the National Association of Printing Ink Manufacturers (NAPIM) for more information.[6]

SUMMARY

The EDS is well suited as a qualitative analytical instrument for wood fibers and paper samples. The major elements constituting the numerous additives and contaminants present in pre- and post- consumer paper products can be easily identified. Its main advantage is the speed with which a spectrum can be acquired and the chemical composition of the sample evaluated. On the negative side, its relatively poor energy resolution, its inability to clearly resolve peaks (especially at low energy) and the presence of spectral artifacts require a certain amount of knowledge and attention to details on the part of the operator.

Non-conductive samples such as paper or pulp fibers should be coated with a thin film of evaporated carbon except when carbon is one of the elements sought. When trying to identify light elements, spectra of uncoated samples should be accumulated at low beam energy using an ultra-thin window or windowless x-ray detectors. Analyzing this type of sample requires the acquisition of at least two spectra, one of which should be acquired at high (20 keV) and the other at low (between 5–7 keV) incident beam voltages so that all the peaks of all the elements present are excited. Once the statistically significant peaks are identified a thorough check of all the x-ray lines (K-, L- and M-lines) of a specific element (including escape and sum peaks) should be conducted to avoid misidentification. Special precautions should be taken when comparing spectra. Microscope conditions and sample preparation methods should be exactly the same to ensure comparability. Reporting of peaks that correspond to trace amounts ($\leq$0.1 weight percent) should be done with the utmost caution and only after all the escape and sum peaks of the major elements present have been accounted for.

ACKNOWLEDGEMENTS

We would like to thank several people who made contributions to this chapter. First of all, we would like to thank the person who gave TEC a draft of Table 1 some years ago; we would like to acknowledge you personally, but he no longer has any record of your name or company affiliation. Other contributions to this table have been made by the authors, by Paul Glogowski of Beloit Corporation, by A.J. Schellahmer of Betz Paper-Chem, and by Ron Zarges of Weyerhaeuser Technology Center. Maria Da Rocha of Sun Chemical, Jack Power of Van Son Holland Ink Corporation of America and Rashmi Patel of Monarch Color Corporation graciously assisted with information about various printing inks and pigments. Dr. Bruce Panuska of the Geosciences Department at Mississippi State University helped us with some of the data on mineral composition and impurities. We also thank the Electron Microscopy Center at Mississippi State University for providing the copper spectrum in Figure 1, Tom Kuster of the USDA Forest Products Laboratory for the linescan shown in Figure 3, and Dr. Lisa Detter-Hoskin of the Georgia Tech Research Institute for the x-ray map shown in Figure 4.

REFERENCES

1. Goldstein, J.I., Newbury, D.E., Echlin, P., Joy, D.C., Romig, A.D., Lyman, C.E., Fiori, C. and Lifshin, E. 1992. Scanning Electron Microscopy and X-Ray Microanalysis. Plenum Press, New York.
2. Fitzgerald, R., Keil, K. and Heinrich, K.F.J. 1968. Solid State Energy Dispersive Spectrometer for Electron Probe X-Ray Analysis. *Science* 159:528–529.
3. Echlin, P. 1992. Low Temperature Microscopy and Analysis. Plenum Press, New York.
4. Fiori, C.E. and Newbury, D.E. 1975. Scanning Electron Microscopy, Volume 1. O. Johari, editor. SEM Inc., page 401.
5. de Silveira, G. Unpublished work.
6. Fetsko, J.M. 1983. NPIRI Raw Material Data Handbook, Volume 4: Pigments. Section D: Color and Chemical Composition. National Association of Printing Ink Manufacturers (NAPIM), Hasbrouck Heights, NY.

Table 1
Attribution of Elements Which May Be Identified in Samples

Element Symbol	Element Name	Possible Source [Chemical Formula]	Use	If Present, Other Elements That Must Be Present
Al	Aluminum	Clay [$Al_2Si_2O_5OH_4$]	Filler, coating, ink extender	Silicon
		Papermaker's alum [$Al_2(SO_4)_3$]	Size precipitator, pitch control, etc.	S, may have a trace of iron
		Polyaluminum chloride (PAC) [$Al_2OH_{6x}Cl_2Y \bullet H_2O$]n; x=1 to 4, y=2x, n may be as high as 15	Polymer additive	Chlorine
		X-ray detector (slight peak)	Artifact, sample holder or microscope part	
		Aluminum oxide or hydroxide	Alumina deposits at wet end	Oxygen
		Aluminum foil	Found in some recycled papers	
		Sodium aluminosilicate	Filler	Silicon, Sodium
Ba	Barium	Barium sulfate [$BaSO_4$]	Filler or ink pigment (white)	Sulfur
		Ink pigment	Printing ink (red)	
Ca	Calcium	Calcium carbonate [$CaCO_3$]	Filler, hard water	*May* have some magnesium
			Extender in some printing inks	
		Ink pigment	Printing ink (red)	Sulfur
		Gypsum [$CaSO_4 \bullet 2H_2O$]	Filler, hard water	
Cl	Chlorine	Chlorides	Byproduct of felt damage, etc.	
		Ink pigment	Some green printing inks	Copper
		Water treatment		
		Various salts	Neutralized bleach	
		Ink pigment	Printing ink (yellow)	
Co	Cobalt	Ink component (drier)	Printing ink	

Table 1, Continued
Attribution of Elements Which May Be Identified in Samples

Element Symbol	Element Name	Possible Source [Chemical Formula]	Use	If Present, Other Elements That Must Be Present
Cu	Copper	Ink pigment Brass Bronze	Blue or green printing inks, metallic inks, black ink Doctor blade, microscope parts Doctor blade	Possibly chlorine in green inks; maybe Fe in black; Zinc Tin
F	Fluorine	Fluorocarbon	Teflon	Carbon
Fe	Iron	Talc [$Mg_3Si_4O_{10}(OH)_2$] Ink pigments [FeO, FeCu] Some toners Stainless steel Impurity: filler, water, alum	Pitch control, filler Printing inks	Fe may substitute for Mg in talc; see Mg Maybe copper Chromium, nickel, manganese Various
K	Potassium	Pitch/black liquor		
Mg	Magnesium	Talc [$Mg_3Si_4O_{10}(OH)_2$] Magnesium oxide [MgO] Hard water [$CaCO_3$ or $CaMg(CO_3)_2$]	Pitch control, filler Rubber additive	Si; Al or Ti may substitute for Si in some forms; Usually Zn, S; Calcium
Mn	Manganese	Ink component (drier) Hard water deposits	Printing ink	
P	Phosphorus	Paper brightener Polyphosphates Ink pigment (blue)	Paper brightener Pitch dispersant Printing ink	 Molybdenum or tungsten

Table 1, Continued
Attribution of Elements Which May Be Identified in Samples

Element Symbol	Element Name	Possible Source [Chemical Formula]	Use	If Present, Other Elements That Must Be Present
S	Sulfur	Wool		
		Gypsum [$CaSO_4 \bullet 2H_2O$]	Filler	Calcium
		Alum [$Al_2(SO_4)_3$]	Size precipitator, pitch control, etc.	Aluminum
		Rubber crosslinker	Rubber additive	Usually Zn, Mg
		Sulfites	Pulping /brightening process byproduct	
		Sulfates	Pulping byproduct	Possibly sodium
		Pearl filler [$CaSO_4$]	Filler	Calcium
		Barium sulfate	Filler, ink pigment	Barium
Si	Silicon	Clay [$Al_2Si_2O_5OH_4$]	Filler, coating, ink extender	Aluminum;
		Talc [$Mg_3Si_4O_{10}(OH)_2$]	Pitch control	Magnesium;
		Silicones, silicates, colloidal silica	Size, antifoam, anti-skid additives, etc.	
		Precipitated silicate [SiO_2]	Filler	5% CaO, 78% SiO_2
		Diatomaceous silica [SiO_2]	Filler	92% SiO_2
		Silico aluminate [$Al_2O_3 \bullet SiO_2$]	Filler	Al, Si
		Concrete	Towers, chests	Al, Ca, K, Fe
Sn	Tin	Solder	Sweated pipe fittings	Lead
Ti	Titanium	Titanium dioxide [TiO_2]	Whitener, filler, opaque ink pigment	
Zn	Zinc	Galvanized steel		Iron
		Zinc sulfide [ZnS]	Phosphorescent pigment	Sulfur
		Zinc oxide [ZnO]	Roll cover additive	Usually Mg, S
		Ink pigment	Printing ink (opaque white)	

10

SIMS: Secondary Ion Mass Spectrometry

Lisa D. Detter-Hoskin and Kenneth L. Busch*

Georgia Tech Research Institute, Atlanta, Georgia, U.S.A.
**Georgia Institute of Technology, Atlanta, Georgia, U.S.A.*

INTRODUCTION

Secondary ion mass spectrometry (SIMS) is the most sensitive of all surface analysis techniques. SIMS is employed to solve practical problems for which both superior spatial resolution and/or detailed chemical information are desired. In the SIMS experiment, an energetic beam of ions is focused onto a solid-phase sample and secondary species (ions, neutral atoms, electrons, and clusters) are sputtered from the topmost monolayer(s) of the surface. The resulting ionized species is then analyzed by a mass spectrometer, giving a positive or negative ion SIMS mass spectrum. Secondary electrons and ions emitted from the sample can also be collected to form SIMS images which contain spatial distribution information showing the origin of elements, fragments, and molecular ions on the sample surface in the xy plane. SIMS is also routinely used to study solid-phase organic and biochemical materials that are thermally labile or non-volatile, and it is, therefore, one approach of desorption ionization (DI) mass spectrometry. Other types of DI mass spectrometric techniques (several of which are conceptually similar to SIMS), include fast atom bombardment mass spectrometry (FAB MS), which uses neutral atoms to bombard solid or liquid samples, and laser microprobe mass analysis (LAMMA). This latter acronym was coined to describe a commercial instrument; LD for laser desorption is the more general acronym, replaced more recently with matrix-assisted laser desorption ionization (MALDI).

The distinction between FAB and SIMS seems straightforward: SIMS uses an incident beam of energetic ions, while FAB uses an incident beam of energetic neutral particles. In reality, incident particle beams can consist of mixtures of ions and neutral particles. Many organic and biochemical "FAB" analyses are carried out with primary incident ions and are therefore "SIMS" experiments. From the sample's point of view,

0-8493-8992-5/95/$0.00+$.50

the charged or neutral character of the incident particle is often of secondary importance. FAB often involves dissolution of the organic sample in a vacuum-compatible liquid support matrix (glycerol or meta nitrobenzylalcohol are common), but this same experiment can be carried out as "liquid SIMS." The relevant distinctions between SIMS and FAB are discussed at length in the book by Lyons.[4] Here we concentrate on the measurements rather than the semantics, but when the experiment is described by the investigators as FAB, we will acknowledge their use here with the hybrid term FAB/SIMS.

There are two modes of SIMS analyses: static and dynamic. Static SIMS employs low-energy (100 eV to 10 keV), low-flux density (<5 nA/cm^2) primary ion beams to bombard a specimen. During the time usually required to measure a mass spectrum, this fluence of primary ions sputters less than a monolayer off the material's surface. The importance of static SIMS lies in its ability to determine structural features and elemental information for samples that decompose with energetic particle bombardment. Static SIMS conditions also enable the intact secondary ions to be collected and an image of the surface to be produced, if the transmission of the mass analyzer is sufficiently high. Dynamic SIMS experiments employ higher current density primary beams for faster sputter rates. The increased sputter rate increases the sensitivity of the SIMS experiment. Dynamic SIMS is commonly used for trace elemental analysis (parts per billion range), but can also be used for quantitative depth profiling studies as the incident ion beam actively erodes the surface in a reproducible manner. For a review of the SIMS technique in both classical inorganic as well as organic and biochemical analytical fields, a number of literature reviews can be consulted.[1–4]

A recent breakthrough in SIMS is the use of a time-of-flight (ToF) mass spectrometer as the mass analyzer rather than the quadrupole or sector mass analyzers used in more conventional SIMS instruments. The main advantages of the ToF SIMS include: *i*) greatly improved sensitivity (using a low-dose pulsed beam of primary ions to eject secondary ions) due to the greater transmission of ions through the ToF mass spectrometer; *ii*) mass assignment accuracy (the ability to define the mass of a spectral peak) equivalent to quadrupole instruments, but with a wider mass range and the ability to measure the entire mass spectrum for a single pulse of incident ions without scanning of the mass analyzer; and *iii*) improved lateral resolution (the ability to resolve adjacent features) for greater imaging capability. Preliminary ToF SIMS studies of delicate species, such as lignocellulosic materials, polymers, and biomedical tissues, have been performed with a number of commercial instruments by several research groups, and minimal structural damage was observed while achieving high lateral resolution and sensitivity.[5]

The technical challenges of producing improved paper in the future depend on many factors, including the ability to better characterize the structure and chemistry of the lignocellulose fiber precursors and/or the structure of the finished paper product.[6] Recent successes in using SIMS to study polymer samples indicates that the SIMS technique may also be applicable to the study of the macromolecular structure of lignocellulosic fibers.[7]

A comprehensive literature survey of the use of SIMS to evaluate paper and related materials has shown that only a limited amount of SIMS research has been performed to date. As further discussed in the *Applications* section, SIMS has been successfully used to study the surface and bulk chemistry of paper, to generate ion images for compositional mapping, and to study organic samples on paper substrates there as part of the paper-making process, or as part of the uses of paper (ink and image transfer materials, for example). From its initial success as a paper surface analysis tool, SIMS should be considered to be a mainstream analytical component for future research programs conducted by the paper industry. The remainder of this chapter will discuss SIMS theory, instrumentation, and paper-specific applications.

INSTRUMENTATION FOR SIMS

As noted in the introduction, the defining attribute of SIMS is the sputtering of a surface by an incident (primary) ion beam, with formation of secondary ions that are mass-analyzed. The essentials of this sputtering process, along with a few of the popular variations in the methodology, are shown in Figure 1. The fact that ions released from the surface are created from the topmost 5–10 nm of the surface is the basis of the surface rather than bulk sensitivity of SIMS, as well as its depth profiling capability when the surface erosion is carefully controlled.

SIMS instrumentation consists of hardware to create and manipulate the primary ion beam, sample introduction and positioning devices, and secondary ion extraction, mass analysis, and detection systems. SIMS instruments based on quadrupole and sector mass analyzers have been available since the 1950s, and were used exclusively for inorganic and elemental analysis until the late 1970s. It was at that time that SIMS methods for the examination of organic surfaces and the determination of organic compounds were developed, and more sophisticated instruments such as ion microprobes, ion microscopes, and ToF SIMS instruments were developed. The monograph by Benninghoven *et al.*[1] contains a comprehensive overview of SIMS instrumentation.

The *primary ion source* in SIMS must provide a stable, reproducible, energetic beam of ions directed to a sample surface. Briefly reviewed here are discharge sources, liquid metal ion sources, and thermionic sources. The earliest SIMS source was a discharge source (based either on plasma or filament electron sources) filled with an inert gas such as argon. Interaction of the electrons with argon forms stable positive ions that are focussed, accelerated, and extracted from the ion source. For increased secondary ion yield from some surfaces, an oxygen negative ion beam can be drawn from discharge gas sources and used to bombard the surface. Over the past thirty years, the discharge

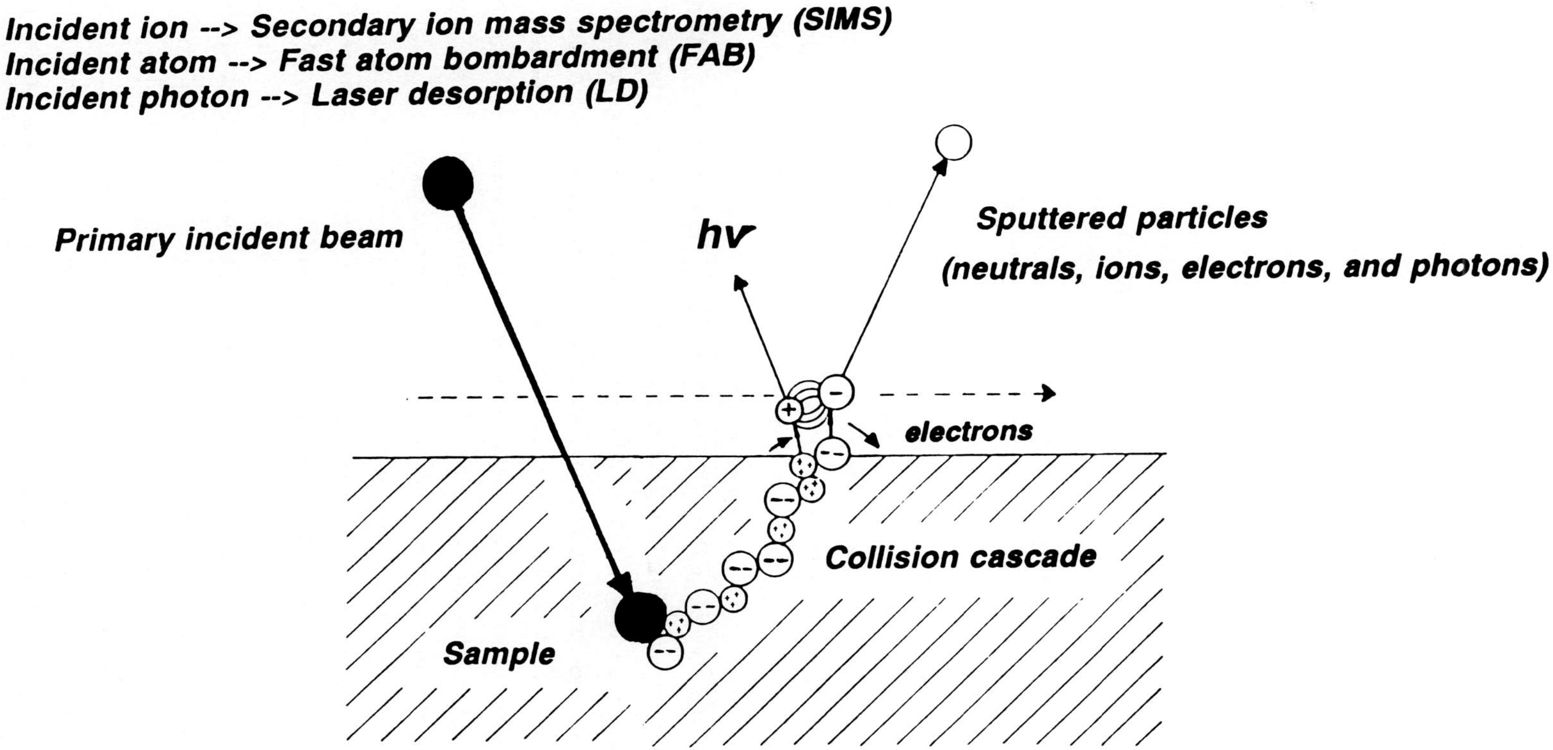

Figure 1. Schematic of the essentials of the sputtering process in SIMS, identified witha few variations of the method.

sources have evolved in several parallel paths. Such discharge sources are available as inexpensive SIMS sources, but can also be configured as highly focused imaging sources for ion microprobes and ion microscopes. Depth profiling through a sample requires a stable ion current of relatively high magnitude; the classical high current ion source is the duoplasmatron ion source. Another source of primary ions is the intense and highly focussed beam of ions that can be extracted from a liquid metal ion source. Such a source operates on the principle of field ionization of metal atoms held in a thin film in a very high potential gradient. Liquid metal ion guns provide different configurations of source heaters and emitter tips to match the different melting points and surface wetting characteristics of the various molten metals; gallium (a low melting point metal) is often used, but many other source metals are available. Finally, a thermionic source has become very popular recently; it provides a stable current of alkali ions (usually cesium, with its low ionization potential) of variable energy, is simple and rugged in construction, and provides no gas load to the high vacuum system. There are many variations, but in general, an alkali salt solution is doped into an aluminosilicate zeolite bead, which is then resistively heated to a very high temperature. The thermal emission of cesium ions from a heated surface will commence at a temperature of only 1300 °C. Since there is no discharge to manage, the cesium ion beam can be accelerated to a very high kinetic energy, improving SIMS sensitivity. Cesium ion sources on organic SIMS instruments routinely operate at voltages of 35 to 50 keV; the higher primary ion energies increase the sputter yield for most organic molecules sputtered from surfaces or from solutions.

If secondary ions are to be sputtered from insulating surfaces, a residual charge can accumulate on the surface that prevents further ion sputtering. Charge compensation is accomplished by using a low energy electron (< 10 eV) flood gun placed near the surface. An essentially neutral surface is maintained through a balance of the arrival of positive ions, the creation and capture of low energy electrons, and the departure from the surface of sputtered ions and electrons. The ion extraction optics of a SIMS instrument vary with the intended use and resolution required. Quadrupole-based SIMS instruments provide low resolution mass spectra, and can achieve moderate spatial resolution. Sector-based instruments can provide higher resolution mass spectral data, and dynamic emittance matching is used in a sophisticated ion optical system to maintain congruous transmission of ions into the mass analyzer from a small surface area. In an ion microscope, secondary ions from a relatively large irradiated surface area pass through the instrument optics in such a way that the image of the surface is preserved. An ion microprobe uses a scanning approach in which the position of the primary ion beam on the surface is rastered, and the ions that pass through the extraction optics to the detection system at any specified time are referenced to the surface position of the rastered beam.

The *mass analyzer* may take any one of several forms. Low-cost SIMS instruments are usually based on quadrupole mass analyzers. The quadrupole provides only low-resolution mass spectral data (typically unit mass resolution), but has forgiving tolerances in terms of the energy and trajectory apertures for secondary ions. A combination of radio-frequency and DC potentials is applied to adjacent rod pairs, and the combination of rf and DC is set such that, at any instant, ions of only one m/z value follow a stable trajectory through the rod structure. The rf/DC ratio is scanned to pass ions of different m/z through the rod assembly. Relative to other forms of mass analyzers, quadrupoles exhibit relatively poor overall ion transmission, and are limited to ions with masses of less than 2000 Daltons. In magnetic sector mass analyzers, the sputtered ions are accelerated and then pass through a curved path under the influence of the magnetic field. The ions that pass through the magnetic sector are described by the equation $m/z = B^2r^2/2V$, where r is the fixed radius of the electromagnet, B is the magnetic field strength and V is the accelerating voltage to which the ions are subjected. Scanning B passes ions of different masses through to the detector of the instrument. The accelerating voltage is typically on the order of a few thousand volts, compared to the very low extraction energies used in quadrupole instruments, and this difference can complicate the design of the secondary ion extraction optics. The ion transmission through the magnetic sector is generally higher, especially at higher masses. In addition, with the incorporation of an electric sector in the proper geometry, a double focusing mass spectrometer provides higher resolution mass measurements, and mass doublets can be resolved. Common examples in which such resolution is desirable include the separation of inorganic from organic species (*viz.* Al^+ from $C_2H_3^+$, or Zn^+ from TiO^+). In some instances, the proper observation of isotopic abundances may avoid the need for measurements at higher mass resolution. Together, sector- and quadrupole-based instruments constitute the majority of commercial SIMS instruments.

Time-of-flight mass analyzers have become much more popular over the past decade. In ToF instruments, mass analysis is accomplished via measurement of the time required for ions of constant kinetic energy to traverse the flight tube of the instrument. That time is given by the relationship $t = L(m/2zV)^{1/2}$, in which t is the flight time, L is the length of the flight tube, and V is the accelerating voltage in the source. At a constant kinetic energy, light ions have a higher velocity than more massive ions. Although the mass resolution achieved with time-of-flight mass analyzers is generally low, it can be improved with the simultaneous use of electric sectors to narrow the energy spread of the sputtered ions. A tremendous advantage of a SIMS ToF system is that the complete mass spectrum can be recorded for every pulse of the primary ion source. The very high transmission efficiency (50–100%) of the time-of-flight mass analyzer means that a very high surface sensitivity can be achieved, although the restrictions imposed by the low duty cycle conspire against the use of the time-of-flight mass analyzer in depth profiling studies. Finally, the Fourier transform ion cyclotron resonance mass spectrometer offers substantial advantages in terms of ultra high mass resolution and very high ion sensitivity, along with the ability to complete MS/MS experiments. This particular incarnation of mass analyzer is not generally considered in combination with SIMS, but several instruments in research laboratories have demonstrated outstanding results, and this ionization/mass analyzer combination will be a large part of future SIMS work.

BASICS OF THE SPUTTERING AND IONIZATION PROCESS

The phenomenon of sputtering was first noted by J. J. Thomson in 1910, who observed the emission of secondary ions from a bombarded metal surface. However, it was not until 1950 that the first SIMS spectra were deliberately recorded for metal and metal oxide surfaces. In the 1960s, commercial instruments for static and dynamic SIMS, depth profiling and imaging SIMS became widely available. Sputtering, as a result, has been the focus of a number of theoretical and experimental studies. Detailed sputtering models are described in several specialized references; only the more salient points are emphasized here. The first of these points is the fact that SIMS is a surface-sensitive technique. Theory and experiment have shown that sputtered ions for any individual primary ion impact originate from the top 2.5–3.0 nanometers of the surface. This result is based on the fact that the energy loss mechanisms for a primary ion impact with a condensed phase is a problem of molecular dynamics (Figure 2). The first energy loss is based on a nuclear stopping process. Essentially, particles first interact with each other in direct "billiard-ball" collisions over a time of 10^{-15} to 10^{-14} s in which the incident particle interacts with the particles of the specimen (direct knock-on collisions). Within 10^{-14} to 10^{-12} s, particles in the specimen originally set in motion by the primary ion impact continue to interact in a linear cascade regime. Finally, as the energy density decreases, the thermal spike or spike regime (10^{-12} s or greater) describes a microvolume in which all of the particles are in motion, and some of them have the proper velocity and orientation to leave the surface. These sequential processes can be theoretically modeled, and studies consistently show that only particles within the microvolume energized in the spike regime can leave the specimen to form secondary ions.

This process exists as the result of the arrival of a single incident ion particle. With a large number of incident particles, the specimen is damaged and the surface is eroded by

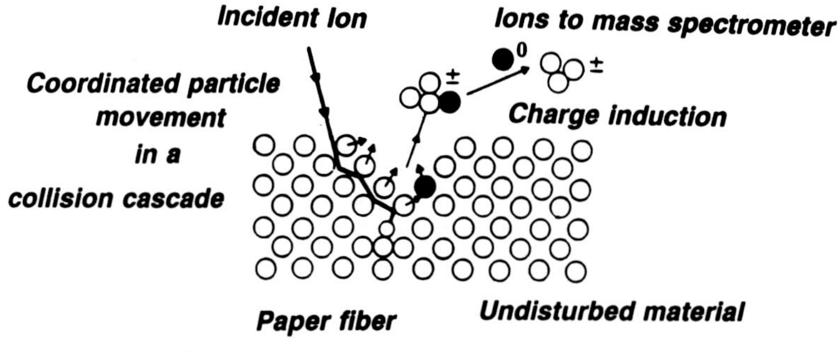

Figure 2. Molecular dynamics detail of the knock-on process in sputtering in secondary ion mass spectrometry.

the primary ion beam. With control of the primary ion source (see previous section), the surface can be eroded in a controlled manner, exposing successively greater depths of the specimen. This is the depth profiling experiment. The conditions that describe a depth profiling experiment will differ from those that characterize a surface coverage experiment. This difference is conceptualized in the distinction between dynamic SIMS and static SIMS. The boundary is sometimes rather arbitrarily defined as a primary ion current of 10^{-9} A/cm^2, but more practical considerations define a sputtering rate (the rate at which the primary beam erodes the surface expressed in terms of the depth dimension z), or the lifetime of a monolayer (several thousand seconds under low fluence conditions to a few seconds under rigorous bombardment). Molecular dynamics that describe sputtering must be integrated with a model of ionization.

What affects ionization? Certainly the nature of the specimen itself will affect the ionization process; relevant factors here include the chemical and physical nature of the specimen (its ionic or covalent nature, crystal structure, morphology, etc.), the nature of the primary particle itself (significant differences exist in secondary ion yields for metals sputtered with Ar^+ ions or O^- ions) and its energy and flux, and the details of the secondary ion extraction and mass analysis (different time windows for viewing the surface). The molecular dynamics determine the ***sputtering rate***, that is, the rate at which material is physically removed from a surface. The ***ionization probability*** is convoluted with the sputtering rate to determine the rate of production of secondary ions. Ionization potentials for atoms follow clearly defined trends across the periodic table, and, for otherwise constant conditions, the ionization probability can vary by several orders of magnitude for different atomic elements. Additionally, the ionization probability depends on the chemical nature of the target. If a metal surface is oxidized, then (in general) the ionization probability of the positive metal ion increases, and the observed yield increases. This increase is not constant for each element, and the individual calibrations that are required can become tedious. The ionization probability changes when a reactive primary particle such as O^- is used, which leads to an oxidation of the metal atoms in the vicinity of the ion impact, and an increase in the ionization probability. For analysis of organic samples by SIMS, the ionization probabilities of individual compounds are similarly variable. Although the ionization potentials of organic molecules fall within a narrower energy range, the chance that a molecular ion will dissociate must also be factored into the probability of observation of the molecular ion. For organic analysis, the nature of the primary particle does not exert a drastic effect on the ionization probability. Rather, it is the chemical nature of the sample that exerts the dominant effect. Generally, organic compounds that exist as preformed ions at the surface, or can be readily induced to form ions via acid/base reactions, exhibit high ion yields. Organic samples that are more difficult to ionize or fragment readily do not produce high quality secondary ion mass spectra. It is ironic to note that paper as a sample or as a support for organic compounds is a relatively simple sample compared to the diversity of other samples analyzed by SIMS, and that paper samples require only minimal sample preparation, if any, prior to SIMS analysis.

SIMS spectral interpretation can proceed on several levels. Atomic ions (and perhaps oxides) are found in positive and negative ion SIMS mass spectra for inorganic compounds. Any ambiguity in identification that involve mass overlaps between ions can often be solved with moderately higher mass resolution, or with a deconvolution program that uses the measured isotopic intensities to correct for signals at overlapping masses. For organic compounds sputtered by SIMS, and for paper, SIMS mass spectra can be far more complex. Most often, the positive ion SIMS mass spectrum of an organic compound contains an ion corresponding to $(M+H)^+$ and the negative ion SIMS mass spectrum contains the $(M\text{-}H)^-$ ion. The prevalence of these ions reflects the fact that acid/base reactions of the organic molecule with protons and (presumably) hydroxide ions formed by sputtering of a "dirty and wet" surface are the lowest energy ionization routes. The SIMS mass spectra also contain ions formed by dissociation of the molecule or the molecular ion, and the dissociations from protonated molecules $(M+H)^+$ in SIMS seem to parallel those found in positive ion chemical ionization. Unfortunately, there is no library of chemical ionization mass spectra (or SIMS spectra) to match the extensive computer database of library mass spectra commonly available for electron ionization mass spectra. Interpretation therefore relies on intuition and rationalization, and on the comparison of mass spectra of unknown and closely related standard compounds. A handbook of secondary ion mass spectra for commonly encountered inorganic and polymer surfaces has been published by Briggs *et al.*[8] that eases the task of spectral interpretation. Library SIMS spectra include those of a number of commonly encountered polymers, as well as common contaminants (fatty acids and their sodium salts, and cationic and nonionic surfactants). The handbook also contains positive and negative ion SIMS mass spectra for a number of metal oxides and halides. Unfortunately, there are no examples in this handbook of SIMS analysis of paper substrates or materials.

APPLICATIONS

Perhaps the greatest challenge to paper makers is learning how to identify and control variation in paper. In order to reduce variations, the paper maker must find a means to detect and to control both the machinery as well as the chemical processes used to manufacture the paper. In this section, we will talk about the use of SIMS to gain a better understanding of the bulk and surface information of the cellulose fibers and the formed paper sheet. SIMS is now being used to study the surface chemistry and the spatial distribution of raw and chemically treated cellulose fibers. This provides necessary information to optimize wet-end papermaking processes. SIMS is also being used to study dry-end paper problems including deficiencies with printing (print mottle), gluing, coating, self-sizing, sizing, and other production problems. The use of SIMS in combination with other surface analysis techniques will greatly enhance the papermaker's ability to solve problems, enhance production control, and assist in research and development.

Recent advances in SIMS instrumentation (such as charge neutralization, pulsed ion sources, and ToF mass spectrometers) have enabled the morphology of delicate materials (biomolecules, polymers, and plant fibers) to be studied. A review of the literature indicates that most of the SIMS studies of paper (and its fillers, binders, and other components) have occurred during the last three years. In the following section, several SIMS paper studies will be highlighted. The types of characterizations to be presented include the use of SIMS to identify and characterize trace elemental constituents in the bulk material and the surfaces of paper, to identify the molecular composition or functional groups of materials associated with the paper, and to identify the spatial distribution of the elements and the compounds associated with the paper products.

SIMS BULK COMPOSITIONAL ANALYSIS

Several research groups have reported the results of SIMS techniques for examining the bulk elemental composition of paper products. Klein and Bauch published three papers which discussed the use of LD for the study of the distribution of inorganic wood preservatives in the wood cell wall.[9–11] Similarly, Schmidt and Brinkmann used LD to evaluate the absorption of an indigo based dye on cellulosic cotton fibers.[12] In their study of photographic paper, Saler *et al.* used dynamic SIMS experiments to evaluate the bulk elemental composition of photographic paper.[13]

In the production of photographic paper, the untreated formed paper is coated with white-reflecting materials. These reflecting materials are suspensions of inorganic compounds (such as titanium dioxide (TiO_2) or aluminum oxide (Al_2O_3), and sodium (Na)) in organic gelatins. The organic materials contribute to the visual properties to the paper, but the overall quality of the paper is determined by the distribution of the inorganic compounds throughout the fiber matrix. This SIMS study was therefore limited to information of inorganic positive secondary ions.

Dynamic SIMS was used to make the appropriate spectral measurements. The primary Ar^+ beam sputtered deep into the wood fiber matrix, allowing the sub-surface chemistry of the photographic sheets to be monitored. The procedure used by Saler *et al.* involved examining the peak intensities of Ti^+, Na^+, Al^+, and TiO^+. Each of these ionic constituents exhibited a similar sensitivity in the SIMS experiment. It was noted that trace iron (Fe^+) peaks were also observed in the SIMS mass spectra and were likely introduced to the paper as contamination by production and cutting equipment. By taking a ratio of the peak heights (in counts) of the elements with similar sensitivity, compositional differences in the papers were identified. For example, the peak height ratios of Al^+ in two of the samples suggested a higher Al^+ concentration near the surface in some sheets, while more Ti^+ was found deeper in the sheets. Other samples appeared to contain similar concentrations of Al^+ and Ti^+ throughout. The Na^+ yield changes with depth were constant in all of the photographic sheets, and in agreement with the expected yields of the paper production process. Therefore, the intensities of the Na^+ peaks served as baselines in this study. In addition to evaluating the homogeneity of the desired inorganic compounds, this SIMS experiment identified a trace level of iron in the paper, an undesirable contaminant. This trace level of Fe^+ in the paper may not have been detected by conventional analytical techniques with inferior surface sensitivity.

SIMS SURFACE COMPOSITIONAL ANALYSIS

FAB/SIMS can also be used to evaluate the surfaces of cellulosic materials (with similar properties to lignocellulose) for molecular weight and functional group information. A study of the surfaces of cellulose nitrates by Fowler and Munro illustrates this FAB/SIMS capability.[14] The objective of this study was to evaluate the chemical and physical changes of the cellulose nitrates after exposure to heat and light. Peak intensity differences observed between the FAB/SIMS spectra of a pristine control sample of cellulose nitrate, a thermally degraded sample (17 hours at 130°C), and an x-ray-degraded sample (12 h, Ti kα) were used to study the extent of thermal and light degradation. The peaks of importance included m/z 30 (NO^+) and m/z 46 (NO_2^+). After the cellulose nitrate was exposed to heat, the relative peak intensities at m/z 30 and m/z 46 decreased. These peak intensities were reduced even more when the cellulose nitrate was exposed to x-rays. By comparing the spectral fingerprints of the pristine cellulose nitrate and the degraded samples, it was possible to qualitatively assess the extent of thermal and light degradation.

IMAGING SIMS AND COMPOSITIONAL ANALYSIS

Perhaps the greatest potential use of SIMS in paper science will be its surface imaging capabilities. In the imaging mode, SIMS can be used to identify the composition *and* the spatial distribution of chemical species on paper, even with respect to the individual fibers. Under optimum operating conditions, SIMS can detect picogram levels of material, while maintaining sub-micrometer lateral resolution. Furthermore, SIMS possesses the ability to detect ions across a wide mass range extending from H^+ to intact molecular ions of organic compounds or polymers with masses of several thousand Daltons. Several practical examples of the use of imaging SIMS to study cellulose fibers and paper surfaces follow.

Imaging SIMS has been used to show the spatial distribution of inextractable organochlorine in fully-bleached softwood kraft fibers.[15] From the SIMS data, inextractable, pulp-bound chlorine was observed to cover the entire exterior surface of fibers. A heterogenous distribution of chlorine was found in the interior of the pulp fibers. Knowledge of the physical location and distribution of pulp-bound organochlorine may provide insight into its behavior during chlorine use in papermaking, its behavior in the environment, and basic information for manufacturing process improvement. Of particular interest is the bleaching of chemical pulps with chlorine or chlorine dioxide, and the ability to track chlorine with changes in the manufacturing process.

There are three types of organochlorine species formed in pulp during its chlorine bleaching; a water-extractable fraction, which can be separated, an organic solvent-extractable portion, which can be removed, and an inextractable (pulp-bound) portion. The concentration of the inextractable organochlorine species ranges from 50 μg Cl/g pulp to 500 μg Cl/g pulp, depending on the pulp type and its processing history.[15] Of interest to the pulp and paper industry is the physical distribution of this residual chlorine. Knowing the location allows user and environmental risks to be determined. The difficulty in identifying and measuring residual chlorine and lignin in semi-bleached or bleached pulp fibers is due to the inherent limitations (sensitivity or lateral and spatial resolution) of most analytical instrumental techniques. Many studies on pulp chlorination kinetics have been performed, and these kinetics studies determine the accessibility of lignin and its removal from the cell walls of fibers. However, kinetic studies do not provide spatial information about the distribution of chlorine within individual fibers. Hence, only inferences about the location of residual lignin or chlorine are possible from kinetic experiments. The terminal objective of the research by Tan and Reeve was to use different micro-analytical techniques to evaluate the spatial distribution of chlorine, organochlorine species, and lignin in fully-bleached kraft pulp fibers.

A CAMECA 3f ion microscope located at the Surface Science Western Laboratory at the University of Western Ontario was used in this work. The chlorine analysis imaging conditions included a Cs^+ primary beam accelerated to 10.0 kV with a total 3–5 nA current. The pulp consisted of a fully bleached, commercial softwood kraft pulp (85% jack pine and 15% spruce). The pulp was bleached with the O(DC)(E+O)HD sequence. Both longitudinal and cross-sectional views of fibers were prepared for SIMS study. The chlorine distribution was monitored by evaluation of the $^{35}Cl^-$ and $^{37}Cl^-$ negative ions; the known isotopic ratios of these ions provided a valuable internal standard. The imaging SIMS data indicated that chlorine ($^{35}Cl^-$ ion image) was present and uniformly distributed over the entire fiber surfaces. There were small dark-colored regions on the fibers where chlorine was not observed. SIMS elemental linescans were performed across these areas in order to qualitatively assess their chemistry. The linescan data indicated that the chlorine as well as the carbon ion intensities decreased across these dark-colored regions, and suggested that the outer surface of the fibers contained pits or voids.

SIMS images of fiber cross-sections enabled the spatial distribution of chlorine within the fiber to be determined. This data showed that chlorine was present throughout the fiber, but was unevenly distributed. Of particular importance were the chlorine-rich areas observed in the middle of the secondary fiber walls. Based on previous literature and the SIMS information, the authors proposed that the residual chlorine is pulp-bound organochlorine. The organochlorine is predicted to be attached to lignin fragments. Tan and Reeve found no evidence in the literature that the chlorine attaches to hemicellulose or cellulose molecules. To further confirm their hypothesis, the authors regenerated air-dried pulp samples in sulfur dioxide/diethylamine/dimethyl sulfoxide and N-methylmorpholine N-oxide monohydrate solutions. Neutron activation measurements of chlorine indicated that 60% of the chlorine was still present in the pulp. This provides additional evidence that the non-extractable chlorine is bound to the lignin (which in turn is linked to carbohydrate). If the pulp-bound organochlorine is covalently bound to large carbohydrate molecules deep within the fiber walls, it is not likely to diffuse out of the fiber. Therefore, the pulp-bound chlorinated species should not be a risk to humans who manufacture or use paper products, or a risk to the environment.

Since the early 1990s, Brinen *et al.* of Cytec Industries, formerly a business unit of the American Cyanamid Company, have been actively using SIMS to gain an understanding of the chemistry of papers treated with various additives and surfactants. Prior to SIMS, Brinen and co-workers used ESCA (Electron Spectroscopy for Chemical Analysis) [also known as XPS (X-ray Photoelectron Spectroscopy)] to study paper. While ESCA has the advantage of being a non-destructive technique and can provide elemental and chemical bonding information of the surface layers of paper, it suffers from limited lateral resolution (~10 micrometers) and relatively limited sensitivity. However, the analytical capabilities of ESCA are complemented by the abilities of SIMS, and when these techniques are used in combination, the adsorption properties and distribution of chemical additives in paper can be very thoroughly evaluated. The subject of ESCA will be discussed in the following chapter. The remainder of this section will focus on the use of imaging SIMS for paper analysis.

The first SIMS work published by Brinen and associates discussed the use of SIMS (and ESCA) to identify and map the location of inorganic additives used in various papermaking processes.[16] The SIMS instrumentation included a modified VG MICROLAB Mark II Surface Spectrometer (VG Scientific, East Grinstead, UK). The SIMS analyzer was an MM12-12 quadrupole mass spectrometer (mass range 0–800 Daltons), with charge neutralization capabilities, and the instrument was equipped with a gallium ($^{69}Ga^+$) liquid metal ion gun. The gallium ion gun was operated at 20 or 25 kV with beam currents between 0.5 and 1.5 nA. For the production of the ion images, the positive ions chosen were those with the highest intensities, including m/z 40 (calcium), m/z 23 (sodium), m/z 27 (aluminum or the organic fragment $C_2H_3^+$), and m/z 57 (possibly a calcium hydroxide or the stable $C_4H_9^+$ hydrocarbon fragment). The negative ion fragments monitored were m/z 24 (C_2^-) and m/z 16 (O^-).

The objectives of this study were twofold: *i*) evaluate the compositional and structural variation between papers treated with different wet strength agents and sizing resins, and *ii*) investigate the utility of SIMS to identify and characterize surface differences. The test vehicles for the wet strength study were handsheets prepared from a blend of equal amounts of hardwood and softwood bleached kraft pulp. The consistency of the pulp was 2% and the pH was adjusted to 6.5. Sodium sulfate (0.2 g SO_4^{2-}/ kg pulp) and calcium chloride (0.05 g Ca^{2+}/kg pulp) were added to the pulp and all water used for the handsheet forming. The pulp was diluted to 0.6% consistency before use. For the acid paper, alum ($Al_2(SO_4)_3$ $16H_2O$) was used (1% w/w based on dry weight of pulp) and the pH readjusted to 6.5. The handsheets for the alkaline sizing work were prepared in a similar manner except the pH was adjusted to 8.0 after addition of the sodium sulfate and calcium chloride. Blanks were prepared for both of the studies. The starch used was ACCOSIZE 80* (a product of Cytec Industries) and the precipitated calcium carbonate used was a dispersant-free product made by Pfizer, Inc.

Initial static SIMS elemental analysis of the wet strength blank indicated that the surface was rich in hydrocarbon material. The surfaces of these handsheets were continuously sputtered to remove the outermost layers of the paper and to expose the presence of the inorganic species added to the pulps. With time, the peak intensities associated with various hydrocarbon fragment ions lessened in the SIMS spectra, and the intensities of the inorganic ion signals increased. The prominent peaks observed in the positive ion SIMS mass spectrum were m/z 23 (Na^+), m/z 27 (Al^+), m/z 40 (Ca^+), m/z 57 (possibly $CaOH^+$) and m/z 69 and m/z 71 (certainly Ga^+ from the incident ion beam). Of surprise in the blank's mass spectra was the presence of aluminum which was not added to the blank's pulp slurry. Chemical bulk analysis confirmed that the Al was present at 240 μg/g).

SIMS images of the inorganic ions were collected to determine their distribution in the blank handsheets. The Ca and Na ion images indicated that both of these elements were uniformly distributed over the fiber surfaces. The map of the Al ions showed that isolated discrete regions of Al were present both on and off the fiber network. Since the Al distribution was not confined to the fibers, handsheets treated with alum (with a bulk Al concentration of 600 μg/g) were prepared and measured with imaging SIMS. The distribution of the ions for Ca, Na, and the Al are along the fiber network.

The blank and the alkaline-sized handsheets displayed similar SIMS mass spectra in the alkaline sizing experiment, hence only the data from the blank was reported by Brinen *et al.* The alkaline size used in this study was alkenyl succinic anhydride (ASA). The blank's SIMS spectrum showed the same elements as observed in the wet strength handsheets except that the calcium peak was dominant. The ion images agreed with observations of the wet strength sheets. However, the intensity of the Al ion was too weak to obtain a good image.

The results of Brinen's preliminary study indicated that SIMS can be used to study the composition of paper. While the emphasis in this study was the evaluation of inorganic species, the authors proposed that SIMS should hold promise for the monitoring of organic additives in paper as well. Among factors to be considered in designing such experiments are the vacuum and ion beam stability of the organic additives to be evaluated. Two subsequent publications of Brinen *et al.* discuss the development of surface spectroscopic (SIMS and ESCA) methods to identify and evaluate the distribution of organic additives and surfactants on and into paper.

In the first of these studies, Brinen and Proverb used conventional quadrupole and ToF SIMS instrumentation to demonstrate that a fluorinated organic surfactant could be detected and mapped on the surface of paper fibers.[17] Fluorine-containing surfactants are used in paper manufacturing when low-energy surfaces are required. Fluorinated surfactants change the wetting ability of paper in contact with various solvents and oil. The presence of the surfactant layer also inhibits both charge and color development.[17]

The paper model systems in this study were handsheets that were made on a commercial machine, and which had not been surface treated. The pulp fiber composition was 60/40 hardwood/softwood with 30% broke. The paper is alkaline, containing ASA as the internal size and calcium carbonate as the filler. The fluorinated surfactants were applied to the paper surfaces by "tub-sizing." A 3% starch solution was used as the carrier. From determination of the starch "pick-up" by the virgin handsheets, the desired concentration of the fluorinated surfactants to be incorporated into the paper could be estimated. The three fluorinated systems examined included: *i*) du Pont's Zonyl TBS fluorosurfactant, a perfluoroalkyl sulfonic acid, ammonium salt; *ii*) 3M's Fluorad FC-170 fluorochemical surfactant; and *iii*) 3M's Fluorad FC-170-C fluorochemical surfactant. After the handsheets were treated with a particular surfactant, they were evaluated with ESCA to determine the level of organofluorine species on the surface of each paper. From ESCA data (using the intensity of the F 1s line) of the treated papers, the highest level surfactant incorporation was observed for the Zonyl TBS fluorosurfactant. Hence, Brinen and Proverb chose the Zonyl TBS fluorosurfactant paper system for in-depth SIMS study.

The quadrupole SIMS instrument was discussed in the previous summary. Two ToF SIMS instruments were also used in this work: a Kratos Analytical Prism and a Charles Evans and Associates™ TFS. The Kratos instrument was used with 25 kV ^{69}Ga ions in the microprobe mode, while the Charles Evans and Associates instruments used Cs^+ ion primary ion bombardment in the microscope mode and Ga^+ ion primary ion bombardment (25 kV) in the microprobe mode. To help prevent charge accumulation on the surface of the poorly conducting paper surfaces, a low-energy electron flood gun near the sample surface was pulsed out of phase with the ion source.

The authors first evaluated the outermost monolayers of the treated papers under static SIMS conditions using a Xe ion source. A typical positive ion mass spectrum for the 0.2% Zonyl fluorosurfactant contained the following peaks attributable to the surfactant: m/z 31 (CF^+), m/z 69 (CF_3^+) and m/z 93 ($C_3F_3^+$) Daltons. Other characteristic peaks at m/z 119 ($C_2F_5^+$) and m/z 131 ($C_3F_5^+$) Daltons were also noted. A strong 19 (F^-) peak was observed in the negative ion SIMS mass spectrum. The quadrupole SIMS experimental parameters were then changed to perform imaging experiments; gallium ion excitation was employed. Higher ion current dose rates were required to achieve the desired lateral resolution of the fluorosurfactant on the paper fibers. Under these dynamic SIMS conditions, many of the organofluorine fragments were not sufficiently stable for positive ion SIMS analysis or mapping. However, the F^- peak (a highly stable negative ion with no means of dissociation) was persistent in the negative ion SIMS mass spectrum, and was used to determine the distribution of the surfactant. The m/z 16 (O^-) and the m/z 25 (C_2H^-) peaks were used to define the fiber network. The SIMS maps of the F^-, O^-, and C_2H^- ions indicated that the fluorinated surfactant was isolated as islands along the paper fibers. These islands ranged in size from 1–50 μm. There was no evidence that the surfactant was agglomerated at the points of intersection between fibers.

To more fully evaluate the distribution of the fluorinated surfactant, ToF SIMS instrumentation was employed. The advantage of using ToF SIMS is the much higher ion collection efficiency and the simultaneous mass detection achievable at much lower ion currents. The typical ToF ion doses are at least an order of magnitude below static SIMS conditions. Theoretically, these more gentle desorption ionization conditions should enable larger characteristic fragments of the parent ion to be measured. The positive SIMS mass spectrum of the treated paper, collected with both of the ToF SIMS instruments using gallium ion excitation, shows more intense peaks for all of the F-containing fragments noted in the static SIMS spectra, as well as the presence of a higher organofluorine fragment at 169 Daltons. The F^- peak was also larger in the ToF negative ion SIMS mass spectrum. The increase in sensitivity gained through use of the ToF SIMS enabled thin films of the fluorinated surfactant to be observed along the fibers. The discrete islands of surfactant were also noted. The films and the islands were present in both the positive (CF^+, CF_3^+, and $C_3F_5^+$) and negative (F^-) ion images.

The treated paper samples were also analyzed on the Charles Evans and Associates ToF SIMS using a pulsed cesium ion source. As with the case of the gallium-generated spectra, higher intensity peaks were observed for the organofluorine fragments, along with the higher mass ion at m/z 169. A higher mass resolution was achieved with the Charles Evans and Associates instrument (which is a multiple electric sector/ToF combination instrument), which enabled the peak CF_3^+ at 68.9952 Daltons to be differentiated from the $C_5H_9^+$ peak at 69.0704 Daltons, and these positive ions could be imaged independently. As before, all the images acquired indicated that the Zonyl fluorosurfactant was present as a thin film along the fibers and as discrete islands. Linescans across different regions of the treated papers also agreed with the image data.

The most recent study published by Brinen evaluated the use of surface spectroscopies (ESCA and SIMS) to monitor the distribution of both commercial and experimental sizes (internal and external) that had been applied to paper.[18] Since sizes are typically added in small amounts to paper (*e.g.* 2 to 4 lb per ton), conventional spectroscopies lack the sensitivity to evaluate them. The use of ESCA and SIMS (particularly ToF SIMS) enable the size to be detected, and the spatial distribution determined. ESCA is suited to monitor the long aliphatic functional groups typical of most sizes. An ESCA C 1s spectrum would be collected from an untreated paper sample, and the location and shape of the C 1s peak(s) would be evaluated. A C 1s spectrum of a sized paper would also be collected and compared to the spectrum obtained from the control. If size was present, the intensity of the C 1s peaks would increase, shifts in the eV-position of the C peaks would be expected, and the shape of the C 1s peak(s) might change. If heteroatoms are associated with the size, large shifts in the C 1s peak are likely. While ESCA can give insight to the chemical environment of certain elements, its disadvantage is that it cannot provide direct structural information about such materials. However, SIMS can be used to identify size in paper and its distribution. From the presence of molecular ions (M^+) (or protonated $(M+H)^+$ or deprotonated $(M\text{-}H)^-$ molecular ions, or perhaps even derivatized molecular ions) in the SIMS spectrum (along with characteristic fragmentation patterns), it is very possible to identify the exact chemical nature of the size. In this study, Brinen observed the molecular ions of several sizes. The internal sizes examined included stearic acid, stearic anhydride, N-Cl-stearamide, alkyl ketene dimer (AKD), and ASA. The surface sizes evaluated consisted of styrene acrylic type resin and styrene/fluorinated-acrylic type resin.

Two different paper preparations were used in this study. The first was used to produce internal sized paper and the second was used to prepare surface sized sheets. A pilot paper machine was used in the surface size experiments, and a hand sheet mold was used for the internal size study. Both sample preparation methods used *i*) higher than normal concentrations of sizing additives and *ii*) a high solubility solvent (toluene) to carry the organic sizes into the paper.

The SIMS data in this study were collected on a ToF SIMS instrument at Charles Evans and Associates, Redwood City, CA. A 25 kV ^{69}Ga ion beam, operated in the microprobe mode, was employed to sputter the samples. Spectrometer conditions were selected for optimized lateral resolution at the expense of optimum mass resolution. An electron flood gun was pulsed out of phase with the ^{69}Ga ion gun to help dissipate positive charge accumulation on the paper surfaces. The ToF SIMS images and spectra were collected under static SIMS conditions. The ESCA experiments were performed with an HP ESCA spectrometer that used Al k alpha radiation to excite the paper samples.

In the internal size studies, papers treated with stearic acid, stearic anhydride, N-Cl-stearamide, and AKD were successfully studied with the ToF SIMS methods. For each size, either characteristic molecular ions (or derivative ions) and/or unique fragments were observed in their SIMS spectra. While the intensities of the ions originating from size were often times very weak, Brinen was able to sum together signals that originated from the size to generate ion images. From the ion images, the internal sizes appeared to be uniformly distributed on the fiber network of the papers, suggesting that preferential areas of size adsorption on the fibers were not present.

In the second part of this study, Brinen used ESCA and SIMS to evaluate papers that were internally sized with ASA and also surface sized (via a size press or tub sizing). He initially analyzed the ASA/starch internally sized paper, applied the surface size, and then performed the second surface analysis. The ESCA data from both the internal and the surface size papers showed the expected increase in C 1s peaks from the addition of hydrocarbon moieties.

Figure 3 shows the SIMS spectrum collected from paper internally sized with ASA. Comparison of this mass spectrum to the mass spectrum from a control paper sample (not treated with a sizing additives) shows the presence of a peak at m/z 323, which corresponds to the parent molecule of the ASA less a carbonyl (CO) group. The positive SIMS images from the ASA sized sample are given in Figure 4. The top image was formed by collecting all the total positive ions emitted from the paper's surface. The middle image is from the m/z 41 ($C_3H_5^+$) ion and the lower image is from the m/z 40 (Ca^+) ion. Compare these images with the ion images collected from the control paper

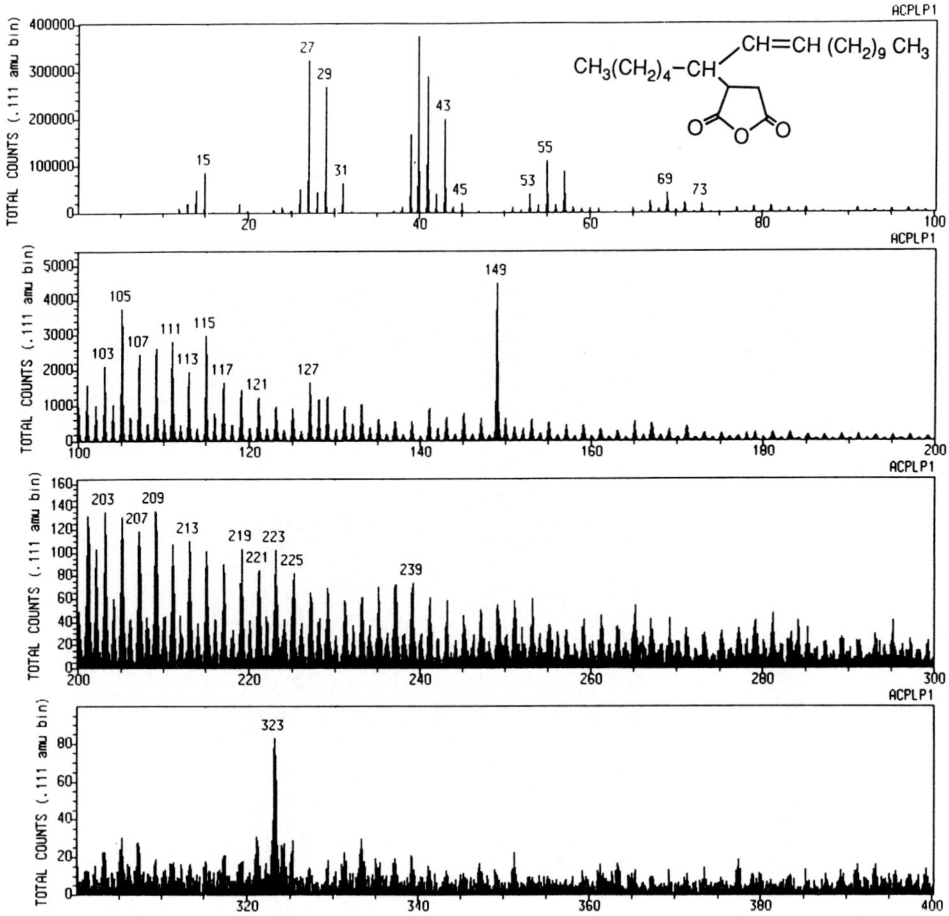

Figure 3. Positive ion ToF SIMS spectrum of pilot paper machine paper containing ASA/starch internal size.[17] Copyright 1993 The American Cyanamid Company. All rights reserved and reprinted with permission.

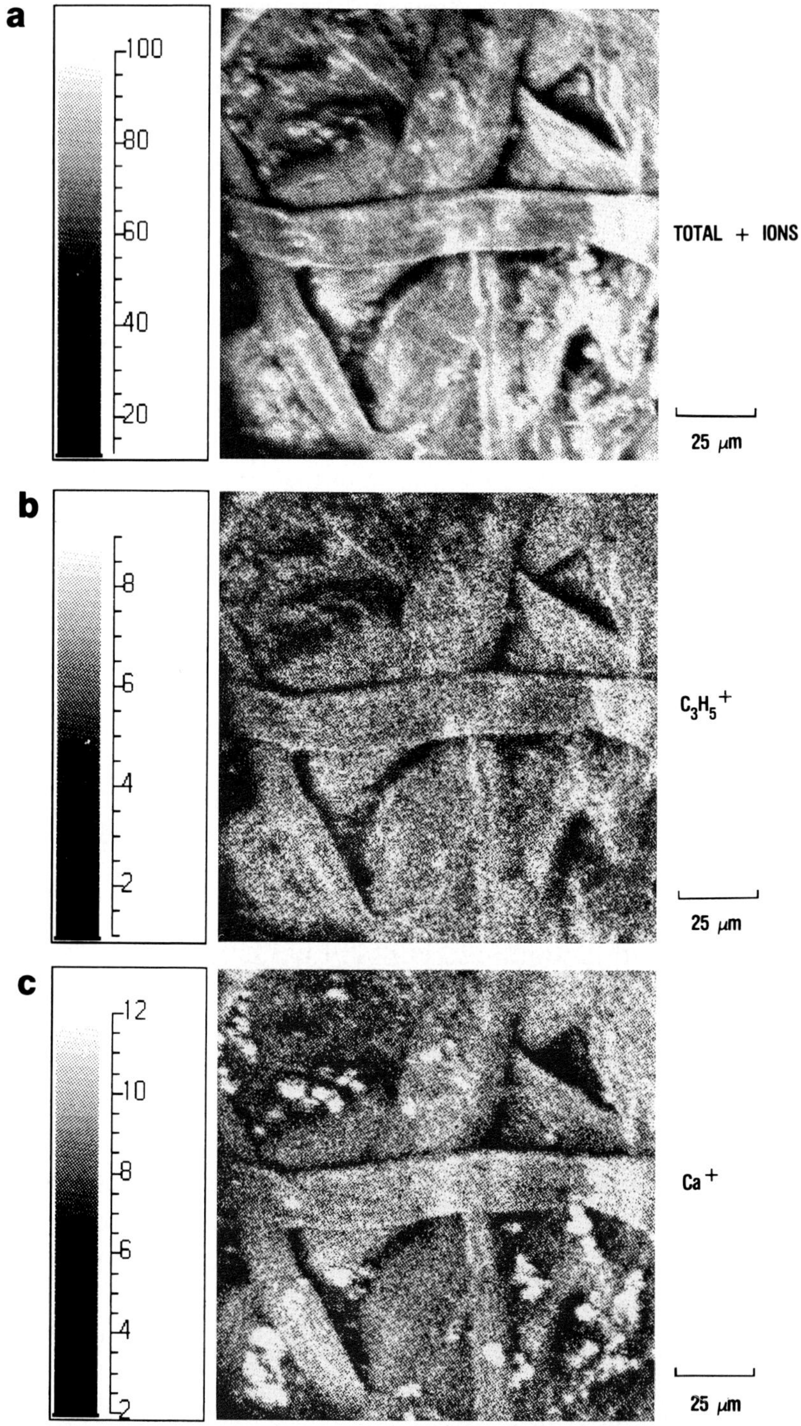

Figure 4. Static SIMS positive ion images from pilot paper machine containing ASA/starch internal size. (a) Total positive ion image; (b) Image of mass 41 Daltons, corresponding to $C_3H_5^+$ and (c) Ca^+ image at 40 Daltons.[17]

sample in Figure 5. As with the other internal size studies, the ASA size appears to be well distributed across the paper surface. Furthermore, comparison of the Ca^+ image collected from the control sample to the Ca^+ image from the sized sheet suggests that the ASA/starch application has dispersed the calcium particles. In the control sheet, larger agglomerates of calcium particles were noted and they were not associated with the paper fibers. In the ASA/starch treated papers, the Ca-containing particles appear smaller and the fiber network is homogeneously coated with calcium.

The ASA/starch internal sized papers were next surface sized with a styrene acrylic-type resin or a styrene/fluorinated acrylic-type resin. Both positive and negative ion SIMS mass spectra were collected. The positive ion mass spectra taken from the papers treated with the surface size resins showed peaks characteristic of polystyrene. These peaks were observed at m/z 91, m/z 115, and m/z 128, as shown in Figure 6a. The most prominent peak in the SIMS mass spectra was from the Ca^+ ion. The very intense Ca^+ signal indicates that the surface size solution further dispersed the Ca particles in the sheet resulting in additional coating on the fibers. A Ca^+ ion image would verify this supposition. The most prominent peaks in the negative ion SIMS mass spectrum of the fluorine-containing surface size solution were from the m/z 16 (O^-) ion and the m/z 19 (F^-) ion (refer back to Figure 6b). The total negative ion and the F^- ion images of the paper with the styrene/fluorinated acrylic-type resin are shown in Figure 7. The F^- ion image suggests that the surface size has uniformly covered most of the fiber network in the paper. However, some of the fibers did not seem to be covered with the F-containing surface size. After investigating the source of the void regions, Brinen noted high concentrations of $C_3H_7^+$ ions in these regions. He suggested that a hydrocarbon-like material had been absorbed onto the fibers in these regions, as shown in Figure 8. He further verified these observations with linescans.

Zimmerman *et al.* report the successful application of ToF-SIMS to study two common problems in the paper industry: *i*) poor print quality (mottling) and *ii*) contamination leading to discoloration and spotting on paper.[19] The instrument used in their experimentation was a ToF-SIMS IV developed at the University of Munster. It was equipped with a 30-keV liquid metal ion ($^{69}Ga^+$) source operating at a current of 1.5 nA. A 40–100 ns pulse length was used to obtain both the spectra and the images and, the primary ion beam was defocused to ca. 1 micrometer diameter. The spectra and images were accumulated in the ion-counting mode (time resolution of 472 ps) and a total time range of 160 microseconds. A 256 x 256 or 128 x 128 digital raster was used to generate the images. A 10 eV electron beam pulsed out of phase with the extraction lens and primary ion beam was used to compensate for charge build up on the papers.

In their first study, Zimmerman *et al.* collected static ToF-SIMS spectra and images from known high-quality and substandard coated papers. The quality of the paper was verified using a backtrap mottle test on a Prufbau backtrap tester. These papers were

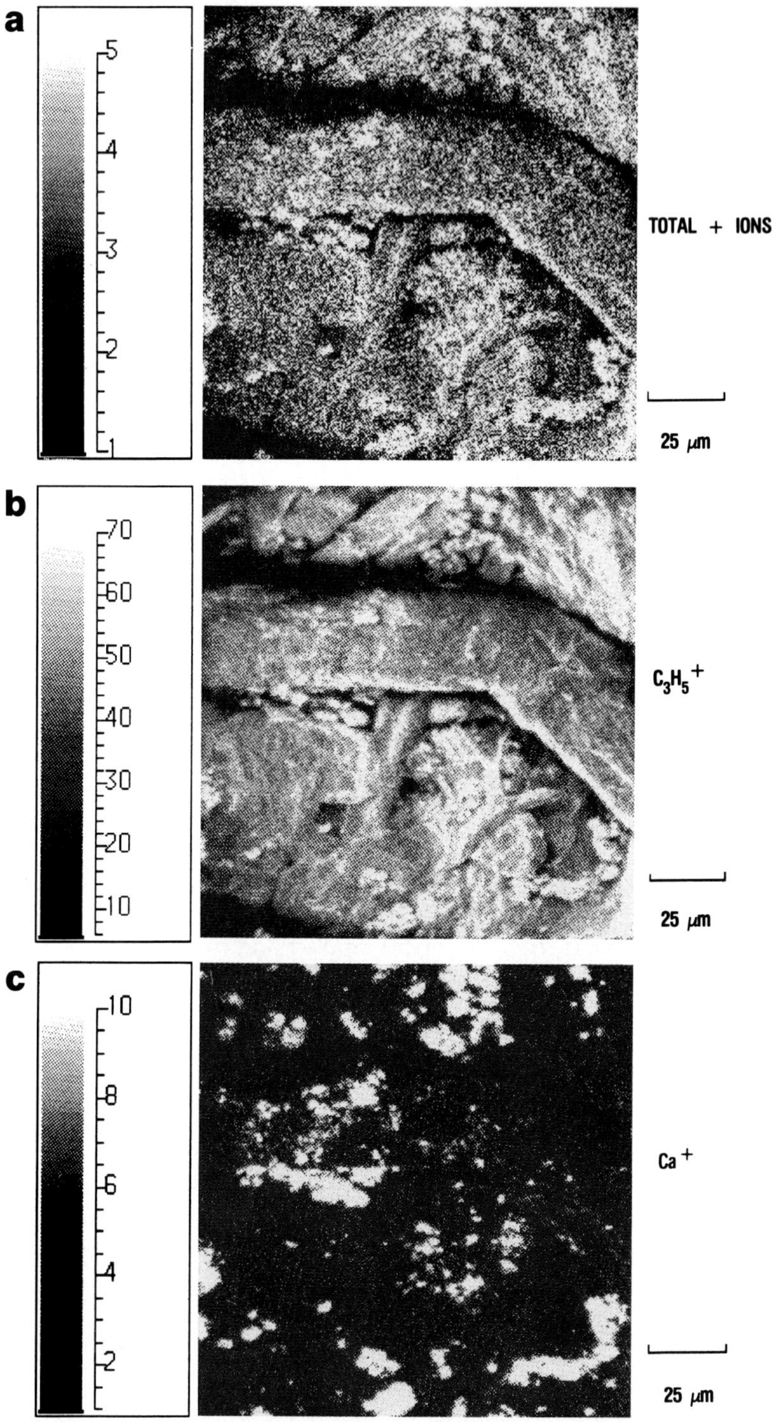

Figure 5. Static SIMS positive ion images from a control (not treated with sizing additives) from pilot paper machine. (a) Total positive ion image; (b) Image of mass of 41 Daltons, corresponding to $C_3H_5^+$ and (c) Ca^+ image at 40 Daltons.[17] Copyright The American Cyanamid Company. All rights reserved and reprinted with permission.

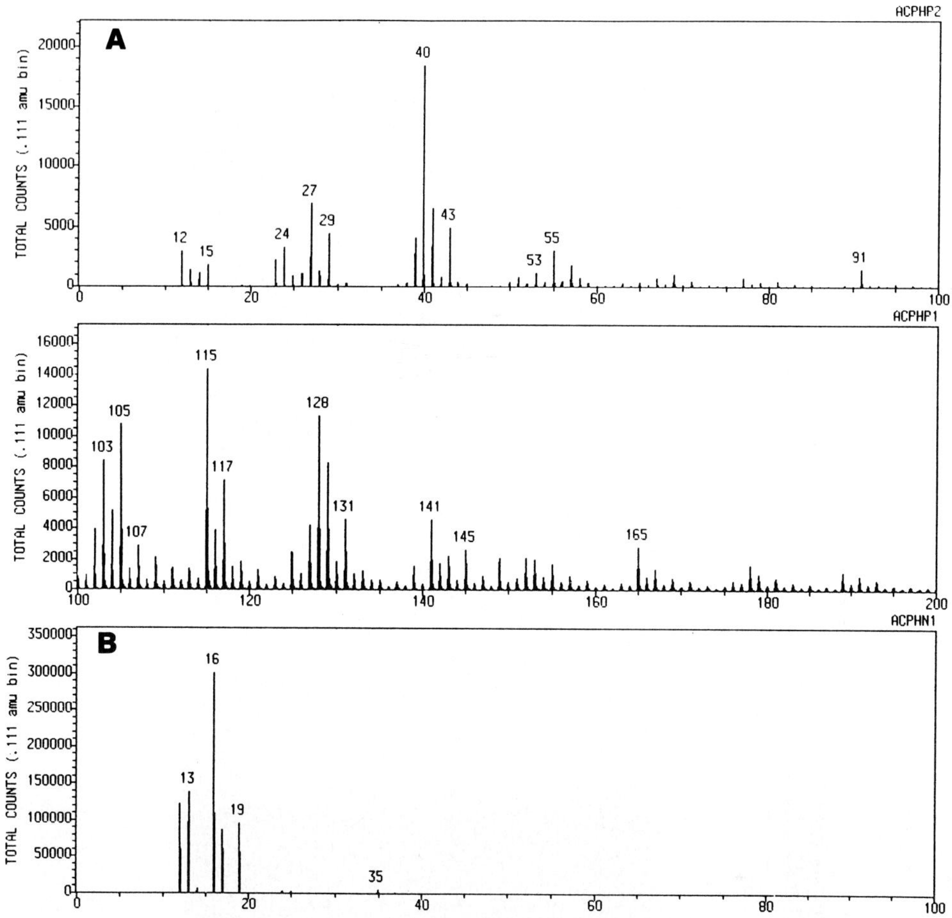

Figure 6. Positive (a) and negative (b) ion ToF SIMS spectra from pilot paper machine containing ASA as internal size and a surface size (tub sized) of a styrene/fluorinated acrylic-type resin.[17]

coated with a latex/$CaCO_3$ mixture. The latex coating on the paper was a cross-linked styrene-butadiene co-polymer.

The authors observed the same ions (Ca^+, $C_3H_5{}^+$, $C_4H_9{}^+$, $C_5H_9{}^+$, $C_6H_5{}^+$, $C_7H_7{}^+$, and $C_9H_7{}^+$) in the spectra of both the high-quality and substandard coated papers. The relative intensities of these ions were similar from six analyzed areas on the high-quality paper (*i.e.*, independent of location). However, the spectra obtained from the poor-quality paper showed variability in absolute intensities from spot to spot, indicating an non-homogeneous paper surface. The most obvious variability in the poor quality papers was the relative intensity of the Ca^+ ion from spot to spot.

To better understand the spatial distribution of the Ca^+ and other ions on the good and poor-quality papers, ToF-SIMS ion images were collected. The good-quality paper

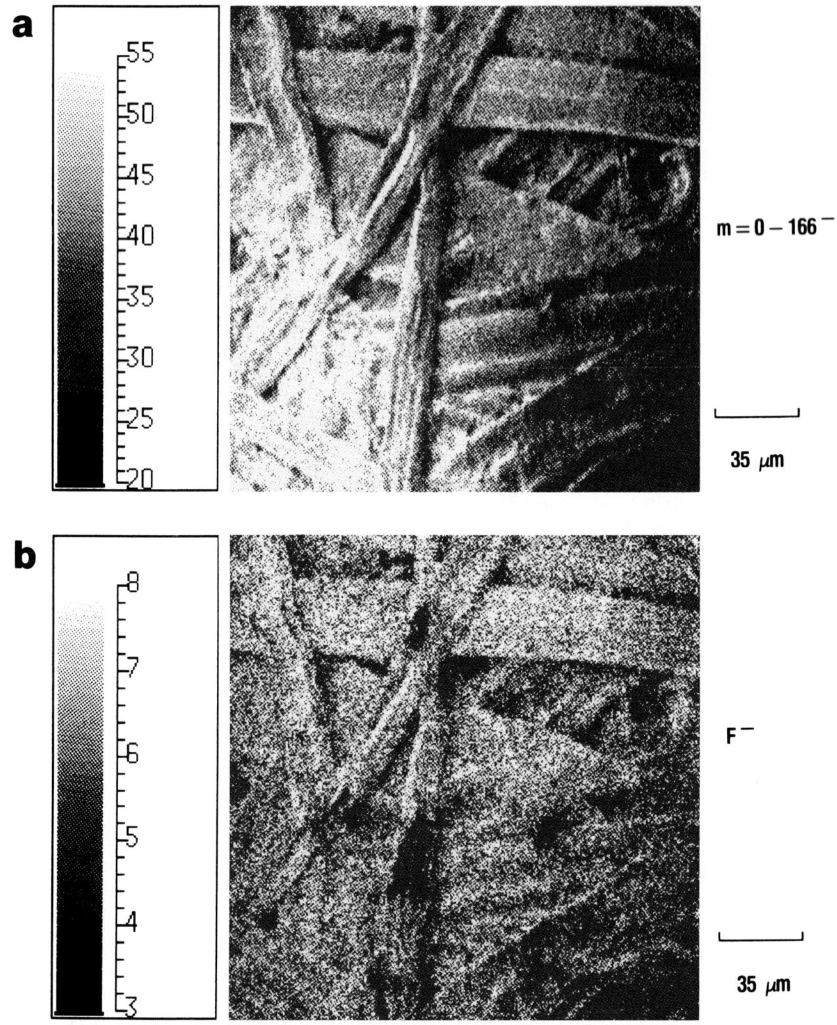

Figure 7. Static SIMS negative ion images from pilot paper machine containing ASA as internal size and a surface size (tub sized) of styrene/fluorinated acrylic-type resin. (a) Total negative ion image and (b) Fluorine image at 19 Daltons.[17] Copyright 1993 The American Cyanamid Company. All rights reserved and reprinted with permission.

showed a fairly homogeneous distribution of all the ions in the 500-μm^2 area analyzed. The authors did note that localized regions of sodium were identified in the good paper, but they attributed these to sublayer domains since there was no correlation between Na^+ and any of the previously identified organic ions. Ion maps were likewise collected from the paper that gave mottled images when printed. It was apparent that these papers contained regions of high concentration, localized calcium and these bright Ca-rich regions were complemented with dark regions in the other ion maps (organic fragments) from the latex. Furthermore, all of the ion images suggested that the poorer quality paper was less homogeneous when compared to the good quality paper. Since a homogeneous

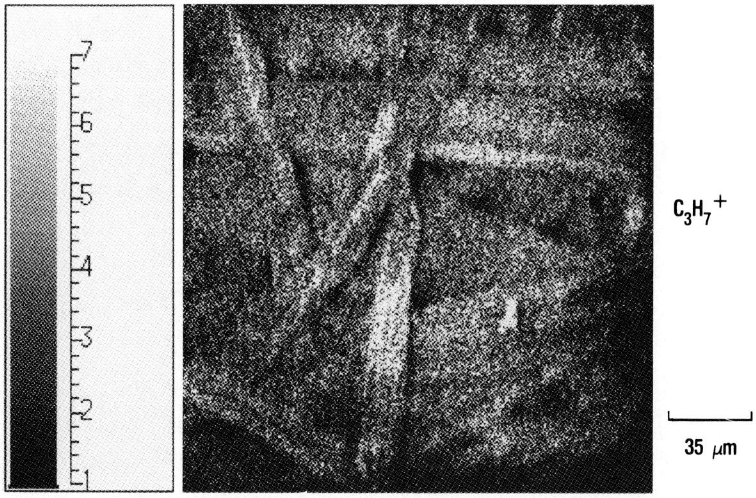

Figure 8. Static positive ion SIMS image at 43 Daltons ($C_3H_7^+$) for the same sample and field as shown in Figure 5.2.[17] Copyright 1993 The American Cyanamid Company. All rights reserved and reprinted with permission.

distribution of latex and calcium is needed for uniform printing, the presence of localized species, such as Ca, would lead to mottling and poor print quality.

In their second study, Zimmerman *et al.* evaluated discolored spots in uncoated adhesive labels. A static ToF-SIMS spectrum was collected from several of these spots and the fragments observed included ions at mass/charge ratios of 73, 147, 207, 221, 281, and 355. This fragmentation is typical of polydimethylsiloxanes (PDMS). Images of several siloxane fragments ($C_2H_5OSi^+$ (73 Da), $C_7H_{21}O_4Si_4^+$ (281 Da), and $C_9H_{27}O_9Si_5^+$ (355 Da)) were collected from the spotted regions. These images show higher concentration of siloxanes on the discolored regions of the paper. A portion of the adhesive label with no apparent discoloration was analyzed for comparison. Both the static spectrum and the ion maps showed the presence of PDMS. However, the relative intensities are significantly lower than on the portions with the spots.

The authors performed additional tests on the back-side of the adhesive label to further identify the source of the PDMS. The static spectra and images collected, both displayed lower concentrations of the species originating from PDMS. Hence, the authors concluded that the contamination did not originate from the adhesive, but rather the release sheet which comes into contact with the surface of the label stock as it is rolled or stacked.

SIMS OF ORGANIC COMPOUNDS IMPREGNATED INTO PAPER

A conundrum exists in the SIMS analysis of organic compounds loaded into paper, either as additives in manufacturing, as agents of some process (copying or character formation), or as vehicles in as simple a situation as ink on paper. The dilemma arises in that the conditions generally required for release of these compounds from the paper and sputtering by the primary ion beam generally lead to excessive dissociation of the sample structure so that the intact molecular ion of the sample is not observed in the SIMS mass spectrum. This situation is beginning to change with the use of ToF SIMS instruments with extraordinarily high ion transmission, as will be discussed shortly. Extremely promising experiments have been completed, but the success was achieved under fairly constrained conditions of sample preparation and analysis. A considerable amount of effort remains to be completed over the next few years to establish the various conditions under which successful organic analysis might be completed with secondary ion mass spectrometry.

Direct analysis of organic compounds in papers by SIMS was the goal of some recent research completed by Bartlett and Busch.[20] Control experiments were completed with blank papers to determine the level of background signal for organic compounds that could be extracted from a number of different papers using simple solvent extraction. Paper samples with a diverse geographical origin were analyzed. In each case, the control experiment involved analysis of the extract with positive ion electron ionization mass spectra. Compound identification was accomplished via library search of the measured mass spectra against the standard electron ionization library. To confirm mass assignments, chemical ionization mass spectra were also recorded for these same peaks. For most paper samples, no identifiable organic compounds could be extracted. A few papers were characteristic in the distribution of compounds that could be eluted from them. For instance, paper with one particular geographical origin could be extracted to produce a peak in the gas chromatographic trace that corresponds to any one of several structures (dehydroabietic acid or epidehydroabietic acid). Abietic acid is a common organic acid prepared by isomerization of resin, used in paints and varnishes, and also in paper sizing. Identification of the same compound, but at a different quantitative value, was accomplished in the extract of a second paper produced in a different country. With the exception of the dehydroabietic acid, all of the mass spectral data recorded suggest that no discrete components were extracted from any other paper sample. This is not to conclude that organic compounds *cannot* be extracted at all (supercritical fluid extraction would be an attractive method). However, the desired experiment is the direct SIMS analysis of the paper, and the sole chance to extract the organic compound from the paper is the placement of a small amount of liquid solvent onto the paper to act as the support matrix in the liquid SIMS (FAB) experiment. In support of the difficulty with which even abietic acid was extracted, direct SIMS analysis (quadrupole mass analyzer, cesium ion gun at 35 keV energy) of the papers that could be shown to contain this compound provided no unambiguous mass spectral evidence for its presence. Known organic compounds purposefully impregnated into the paper to participate in image developments (levels of 1–50 ng/cm^2) similarly could not be detected as molecular ions in the direct SIMS analysis of the paper.

Persistent difficulties with organic sample release from the paper substrates led to our investigation of enzymatic digestion of the structure of the paper itself to release the compounds of interest. If the organic compounds are strongly bound to the paper, or otherwise trapped within the structure of the paper fibers, then it was thought that destruction of the cellulosic structure itself might lead to more efficient extraction of the compounds. The cellulase enzyme isolated from the bacteria *Trichoderma viride* was used to break down the cellulosic structure of the paper. This enzyme is available commercially, with a known activity specified against the digestion of filter paper. Conditions were found in the literature for the standard use of this enzyme. A pH of 4.0 was used for the digestion, with a temperature of 50 °C and a digestion time of 24 hours. Four paper samples with different geographical origins were analyzed. Two samples were blank papers, and two samples were loaded with known amounts of a target organic compound. In each case, a 4 cm^2 sample of paper was used, corresponding to about 10 micrograms of each of the organic compounds. The digestion solution was extracted, the sample fractionated, and then concentrated to a final volume of 2 mL. A 10 μL aliquot of this sample is loaded onto the direct insertion probe and analyzed by positive ion electron ionization mass spectrometry. Using these parameters, the loaded samples should contain about 30 ng of sample for the pre-loaded organic compounds, and electron ionization mass spectrometry provided independent confirmation for the target compound, and no identifiable organic compounds for the blank paper substrates. This level of organic compound was very near the detection limit for SIMS established independently, but SIMS analysis of the extract was unsuccessful. It was speculated that other compounds in the extract (particularly trace levels of surfactants) depressed the ion signal for the target organic compound in the positive ion SIMS mass spectrum. The enzymatic digestion procedure thus shows promise for the release of organic compounds, but additional work with more sensitive instrumentation needs to be completed.

Pachuta and Staral[21] have used ToF SIMS for the analysis of colorants (inks) on paper surfaces, surveying a total of twenty-one pen inks and sixteen printed paper specimens. Static SIMS conditions were used with a Charles Evans and Associates TFS instrument with 25 keV gallium ions extracted from a liquid metal ion source. In almost all cases, positive ion SIMS mass spectra were recorded, since they provided more useful structural information. The sensitivity of the SIMS analysis is such that a single fiber of the ink-impregnated paper can be removed for analysis, leaving no visible damage to the specimen. However, the SIMS mass spectra exhibit the selectivity necessary to differentiate brands of ink, and to identify at least some of the organic dyes present in the ink. For example, Figure 9 is the positive ion SIMS mass spectrum of the ink from a Pilot blue ink; the ion at m/z 358 may correspond to the molecular ion of methyl violet, and the ion at m/z 372 may correspond to the molecular ion of crystal violet, although independent confirmation for the presence of these dyes is still required. In addition to the ability to identify the organic dyes that may be present, the SIMS mass spectrum

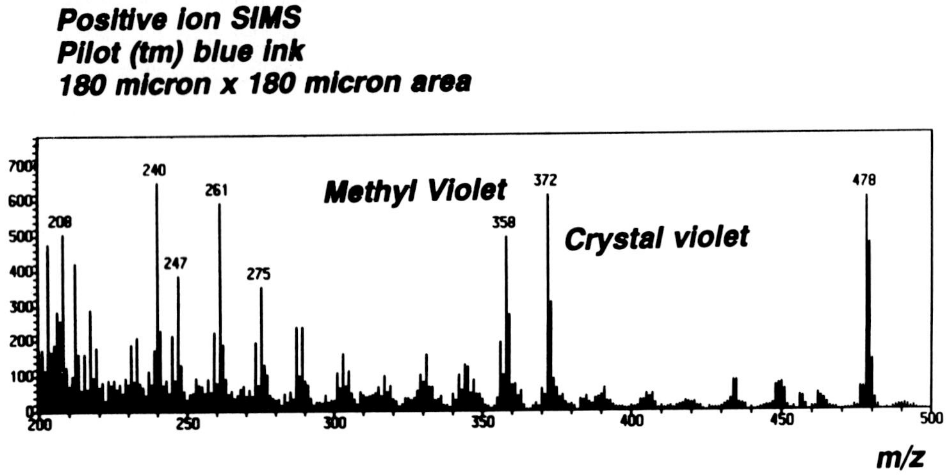

Figure 9. Positive ion SIMS mass spectrum of blue ink on paper; the tentative identification of two dyes is noted on the Figure. Adapted from Pachuta and Staral.[21]

provided direct evidence for various metal salts, such as copper, zinc, and lead found in various inks. Pachuta and Staral noted that ball-point pen inks provided the most intense SIMS mass spectra, attributed to the tendency of these inks to reside on the surface of the paper fibers. Inks from fiber-tip pens and from fountain pens are carried deeper into the paper fibers, and the lowered surface concentration makes detection with SIMS more difficult. It remains to be seen if a "liquid SIMS" experiment can improve this situation.

CONCLUSIONS AND OUTLOOK

Secondary ion mass spectrometry was developed for inorganic ion analysis, and the characteristics of sample preparation, data analysis, and instrumental design all reflect this traditional origin. With few exceptions, however, this paradigm does not lead to success in the direct analysis of paper and compounds in paper by SIMS. Innovative research by a number of groups has nevertheless demonstrated a remarkable potential in both imaging studies and for the analysis of the organic components of paper, and of compounds impregnated into paper. Successes have been catalyzed by the ability of new ToF instrumentation that provides both excellent imaging abilities and extraordinary sensitivity. As Fourier-transform ion cyclotron resonance mass spectrometers become more widespread, another major leap forward in capabilities can be foreseen. Historically, then, advances have been triggered by developments in instrumentation; sample preparation methods for SIMS are still relatively primitive and unexplored. For example, there are a number of well-characterized functional-group-specific derivatization reactions that can be used to treat paper surfaces, and yet there is no movement into this area. Fluorescent tags and labels can be used, and optical spectroscopic data can be correlated with mass spectrometric data, but so far the only advance in this direction has been in conjunction with thin-layer and paper chromatography, with SIMS detection. New means of organic compound extraction can be envisioned, especially using supercritical fluid extraction or enzymatic digestion, and only requires concerted attention from those expert in paper chemistry. The complexity of a paper sample is still relatively unexposed by rigorous and multi-faceted SIMS analysis.

REFERENCES

1. Benninghoven, A., Rudenaur, F. G., and Werner, H. W. 1987. Secondary Ion Mass Spectrometry: Basic Concepts, Instrumental Aspects, Applications, and Trends, John Wiley, New York. 1227 pp.
2. Morris, H.R. 1981. Soft ionization biological mass spectrometry. *In*: Proceedings of the Chemical Society Symposium on Advances in Mass Spectrometry Soft Ionization Methods. Heydon, London. pp. 1–256.
3. Miller, J.M. 1984. Fast atom bombardment mass spectrometry and related techniques. *Adv. Inorg. Chem. Radiochem.* 28:1–27.
4. Lyon, P.A. 1985. Desorption mass spectrometry: Are SIMS and FAB the Same? *In*: American Chemical Society Symposium Series, Washington, D.C. pp. 1–248.
5. Schueler, B., Sander, P., and Reed, D. A. 1990. *Vacuum*, 41:1661–1668.
6. MacGregor, M.A. 1993. The technical challenges of making better paper: Preparing for the 21st century. *TAPPI Journal* 76(6):143–154.
7. Hua, X., Kaliaguine, S. and Kokta, B.V. 1993. Application of SIMS in Polymers and Lignocellulosic Materials. *Journal of Applied Polymer Science* 48:1–12.
8. Briggs, D., Brown, A., and Vickerman, J. C. 1989. Handbook of Static Secondary Ion Mass Spectrometry (SIMS), John Wiley, New York. pp. 1–156.
9. Klein, P. and Bauch, J. 1979. On the localization of ions in the cell wall layers of treated wood based on laser-microprobe-mass-analyzer (LAMMA). *Holzforschung* 33(1):1–6.
10. Klein, P. and Bauch, J. 1981. Studies concerning the distribution of inorganic wood preservatives in cell wall layers based on LAMMA. *Fresenius Z. Anal. Chem.* 308(3): 283–286.
11. Klein, P. and Bauch, J. 1981. On the determination of wood preservatives in the cell wall of treated pine poles using a laser-microprobe-mass-analyzer (LAMMA). *Wood and Fiber* 13(4):226–236.

12. Schmidt, P.F. and Brinkmann, B. 1989. Investigations of the identification and discrimination of indigo-dyed cotton fibers by means of a laser microprobe mass analyzer. *Microbeam Anal.* 24:330.
13. Saler, J., Bremer, T., and Koschmeider, H. 1989. A study of photographic paper by secondary-ion-mass-spectrometry. *Fresenius Z. Anal. Chem.* 333:426–427.
14. Fowler, A.H.K. and Munro, H.S. 1985. Preliminary investigation of the thermal and x-ray degradation of cellulose nitrates by FAB/SIMS and ^{13}C NMR. *Poly. Degrad. Stabil.* 13:21–29.
15. Tan, Z. and Reeve, D.W. 1992. Spatial distribution of organochlorine in fully-bleached kraft pulp fibres. *Nordic Pulp and Paper Research Journal.* 1(7):30–36.
16. Brinen, J.S., Greenhouse, S. and Dunlop-Jones, N. 1991. SIMS (secondary ion mass spectrometry) imaging: a new approach for studying paper surfaces. *Nordic Pulp and Paper Research Journal.* 2(6):47–52.
17. Brinen, J.S. and Proverb, R.J. 1991. SIMS imaging of paper surfaces. Part 2. Distribution of organic surfactants. *Nordic Pulp and Paper Research Journal.* 4(6): 177–183.
18. Brinen, J.S. 1993. The observation and distribution of organic additives on paper surface using surface spectroscopic techniques. *Nordic Pulp and Paper Research Journal.* 1(8):123–129.
19. Zimmerman, P.A., Hercules, D.M., Rulle, H., Zehnpfenning, J. and Benninghoven, A. 1995. Direct analysis of coated and contaminated paper using time-of-flight secondary ion mass spectrometry. *TAPPI J.* 78(2):180–186.
20. Bartlett, M. G. and Busch, K. L. Unpublished work.
21. Pachuta, S. J. and Staral, J. S. 1994. Nondestructive analysis of colorants on paper by time-of-flight secondary ion mass spectrometry. *Anal. Chem.* 66:276–284.

ABBREVIATIONS LISTING

AKD	Alkyl Ketene Dimer
ASA	Alkenyl Succinic Anhydride
DI	Desorption Ionization
ESCA	Electron Spectroscopy for Chemical Analysis
FAB	Fast Atom Bombardment
LAMMA	Laser Microprobe Mass Analysis
LD	Laser Desorption
MALDI	Matrix-assisted Laser Desorption
SIMS	Secondary Ion Mass Spectrometry
ToF	Time-of-Flight

11

X-Ray Photoelectron Spectroscopy (XPS)

William K. Istone

Champion International Corporation, West Nyack, NY, U.S.A.

INTRODUCTION

Over the past five years, x-ray photoelectron spectroscopy (XPS) has become the workhorse for surface characterization and analysis of polymers and biomedical materials. This technique, also known by an older acronym – ESCA (Electron Spectroscopy for Chemical Analysis), provides a quantitative elemental analysis of the top 1–20 nm of a solid surface. Information about chemical bonding and molecular orientation can also be obtained.

X-ray photoelectron spectroscopy was developed by K. Siegbahn[1] in 1967 (he was awarded the Nobel prize in physics in 1982, primarily for his work on XPS) and commercial instrumentation began to appear in the mid-1970s. Most of the commercial applications of XPS have been in the areas of metallurgy, semi-conductors, and catalysis. As XPS requires ultra-high vacuum (1×10^{-8} torr or better) and the use of an x-ray source, it was initially considered unsuitable for the surface analysis of organic materials. However, recent advances in instrument design involving higher-capacity vacuum systems, faster and more sensitive detector designs, and the introduction of incident-beam monochromators for x-ray sources have permitted the technique to be successfully extended to the characterization of organic materials. It has become the main tool for the surface characterization of polymers and biomedical materials, and samples such as teeth, synthetic body parts, cheese, etc. are routinely analyzed by XPS.[2] The problem-solving capabilities of XPS in the surface analysis of organic materials can be extended further when XPS is used in conjunction with other surface-sensitive techniques such as secondary ion mass spectrometry (SIMS), attenuated total reflectance Fourier transform infrared spectroscopy (ATR–FTIR), and scanning electron microscopy (SEM). A comparison of the type of information yielded by these techniques is shown in Table 1.[3]

0-8493-8992-5/95/$0.00+$.50

Table 1
Comparison of Techniques

Parameter	SEM	EDX	XPS	FTIR (ATR)
Incident Radiation	Electron	Electron	X-Ray	Infrared
Analysis Depth	None	0.5–5.0 μm	0.5–10 nm	0.5–5.0 μm
Spatial Resolution	10 nm	0.5–5.0 μm	5–75 μm	30 μm
Detection Limit	NA	0.10 %	0.10 %	1.00 %
Detectable Elements	NA	B–U	Li–U	NA
Depth Profile	No	No	Yes	No
Chemical Information	No	No	Yes	Yes
Quantification	No	Semi-quant	2–10 % Rel.	10 % Rel.
Imaging	Yes	No	Yes	No
Mapping	No	Yes	Yes	Yes

NA = Not Applicable. (Data from Brundle *et al.*[3])

Given the chemical nature of paper and the multitude of surface-related phenomena associated with its manufacture and use, it is only natural that XPS should be applied to chemical characterization of paper surfaces. Indeed, the vast majority of papers published in recent years concerning the chemical characterization of paper surfaces focuses on analysis by XPS. While significant work has been done on XPS analysis of paper at universities and research institutes, the paper industry has been slow to make XPS a routinely-used tool at corporate research and development centers. At this writing, only three paper companies world-wide have on-site XPS capabilities, although many companies are engaged in utilizing this technique through contract laboratories and universities.

The objective of this chapter is to give the reader a basic understanding of x-ray photoelectron spectroscopy and to provide enough information that someone knowledgeable about the paper industry could evaluate the utility of employing XPS in solving a particular type of paper manufacture or use problem. This chapter is not intended to be an exhaustive treatise on the subject, but suitable references will be provided for those desiring more information on the technique and applications.

THEORY AND INSTRUMENTATION

THE PHOTOELECTRIC EFFECT

Albert Einstein was awarded the Nobel Prize in Physics, not for his discovery of relativity as is commonly believed, but rather for his discovery of the photoelectric effect.[4] When a material is exposed to a photon source, electrons contained in the material can be ejected if the energy of the incoming photons is greater than the binding energy holding the electrons within their orbitals – the photoelectric effect. Thus:

$$KE_e = 0.5\,m v^2 = h\nu - KE_B \qquad (1a)$$

where KE_e is the kinetic energy of a photoelectron, m is the mass of the electron, v the velocity of the electron, $h\nu$ the energy of the incident photon beam, and KE_B the binding energy of the electron in its orbital. If the source of the incident photons is an x-ray source and we measure the kinetic energy of the ejected photoelectrons then we have x-ray photoelectron spectroscopy (XPS).

The electrons generated by the photoelectric effect have a relatively low kinetic energy, typically 0 to 1500 eV (electron volts); because of this low kinetic energy the photoelectrons cannot travel large distances in matter. This has two implications for XPS: *i*) high vacuum is required as air molecules will scatter the photoelectrons; and *ii*) XPS is surface sensitive – the photoelectrons are only capable of traveling a few tens of nanometers in condensed matter. Only those photoelectrons which emerge from the sample without any loss in kinetic energy comprise peaks in the XPS spectrum; the scattered electrons comprise the background.

Einstein's paper on the photoelectron effect was published in 1905 and the first demonstrations of XPS were conducted in the 1920s. Unfortunately, the vacuum technology and electronics of the 1920s were insufficiently developed to permit the construction of a practical x-ray photoelectron spectrometer and hence further work on the technique waited over forty years until Siegbahn revived interest.

INSTRUMENTAL COMPONENTS

Five basic components are required to perform a XPS analysis: *i*) an x-ray source; *ii*) a sample holder; *iii*) a kinetic energy analyzer; *iv*) an electron detector; and *v*) an ultra-high vacuum system. These components are shown as a block diagram in Figure 1. Each of these components will be discussed in more detail along with other elements of XPS instrumentation, but the basic requirements are as follows:

- **X-ray source:** The source must be capable of emitting monochromatic x-ray photons at constant potential and current. Magnesium and aluminum x-ray sources are most commonly used.

- **Sample holder:** In addition to providing a stable platform on which the sample is mounted, the sample holder must also provide the analyst with a means of introducing the sample into the vacuum chamber and manipulating the sample while under vacuum. Most modern instruments have motor-controlled sample mounts.

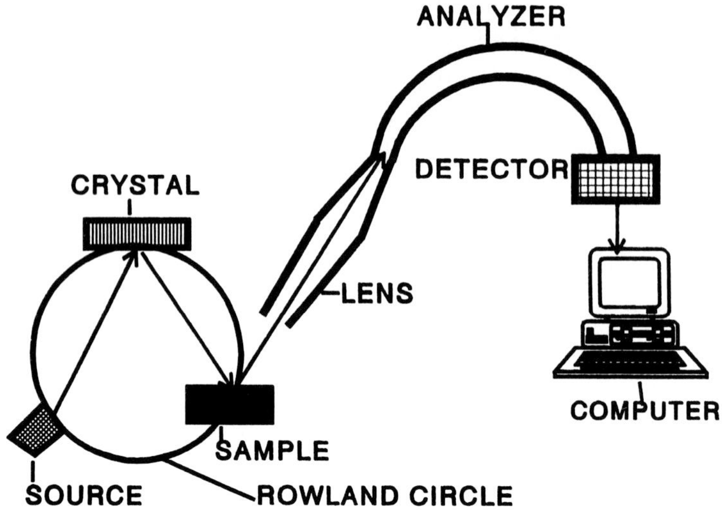

Figure 1. Diagram of a modern x-ray photoelectron spectrometer employing a spherical capacitor analyzer (SCA) and a monochromatic x-ray source.

- **Analyzer:** The purpose of the analyzer is to separate the photoelectrons and Auger electrons by energy much as a prism separates photons by wavelength. Several different analyzer designs have been used over the years, but most modern instruments employ a spherical capacitor analyzer (SCA).

- **Detector:** The detector counts the photoelectrons (as discrete events) as they emerge from the analyzer. An electron multiplier is used for this purpose. Modern instruments use position-sensitive detectors or multi-channel detectors to permit more rapid analysis.

- **Vacuum system:** X-ray photoelectron spectroscopy must be performed under ultrahigh vacuum conditions (greater than 1×10^{-8} torr) in order to prevent scatter of photoelectrons and to minimize contamination of the sample surface. Most instruments use cryogenic or ion diffusion pumps to achieve and maintain these high vacuums.

X-Ray Sources: In XPS instrumentation, the excitation energy is produced by an x-ray source or sources mounted within the high vacuum chamber (or bell). In an x-ray source, electrons are produced by a heated filament and are then accelerated by high voltage to impact a metal anode. Interaction of these electrons with the anode material results in the emission of characteristic x-rays which are focused on the sample. Aluminum and magnesium are the two commonly used anodes for XPS analysis,[5] although occasionally other anode materials (titanium, chromium) may be used for specialty applications.[2] The resultant x-ray beam consists of a primary line (x-rays of a discrete wavelength) referred to as the $K\alpha_{1,2}$ line. Minor lines such as the $K\alpha_{3,4}$ line and $K\beta$ lines are also produced along with high energy white radiation (bremsstrahlung radiation). These secondary lines and white radiation contribute little to the analytical signal, but contribute greatly to spectral noise (background and satellite peaks) and to sample damage, both radiation- and thermally-induced.

An incident beam monochromator is a device that filters the incident x-ray beam and permits only x-rays of a given wavelength to reach the sample. The monochromator consists of a toroidally-shaped crystal situated between the source and the sample. When properly aligned, only the $K\alpha_{1,2}$ x-ray line is diffracted from the crystal face toward the samples. Other x-ray lines are scattered along with white radiation, impacting the monochromator housing. Two conditions must be met for the monochromator to work properly: *i*) the crystal face must have the proper d-spacing to diffract the incident radiation, and *ii*) the x-ray anode, crystal face, and sample surface all must lie on the circumference of a circle (Rowland circle). Typically, a quartz crystal is used in conjunction with an aluminum x-ray source to produce monochromatic Al $K\alpha_{1,2}$ radiation.

Three major advantages are gained by the use of a monochromator. First, the spectral background is reduced due to the elimination of secondary lines and scattered radiation resulting in an increase in signal/noise ratio and hence enhanced sensitivity. Second, spectral resolution is increased due to a narrower line width and elimination of satellite peaks.[2] This makes it easier to accurately identify peak positions and to separate overlapping peaks. Third, sample damage is reduced as less radiation and heat reach the sample. ***The use of an incident beam monochromator is absolutely essential for XPS analysis of paper and related materials.*** Cellulose degrades in the presence of x-rays, converting to an enone form.[6] Other organic materials used in the manufacture of paper are also susceptible to x-ray or thermal damage. It is also necessary to have high spectral resolution in order to be able to successfully separate multiple carbon chemistries in the XPS spectrum.

There is one disadvantage to the use of the x-ray monochromator: charging. As photoelectrons are ejected from the sample surface, the surface becomes more positively charged thus making it more difficult (increasing the binding energy) for remaining electrons to be ejected. In a conducting sample (a piece of metal) the problem is solved by simply grounding the sample and the analyzer/detector to a common point–providing a closed circuit through which the electrons can flow. This is not possible with an insulating sample like paper or a polymer. With a non-monochromatic x-ray source charging is minimized as electrons due to white radiation serve to neutralize the charge on the sample's surface. An insulating sample with a monochromatic source presents a problem. The most common solution is to use a neutralizer or "flood gun" to control charging. The neutralizer is simply a filament that produces a flow of low-energy electrons directed toward the sample. The effectiveness of a neutralizer can be enhanced by providing a conductive path between the sample surface and a common ground. This is accomplished by using metal masks which cover the sample surface, exposing only the area of analytical interest. The masks are then connected to the common ground.

Sample Holder (The Stage): The stage or sample holder provides a means of introducing the sample into the XPS instrument and maneuvering the sample once inside the instrument. Typically, the stage consists of a flat metal plate (usually stainless steel) in which holes have been drilled and tapped so that the sample and masks can be secured with screws. The underside of the stage is machined so that it can be grasped by a manipulator arm. The stage size varies from instrument to instrument, but typically is 2.5–7.5 cm (about 1–3 inches) in diameter. Once the samples have been mounted, the stage is placed on a manipulator arm (or probe) inside of an airlock. The airlock is then pumped (usually by a turbo-molecular pump) to bring the samples and stage down to a vacuum near that of the instrument. A gate valve is then opened and the stage is inserted into the high vacuum chamber with the probe and secured to a stage mount, after which the probe is withdrawn and the gate valve closed. The stage mount is also connected to a manipulator so that the samples can be precisely positioned within the instrument. On basic instruments, a series of micrometers are used to adjust the sample in the X, Y and Z positions along with tilting and rotating the sample. On more advanced instruments, computer-controlled stepper motors allow the sample to be moved from a joystick or mouse. A fully automated stage is useful in that it permits automated analysis of multiple positions on one sample or multiple samples. It also allows for mapping experiments.

The Analyzer: The function of the analyzer is to separate the photoelectrons emitted by the sample according to kinetic energy much as a monochromator separates light according to wavelength in a spectrophotometer. An XPS analyzer consists of two parts: a lens and a kinetic energy analyzer.

The lens serves a twofold purpose, the first of which is to define the analysis area. Using a combination of electrostatic steering plates and metal apertures, the size of the analysis area can be varied from as little as ten micrometers in diameter to as much as several millimeters in diameter (actual range and dimensions depend upon individual instruments). This arrangement also allows one to vary the acceptance angle. Photoelectrons are emitted from various depths in the sample; by controlling the angle of the sample surface relative to the lens one can set the maximum depth from which photoelectrons are detected (this will be discussed in more detail later in the section on *Spectroscopy*). The acceptance angle determines the degree of angular resolution with respect to the photoelectrons being allowed to enter the lens. A large acceptance angle yields poor angular resolution but high sensitivity; a small acceptance angle yields high angular resolution but low sensitivity.

Some XPS spectrometers have the ability to perform chemical imaging. That is, an SEM-like image of the surface is generated with various elemental or chemical maps overlaid on that image. A common approach to imaging is to use the steering plates in the lens to vary the analysis position. A fixed area of the sample is analyzed, then the analysis spot is shifted to an adjacent area which is then analyzed, etc. The individual spectra for each area are then assembled via computer into an image and corresponding chemical maps – a similar process to the way in which satellites prepare composite images of the earth's surface providing not only visual images, but also other information (*e.g.*, infrared data).

The second function of the lens is to apply a retard voltage to the incoming photoelectrons so that these electrons have the correct energy to enter and be separated by the analyzer. When a spectrum is generated, the spectral range is determined by the retard voltage and the actual spectral sweep is accomplished by ramping the retard voltage while the analyzer is operated at constant potential. This permits one to scan a wide range of kinetic energy while a precise energy analysis is performed under constant resolution conditions.

Two types of analyzers are common in XPS analysis: a cylindrical mirror analyzer (CMA) and a spherical capacitor analyzer (SCA). Cylindrical mirror analyzers are generally found on older instrumentation or on instruments where XPS is combined with scanning Auger spectroscopy (SAM). Most high resolution XPS spectrometers employing monochromatic x-ray sources use a spherical capacitor analyzer. As this is the type of instrument most useful for analyzing paper samples, we will discuss this analyzer. The SCA consists of two concentric metal hemispheres. These hemispheres are made from non-metallic material and are usually gold-plated. A voltage can be applied to each hemisphere. The photoelectrons, after being suitably retarded, enter along one edge of the hemisphere from the lens and exit to the detector at the opposite edge. The values of the voltages applied to the hemispheres determine the range and resolution of the kinetic energies passed onto the analyzer. The energy range is referred to as the *pass energy*. As the electrons pass through the hemispheres, electrons having too high a kinetic energy collide with the outer sphere and those having too low a kinetic energy collide with the inner sphere. The remaining electrons are dispersed radially according to their respective kinetic energy and pass onto the detector.

The Detector: The basic detector consists of an electron multiplier. In older instruments, a single channel multiplier was used which necessitated sweeping the lens and analyzer slowly, stepping the kinetic energy in small increments. Most modern spectrometers use channeltron plates to form multi-channel or position-sensitive electron multipliers which permit the lens and analyzer to sweep in larger steps. This greatly speeds the acquisition of a spectrum. A spectral acquisition which took two hours in the late 1970s now takes ten to fifteen minutes on a modern spectrometer. Not only does this reduce analysis time, but this also reduces sample damage for samples sensitive to radiation or heat. The output of the detector is then transferred to a computer.

Vacuum System: As was stated previously, XPS analysis must be performed under conditions of ultra-high vacuum in order to minimize scatter of photoelectrons and contamination of the samples surface. The central part of any XPS spectrometer, therefore, is a large stainless steel vacuum bell containing the various components of the spectrometer and operating at a pressure of 1 x 10^{-8} to 1 x 10^{-10} torr. Most systems utilize multiple vacuum pumps in order to achieve and maintain this pressure. Usually a mechanical vacuum pump is used in combination with a turbo-molecular pump to "rough" the vacuum chamber (pump it down to approximately 1 x 10^{-6} torr) and to provide vacuum for the sample introduction chamber and various instrumental components (such as an ion gun). An ion diffusion pump or a cryogenic sorption pump is then used to achieve and maintain the ultra-high vacuum in the bell. The use of pumps alone is not sufficient to attain these high vacuums. There must also be a way of "baking the system," that is, heating the system to a temperature in excess of 100 °C in order to desorb water vapor and organic materials from the inner walls of the chamber.

X-RAY PHOTOELECTRON SPECTROSCOPY

THE BASIC XPS SPECTRUM

A great deal of information can be obtained from an XPS experiment. This includes elemental information, chemical state information, variation of composition by depth, variation of chemical composition spatially on the surface, and thickness of layers.

Photoelectrons generated by a given element result in a discrete series of lines, or peaks in the XPS spectrum. Figure 2 shows a wide range or survey XPS spectrum of the element copper; Figure 3 is the spectrum of gold. Observe that photoelectrons are generated for not only each electron orbital, but also for electrons occupying the same orbital but having different spins. The predominant lines in the XPS spectrum are due to electrons ejected from inner shells. Electrons ejected from valence shells have very low energy (typically less than 50 eV) and are referred to as valence band electrons. Lines generated by these valence electrons are highly dependent on the binding state of these atoms. The electrons generated by the inner orbitals are also affected by the binding state of the atom. Changes in the valence electron configuration as well as the type of atom bound to the atom of interest affect the electron cloud through which the inner shell electrons must pass as they are ejected and thus change their kinetic energy. The shift in kinetic energy from the values obtained from pure elements can be used to identify the binding state of the atom.

As electrons are ejected from an atom's inner shell, vacancies are created. Electrons occupying a higher energy orbital can descend to fill this vacancy resulting in a burst of energy. In many cases this energy is sufficient to eject another electron. This second ejected electron is referred to as the Auger electron after the Belgian physicist who discovered the effect. Thus the XPS spectrum also contains peaks due to Auger electrons.

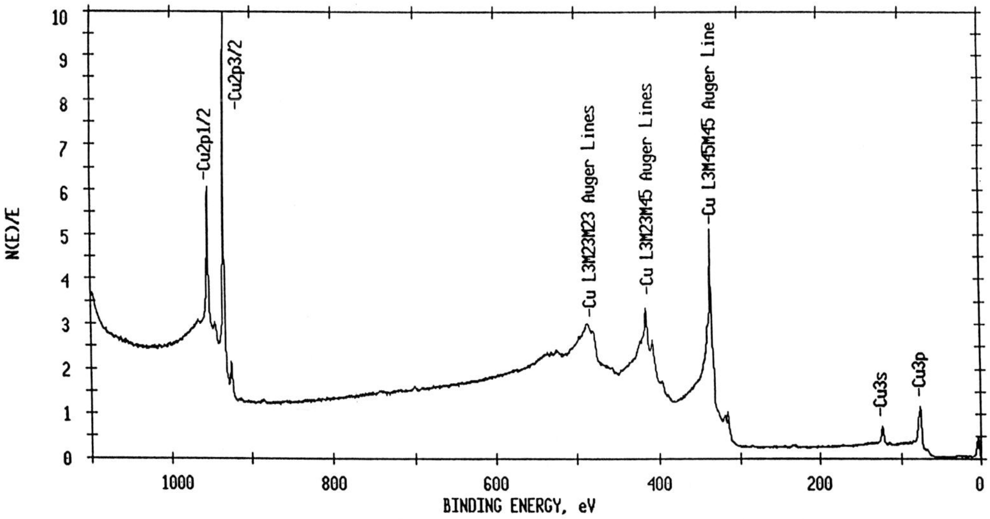

Figure 2. XPS survey spectrum (low resolution, wide energy range) of copper metal showing the major photoelectron and Auger electron lines.

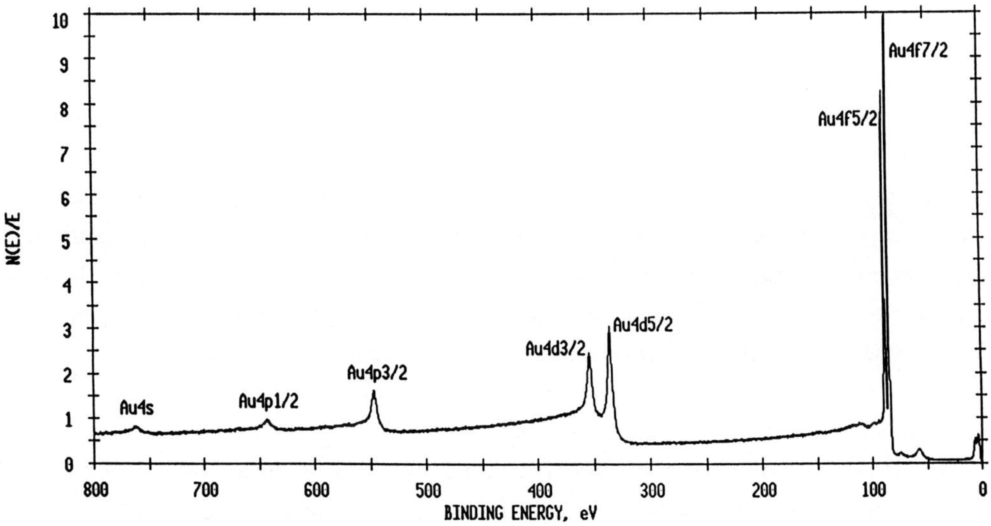

Figure 3. XPS survey spectrum of gold metal showing the singlet photoelectron peak (Au 4s) and the doublet photoelectron peaks (Au $4p_{1/2}$–Au $4p_{3/2}$, Au $4d_{3/2}$–Au $4d_{5/2}$, and Au $4f_{5/2}$–Au $4f_{7/2}$). Note the difference in peak separation as one proceeds from the 4p through to the 4f orbital.

All of the information contained in an XPS spectrum (photoelectron lines, shifts in peak positions, the splitting of a peak due to electrons of different spins, valence band electron lines and Auger lines) can be utilized to provide information about the chemical composition of a surface.

XPS SPECTRA OF CORE ELECTRONS

Peaks due to core electrons constitute the major portion of the XPS spectra. The positions of these peaks are used to identify the elements present on the surface and to provide information about the oxidation states of these elements and what these elements are bound to. The intensities of these peaks are also used for quantification.

The spectrum obtained by the instrument is a function of kinetic energy. Originally, the spectra were displayed with kinetic energy as the X-axis. The current commonly used convention is to display the spectra with binding energy as the X-axis. As the measured binding energy is a function of the energy of the x-ray source and the work function of the spectrometer, the peak positions must be converted as in Equation (1b):

$$KE_e = 0.5 m v^2 = h\nu - KE_B - W \qquad (1b)$$

where KE_e is the kinetic energy of photoelectron, m is the mass of the electron, v is the velocity of the electron, $h\nu$ is the energy of the incident photon beam, W is the spectrometer work function and KE_B is the binding energy of the electron in its orbital.

If the sample is not a conductor (as is the case with paper), then the spectrum must be corrected for charging. The easiest way to do this is to choose a peak in the spectrum for which the correct binding energy is known (the core line of an element having a known or independently verifiable chemistry). The entire spectrum is then shifted so that the reference peak is in the correct position. This procedure works in most cases. The exception is a situation known as differential charging, where the sample surface is not homogeneous and some portions of the surface charge more than others. In the case of polymers or organic materials, the hydrocarbon portion of the Carbon 1s line is usually chosen as the reference peak as most samples of this type normally contain hydrocarbon materials either intentionally or as the result of contamination.[2] One finds two conventions in use today: most spectra are referenced to C1s at 285.0 eV, but the NIST XPS Database[7] is referenced to C1s at 284.8 eV.

The low resolution survey spectrum described above and shown in Figure 2 is useful for determining which elements are present on the sample surface. In order to obtain information about chemistry and to perform quantitative analysis it is necessary to obtain higher resolution spectra. Obtaining high resolution spectra over a wide range similar to the survey spectrum would be very time-consuming, so an alternative approach–the multiplex spectrum–is used. In this approach, a series of narrow (approximately 20 eV) windows referred to as *regions* are scanned at high resolution around each peak of interest. The peak areas or intensities can be used for quantification and the individual peaks can be examined in detail to provide chemical information.

Chemical Shifts: The core lines in the XPS spectra are due to the loss of photoelectrons from the orbitals in the core shells of the atoms at the surface. Atomic orbital notation is used to label these lines. For example: the C1s line is the electron in a carbon atom having a principal quantum number of "1" and an angular momentum quantum number of 0 ("s" orbital). Table 2 gives the binding energies of the major lines for elements of interest in paper and polymer analysis. As XPS lines expressed as binding energies have already been corrected for the kinetic energy of the x-ray source, the same values hold regardless of the source anode used. Complete tables for all elements can be found elsewhere.[2,5]

There is no spin–orbit splitting due to different magnetic spin states in s orbitals. In p, d, and f orbitals, spin–orbit splitting does occur resulting in split peaks. P orbitals have 1/2 and 3/2 spin states producing a split peak in a 1:2 ratio. D orbitals have 3/2 and 5/2 spin states resulting in a 2:3 split while F orbitals have 5/2 and 7/2 spin states resulting in a 3:4 split peak. Figure 3 shows examples of split peaks for p, d, and f orbitals. Differing chemical states can effect the separation of the peaks in these cases and hence this becomes a tool in interpreting the spectra. For example: the split of the Nickel 2p orbital (Ni $2p_{1/2}$, Ni $2p_{3/2}$) is 17.4 eV in the case of nickel metal, but is 18.4 eV in the case of

Table 2
XPS Core Lines of Selected Elements

Element	1s	2s	$2p_{1/2}$	$2p_{3/2}$	3s	$3p_{1/2}$	$3p_{3/2}$	$3d_{3/2}$	$3d_{5/2}$
Carbon	***287***								
Nitrogen	***402***								
Oxygen	***531***	23							
Fluorine	***686***	30							
Sodium	***1072***	64	31						
Magnesium	***1305***	90	51						
Aluminum		119	74	***74***					
Silicon		153	103	***102***					
Phosphorus		191	134	***133***	14				
Sulfur		229	166	***165***	17				
Chlorine		270	201	***199***	17				
Potassium		378	296	***293***	33	17	17		
Calcium		439	350	***347***	44	25	25		
Titanium		565	464	***458***	62	37	37		
Chromium		698	586	***577***	77	46	45		
Manganese		770	652	***641***	83	49	48		
Iron		847	723	***710***	93	56	55		
Nickel		1009	873	***855***	112	69	67		
Copper		1098	954	***934***	124	79	77		
Zinc		1196	1045	***1022***	140	92	89	10	10
Silver					718	604	573	374	***368***
Iodine					1071	930	874	630	***619***

N.B. Most intense line shown in bold italics. (Data from Wagner *et al.*[5])

nickel oxide (NiO). The NIST XPS Database[7] andother sources[5] provide tables of spin–orbit splitting as a function of chemical state. In the case of paper analysis, most of the elements of interest only have s and p core electrons where the splitting is non-existent or very small so that analysis of splitting values is of little use.

Chemical shifts are the most useful information in an XPS spectrum for determining chemical states. Chemical shifts occur when neighboring atoms increase or decrease the electrostatic shielding around the nucleus of the atom being studied. The increase or decrease in shielding results in greater or lesser attraction between the nucleus and the core electrons, resulting in a corresponding increase or decrease in binding energy for the photoelectrons being emitted from that atom. For example, a carbon atom bound to a more electronegative atom such as oxygen would have some of its electrons attracted to the oxygen. This would result in less negative charge to interact with the nucleus and hence a greater attraction between the nucleus and the remaining electrons – hence an increase in binding energy.

Chemical shift data have been compiled from both theoretical models and experimental determinations. There are a number of compilations of chemical shift data available, with one of the most complete being the NIST XPS Database. Beamson and Briggs' book on high resolution XPS of polymers[8] is an excellent source for shift data for organic materials. In Table 3, chemical shift data[8] for functional groups of interest in paper analysis are summarized.

Curve Deconvolution: In analyzing XPS spectra of complex organic molecules or mixtures of materials, the core peaks can have very complicated shapes due to various chemical shifts being involved. Consider the case of a carbon 1s XPS line in the spectrum of a paper coating containing SBR latex, starch, and calcium carbonate. There would be peaks due to the chemical shifts for the following carbon chemistries: C–C or hydrocarbon from the latex, C–O and C=O due to the starch, CO_3 due to the carbonate, and a shake-up peak (discussed later in this section) due to the aromatic character of the latex. Even in a high resolution spectra, these peaks would overlap enough to make accurate assignment of peak position difficult and quantification impossible. The solution to this problem is to use mathematical deconvolution or curve fitting techniques to break the spectra peak down into component parts. Most XPS data processing programs include curve fitting capabilities. Other mathematical and spectral analysis programs that include curve fitting area also available.

In principle, a curve such as an XPS peak would have an infinite number of possible solutions. In curve deconvolution, knowledge of XPS peak shapes and positions and some basic knowledge of the sample chemistry is used to define limits on the curve fitting program so that a unique solution can be arrived at. Limits are placed on allowable peak positions (due to knowledge of the basic sample chemistry), peak shape

Table 3
Selected Chemical Shifts of XPS Core Lines

Functional Group	Carbon 1s	Oxygen 1s	Nitrogen 1s	Fluorine 1s	Silicon $2p_{1/2}$	Sulfur $2p_{1/2}$	Bromine $3d_{3/2}$	Chlorine $2p_{1/2}$
C=C	284.73							
C–C	285.00							
C–O–C	286.45	532.64						
C–OH	286.55	532.89						
*C–O–C=O	286.64							
C=O	287.90	532.33						
O–C–O	287.93	533.15						
C–O–*C=O	288.99							
–C(=O)OH	289.26							
O–C(–O)–O	289.32	532.99						
–C(=O)–O–C(=O)–	289.41	533.91 (532.64)						
Carbonate (CO_3)	290.44	532.38 (533.93)						
C–NO_2	285.76	532.45	405.45					
C–NR_2	295.94		399.77					
C–NR_3	286.11		401.46 (R=H)					
*C–C=N	286.41		399.57					
C=N	286.74		399.57					
C–ONO_2*	287.62	534.70	408.15					
N–C–O	287.78							
N–C=O	288.11	531.87	399.85					
–C(=O)–N–C(=O)–	288.55		400.50					
N–C(=O)–N	288.84		399.89					
N–C(=O)–O	289.60		400.32					
C–Si	284.33				102.35			
C–S	285.37					164.77		
C–SO_2	285.38	531.67				168.82		
C–SO_3	285.16	531.72				169.40		
C–Br	285.74						71.54	
C–Cl	287.02							202.21
–CCL_2	288.56							202.43
C–F	287.91			686.94				
–CF_2	290.90			689.67				
–CF_3	292.69			688.20				

* indicates atom for which binding energy is given. For multiple oxygens, =O is given first, followed by –O in parentheses. All values are binding energies in eV relative to C–C at 285.00 eV. (Data from Beamson and Briggs[8])

(XPS peaks are primarily Gaussian with some Lorentzian character), full width at half maximum–FWHM (this should vary slightly, increasing with increasing binding energy), and the total number of peaks present (once again based on some basic knowledge of sample chemistry). In Figure 4 an example of a curve fit for our coating case is shown. This coating contains carbon from three different sources: styrene-butadiene latex which results in the C–C peak, starch which results in the C–O and C=O peaks, and calcium carbonate which results in the carbonate peak. From the resultant fit, one can determine the peak positions and peak areas (for quantification).

Shake-up Processes: Under certain conditions satellite peaks adjacent to core electron peaks can be formed as a result of "shake-up" processes. The two most common causes of shake-up peaks are paramagnetic states in metals and metallic compounds and unsaturation in organic materials. Shake-up satellite peaks are caused by a portion of the kinetic energy of a photoelectron being used to promote a another electron to a higher orbital thus resulting in a peak shift. As only a portion of the photoelectrons contribute to the shake-up process, the resulting spectrum shows the normal photoelectron peak along with a small satellite at a slightly higher binding energy. This is illustrated is Figure 5. Copper metal, Cu(0), does not undergo shake-up, and hence exhibits a doublet in the 2p spectral region, as shown in Figure 5A. Copper oxide, CuO with copper as Cu(II), does exhibit shake-up resulting in four peaks in the 2p spectral region, as shown in Figure 5B. Shake-up processes in organic materials are commonly caused by π–π^* molecular orbital transitions. Shake-up satellites provide a useful means for determining whether an organic compound on the surface is aromatic or unsaturated.

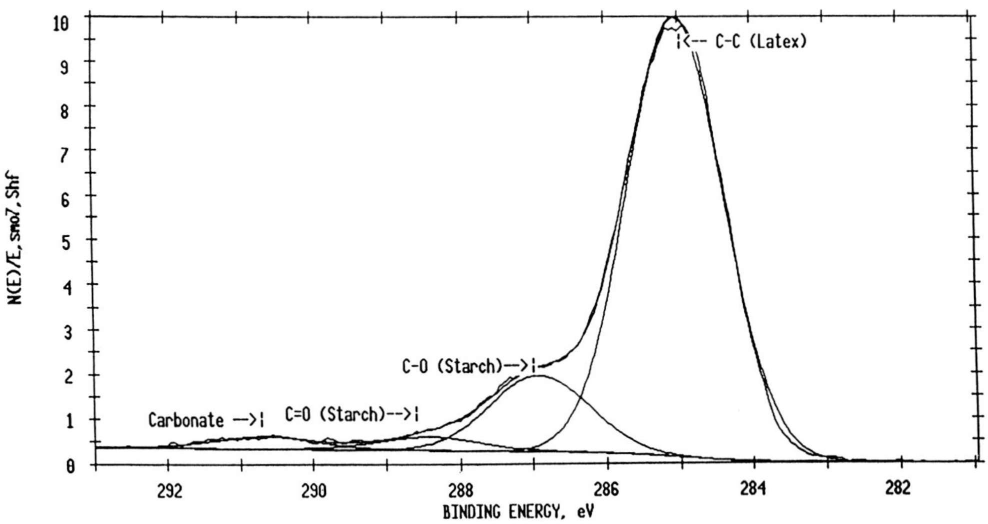

Figure 4. High resolution XPS spectrum of the carbon 1s region for a coated paper surface showing the curve fit peaks for C–C (latex), C–O and C=O (starch), and carbonate.

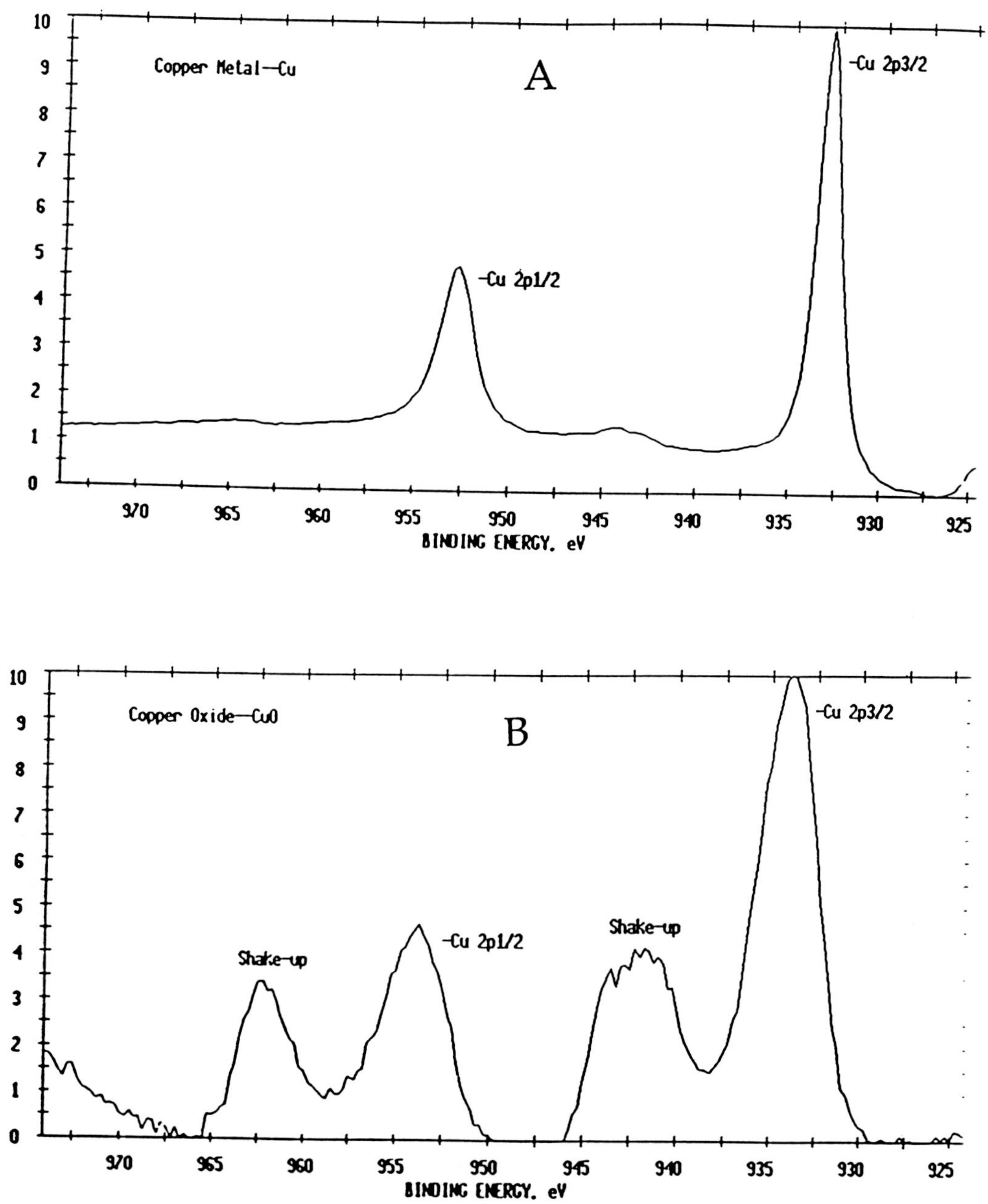

Figure 5. XPS spectra for the 2p region of copper illustrating "shake-up" effects. Spectrum A for copper metal shows no shake-up, whereas spectrum B for copper oxide (CuO) shows significant shake-up peaks.

Quantitative Analysis: There are two commonly used approaches to quantitative analysis for XPS. The easiest approach is to use a fundamental parameters calculation to convert peak areas for the elements of interest into atomic concentrations. (In XPS analysis, atomic concentration or atomic percent is the common unit of concentration. If the sulfur concentration at the surface is thirteen atom percent, then thirteen of every hundred atoms on the surface are sulfur. Atomic percent is equivalent to mole percent). This approach yields quantitative results of ten percent relative or better. The second approach is to prepare suitable standards and construct calibration curves. Using this approach, quantitative results of two percent relative are achievable. Both approaches require that all spectral regions for samples and standards (if applicable) be acquired under identical instrumental settings. An additional requirement for the fundamentals parameters approach is that all elements present on the surface be analyzed for and included in the calcuation. Quantification is almost always done using peak areas of core electron lines.

The basic equation for quantitative XPS analysis is:

$$N_{i,k} = I_0 \rho_i \sigma_{i,k} \lambda_{i,k} T_{i,k} \tag{2}$$

where $N_{i,k}$ is the experimentally determined peak area for atomic shell k of element i; I_0 is the x-ray flux on the sample; ρ_i is the volume density for element i at the surface (concentration); $\sigma_{i,k}$ is the differential photoionization cross section for element *i* shell *k*; $\lambda_{i,k}$ is the mean free path of electrons from the *k* shell of element *i*; and $T_{i,k}$ is the transmission or instrument throughput at the kinetic energy of photoelectrons from shell *k*, atom *i*.

The x-ray flux can assumed to be constant with respect to time for periods as long as several hours. As the calculation of concentrations involves ratioing data for the elements present, the x-ray flux term cancels. The differential photoionization cross section can be calculated from Equations (3) and (4):

$$\sigma_{i,k}(h\nu,\phi) = \sigma_{i,k}(h\nu) \cdot F_{i,k}(\beta,\phi) \tag{3}$$

$$F_{i,k}(\beta,\phi) = 1 + \frac{3\beta}{2\,(2\sin^2\phi - 1)} \tag{4}$$

The first term of Equation (3) is the Scofield total photoionization cross section, whereas the second term, defined in Equation (4), consists of an angular term and a parameter (β) characteristic of the atomic orbital and atomic number. The term ϕ is the angle between the incident x-ray photons and the emitted photoelectrons. This term is a constant for a given instrument. The values for Equations (3) and (4) have been tabulated by Scofield[9] and Reilman.[10] By knowing ϕ for one's instrument and using these tabulated values, one can calculate σ.

The mean free path can be represented by Equation (5):

$$\lambda_{i,k} = CE^m \tag{5}$$

where E is the kinetic energy and C and m are functions of the solid material. It has been shown by Wagner *et al.*[11] that m is approximately 0.75 for inorganic solids and 0.70–1.00 for organic solids for kinetic energies greater that 300 eV. The transmission function, T, is a constant for a given instrument.

In practice, one would determine sensitivity functions for given core lines using the equations above. These sensitivity functions would then be used in conjunction with experimentally determined peak areas in order to calculate concentrations. XPS instruments are generally provided with appropriate algorithms and tabulated values as part of the manufacturer's software package.

The use of calibration curves based upon standards can provide more rigorous quantitative analysis. One must be very careful, however, in the preparation of standards. It cannot be assumed that bulk concentration will translate directly as surface concentrations. Furthermore, contamination of the standards surface by extraneous material (vapors, contact with die or tools surfaces, etc.) will result in poor standards.

AUGER ELECTRON PEAKS

As a core photoelectron is ejected from an inner shell, a vacancy is created. This vacancy can be filled by an electron from a higher energy orbital descending to fill it. Figure 6 illustrates the ejection of an Auger electron. As this process occurs, energy is released. If an appropriate amount of energy is released, a second electron can be ejected as a photoelectron. This secondary photoelectron is referred to as an Auger electron. The Auger process is independent of the energy of incident x-ray radiation. Since XPS spectra are displayed as a function of binding energy corrected for the energy of the x-ray source, Auger peaks will shift position as different x-ray sources are used. Table 4 gives the energy of Auger lines for elements of interest in paper and polymer analysis for aluminum and magnesium sources. More extensive tables are available elsewhere.[5]

Like core photoelectron peaks, Auger peaks can be used for quantification and determination of chemistry. Chemical shifts of Auger lines are available in the NIST XPS Database and elsewhere. Most Auger shift data in the literature are for metals, and are hence of little use in paper analysis at present.

VALENCE BAND SPECTRA

Valence band or molecular orbital spectra refer to those photoelectron lines resulting from outer shell electrons–those electrons involved in bonding. Because of their involvement in bonding, these electrons are far more sensitive to subtle changes in chemistry than are the core electrons. Unfortunately, the photoionization cross section for electrons occupying these outer orbitals is small, resulting in poor sensitivity. Until recently, it was not possible to acquire valence band spectra for organic materials as the long acquisition times required to obtain high resolution spectra for this region often resulted in severe sample damage and led to results of questionable value. With the

Table 4
Auger Lines for Selected Elements

Element	KL_1L_1	KL_1L_{23}	$KL_{23}L_{23}$	$L_3M_{23}M_{23}$	$L_2M_{23}M_{23}$	$L_3M_{23}L_{45}$	$L_3M_{23}M_{45}$	$L_2M_{23}M_{45}$	$L_3M_{45}M_{45}$	$L_2M_{45}M_{45}$	$M_5N_{45}N_{45}$	$M_4N_{45}N_{45}$
Carbon			***1226 (993)***									
Nitrogen			***1108 (875)***									
Oxygen	1012 (779)	997 (764)	***976 (743)***									
Fluorine	878 (645)	859 (626)	***832 (599)***									
Sodium	565 (332)	536 (303)	***497 (264)***									
Magnesium	384	350	***305***									
Aluminum			100									
Sulfur				***1336 (1103)***	***1336 (1103)***							
Chlorine				***1304 (1071)***	***1304 (1071)***							
Potassium				***1238 (1005)***	***1236 (1003)***							
Calcium				***1197 (964)***	***1194 (961)***							
Titanium				1106 (873)	1106 (873)	***1072 (839)***						
Chromium				1000 (767)	1000 (767)	***962 (729)***						
Manganese				948 (715)	948 (715)	***903 (670)***			853 (620)			
Iron				892 (659)	892 (659)	841 (608)			***786 (553)***			
Nickel				781 (548)	775 (542)	715 (482)	709 (476)		***643 (410)***	626 (393)		
Copper				719 (486)	712 (479)	649 (416)	641 (408)	629 (396)	***570 (337)***	550 (317)		
Zinc				662 (429)	655 (422)	585 (352)	576 (343)	562 (329)	***498 (265)***	475 (242)		
Silver											1136 (903)	***1130 (897)***
Iodine											981 (748)	***970 (737)***

Auger line positions in eV (binding energies) for aluminum and (magnesium) anodes. Most intense lines in bold italics. (Data from Wagner *et al.*[5])

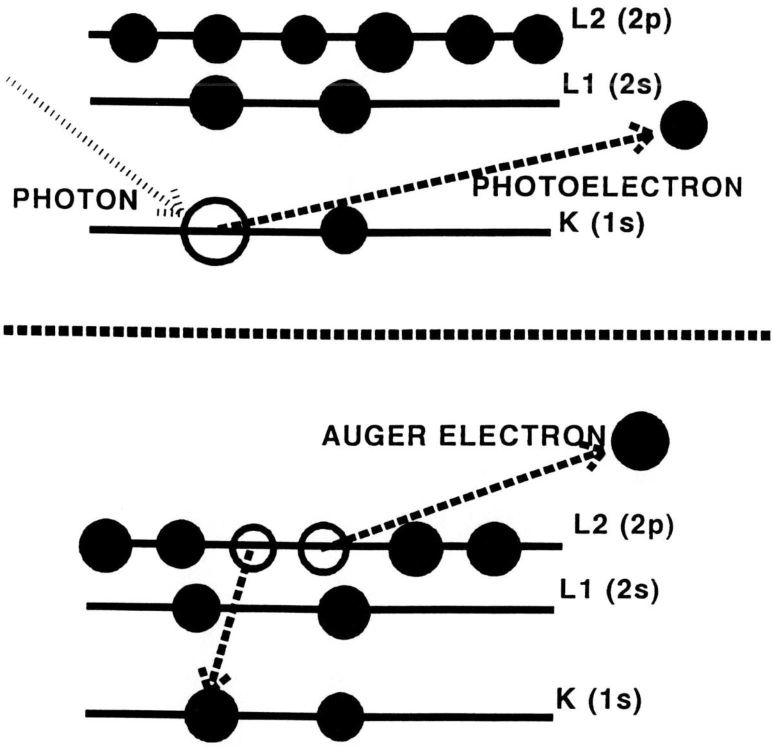

Figure 6. An illustration of the electronic transitions for the photoelectron process (top) and the Auger electron process (bottom). In the photoelectron process, the energy from an incident x-ray photon results in the ejection of a single photoelectron. In the Auger process, an electron from a higher orbital descends to fill the vacancy created by the ejected photoelectron resulting in an energy release, which in turn, ejects a second (Auger) electron.

advent of monochromatic x-ray sources and multi-channel detectors, valence band spectra of organics can now be acquired and used as an additional tool in determining surface chemistries. The spectral region of interest is generally from 0 to 30 eV.

Valence band spectra are capable of providing chemical information that is not obtainable from other forms of XPS data. For example, the polymers polyethylene, polypropylene, poly-1-butene, and polyvinyl methyl ether appear identical when core electron and Auger electrons lines are considered, but are easily distinguished by valence band spectra. This is shown in Figure 7. One can also use valence band spectra to distinguish between different forms of the same compound, such as cis-and trans-polyisoprene or poly-2-vinylpyridine and poly-4-vinylpyridine.

Because of the difficulties associated with obtaining high resolution valence band spectra of organics with older XPS instruments, very little valence band information appears in the literature. Beamson and Briggs' recent book, "High Resolution XPS of Organic Polymers,"[8] is the best source providing both reference spectra and aids in interpretation. Table 5 summarizes the positions of major valence band lines for organics.

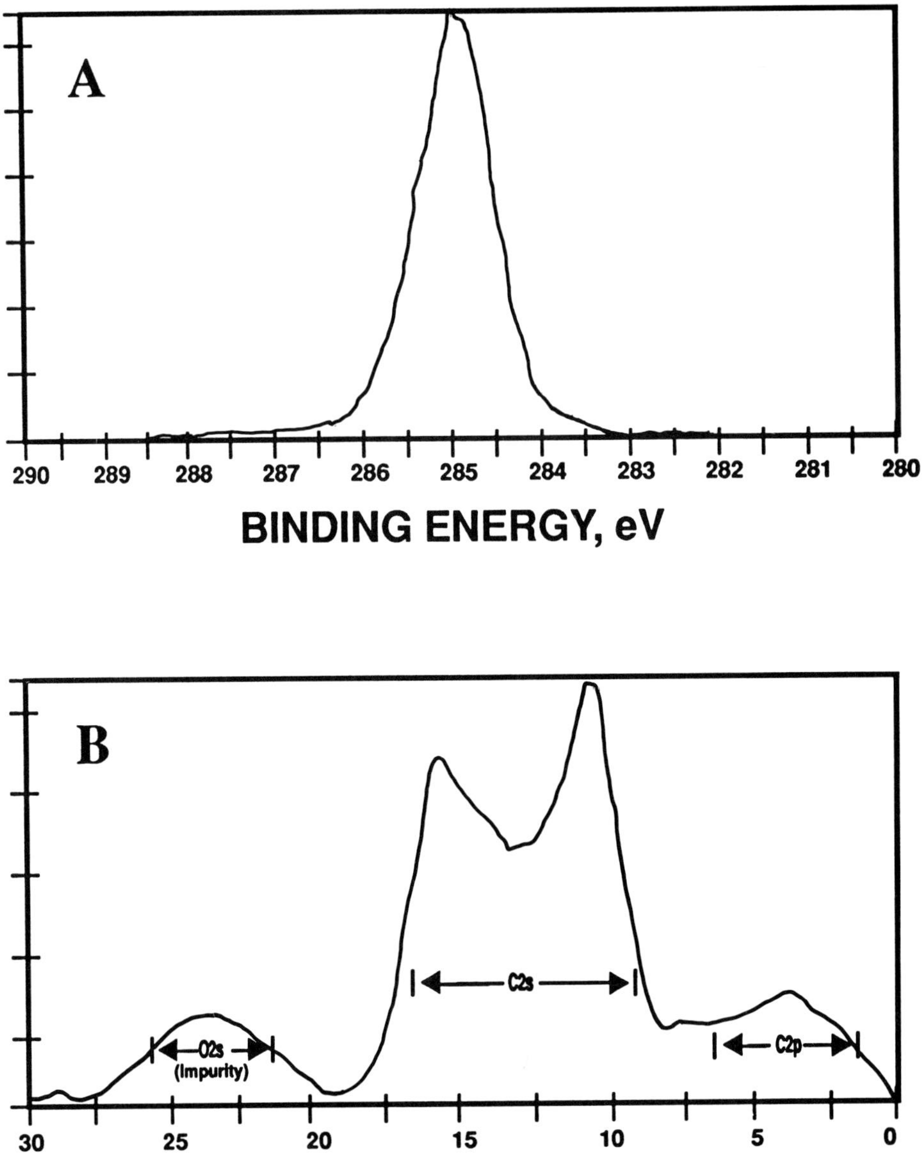

Figure 7. Polyethylene, polypropylene, and poly 1-butene have identical XPS core spectra: a single carbon 1s peak as shown in A. The three polymers have significantly different valence band spectra, however, as shown in B (polyethylene), and C (polypropylene) and D (poly 1-butene) (next page). The major differences occur in the carbon 2s region.

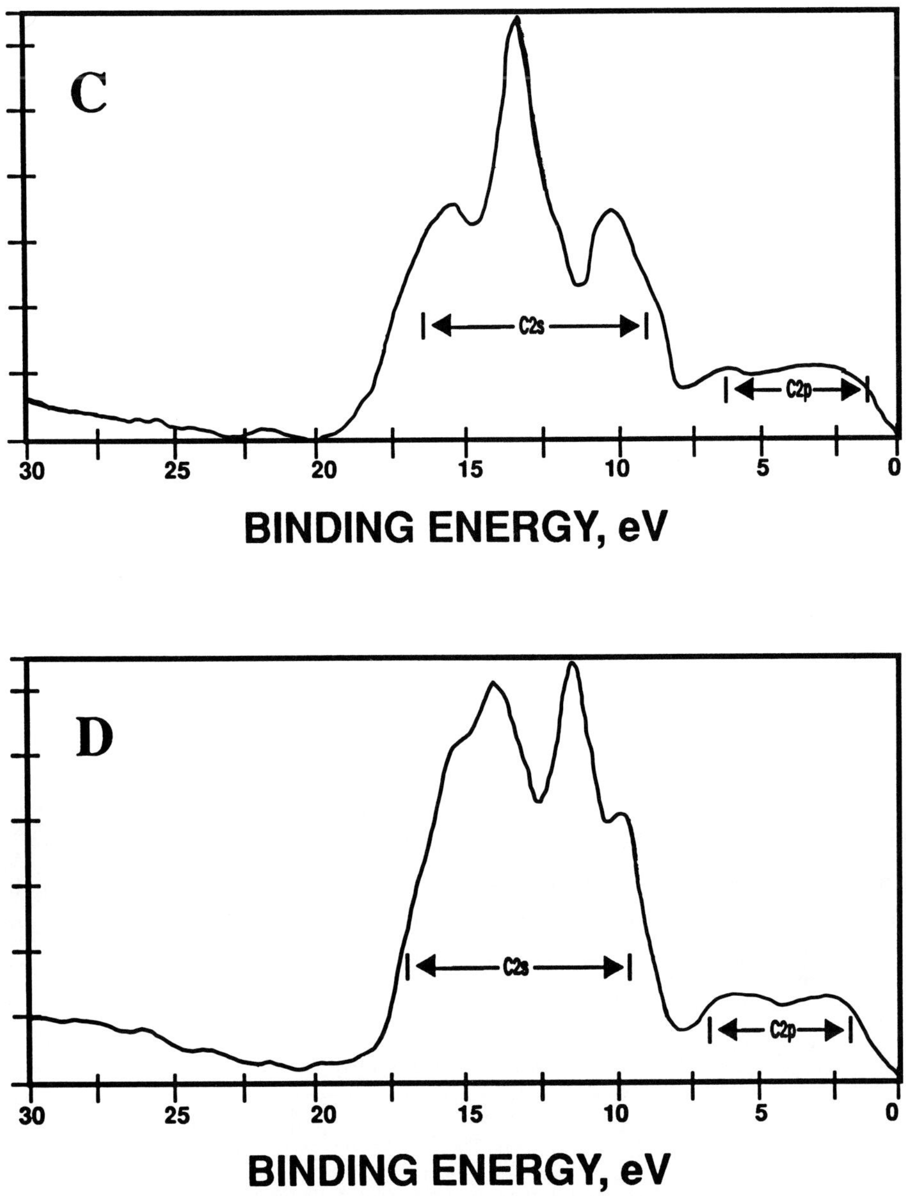

Figure 7, continued. C (polypropylene) and D (poly 1-butene). Compare to (A) and (B) on the previous page.

Table 5
Valence Band Region Peaks

Element	Orbital	Binding Energy Range (eV)
Carbon	2p	4–12
Carbon	2s	13–22
Nitrogen	2s	21–25
Oxygen	2s	26–30
Fluorine	2s	32–36
Sodium	2p	30–32
Chlorine	3p	3–6
Chlorine	3s	14–15
Bromine	4p	4–6

Binding energies relative to C1s=285.00 eV. (Data from Beamson and Briggs [8])

DEPTH PROFILING

It is often of interest to look at surface chemistry as a function of depth, as opposed to a fixed depth. There are two approaches for conducting depth profiles by XPS. For a narrow range of depth, 1–20 nm, angle resolved XPS can be used. For a wider depth range, 20–200 nm, thin layers of material are removed between spectral acquisition.

When the sample surface is exposed to x-ray photons, photoelectrons are ejected from a range of depths in the sample. By setting a narrow acceptance angle for the lens and varying the tilt of the sample surface relative to the lens, θ, it is possible to analyze those photoelectrons being emitted from a given depth. Depth into the sample varies as a function of the sine of the tilt angle, θ, and the escape depth of the photoelectron. While more rigorous equations for the analysis depth are available, a useful approximation for depth (d) dependent upon the incident source energy, hv, and the binding energy (BE) of the line being analyzed is expressed in Equation (6). This equation accounts for the depth down to which 95% of the photoelectrons are emitted.

$$d = 3(h\nu - BE)^{0.5} \sin(\theta) \tag{6}$$

Most instruments offer a range of tilt from ten to ninety degrees giving a range of depths of 1–20 nm. This techniques works well when the sample surface is very smooth. It has been used successfully with coated papers and polymer films extruded or laminated to paper. It does not work particularly well on rough surfaces such as most uncoated papers.

For wider ranges of depth, it becomes necessary to remove thin layers (about 10 nm) between spectral acquisitions. The traditional manner of doing this by XPS is to use an ion gun. An ion gun uses a beam of ionized heavy atoms (*e.g.*, Ar^{++}) at high energy to bombard the sample surface and remove the surface layer–a sort of "atomic sand-blasting." As the ion-gun is mounted in the vacuum bell, depth profiling can be carried out quickly and automatically on modern XPS instrumentation. Ion-gun sputtering has been successfully used in the depth profiling of metals and ceramics. Unfortunately, it does not work very well with organic materials. While it does remove surface layers, it often has the unwanted side-effect of reducing the underlying organic material resulting in the formation of carbides and graphite. This makes it impossible to determine the chemistry of underlying layers.

Other approaches for layer removal in organic samples have been tried with limited success. Mechanical removal of layers does work and has been demonstrated with coated paper; but it is difficult to remove very thin layers reproducibly. Dolezal[12] has demonstrated that plasma etching can be used to depth profile paper coatings without changing the chemistry of the underlying layers. Laser ablation holds some promise in this area as well, although very little has appeared in the literature to date. The difficulty with all of these approaches so far is that they must be used outside of the XPS instrument, making depth profiling very time consuming.

OTHER DATA PROCESSING TECHNIQUES

Data processing techniques and spectral analysis tools specific to XPS analysis of organics have been discussed in some detail above. In addition to these techniques, there are other useful tools which are commonly used in the analysis of spectral data. These include curve smoothing to remove noise, differentiation to detect very small peaks, and normalization, addition, and subtraction techniques to compare spectra. All of these techniques are also commonly employed in XPS analysis. As they are in general use, they will not be discussed further here.

APPLICATIONS

While XPS is being used actively in the surface analysis of paper, very little has been published in the open literature. Salvati published a general application note based upon analysis of a variety of coated and uncoated paper samples.[13] This paper compares results achieved with monochromatic versus standard x-ray sources for newsprint samples. Application of XPS to the analysis of synthetic sizing agents, alkyl ketene dimer (AKD) and allyl succinic anhydride (ASA) is discussed along with general analysis of pigment coated paper samples.

Most of the published literature concerns XPS analysis of paper coatings and general cellulosic materials. These two areas will be discussed in detail.

PAPER COATINGS

A pigment coated paper such as a lightweight coated grade (magazine paper) or coated free sheet grade makes a more ideal surface for XPS analysis than does an uncoated grade. This is due to the coated paper having a significantly smoother and more homogeneous surface. A negative drawback to XPS analysis of coated paper is that the coated surface is less conductive than an uncoated paper surface and hence more prone to charging. If the XPS analysis is performed with a monochromatic x-ray source, charge compensation must be adjusted carefully.

Takao Arai, Hiroshi Tomimasu and colleagues at Mitsubishi Paper Mills, Ltd.[14,15] conducted an extensive study of paper coatings using XPS. The study concentrated on migration of latex in both the Z–direction and the X–Y plane of the coating. Various methods were employed to detect and quantify the latex by XPS. These included: preparation of fluorine-labeled latex by copolymerizing styrene with 1,1,1-trifluoroethylacrylate, butadiene, and acrylic acid ester; labeling of the latex-coated papers with bromine or osmium by exposing the paper surface to bromine or osmium tetroxide vapor; and curve deconvolution techniques. Arai and Tomimasu concluded that analysis of model papers prepared with fluorine-labeled latex provided the best quantitative results, although application of this technique was limited.

Arai and Tomimasu conducted XPS depth profiles of coated papers using mechanical means to remove layers of coating between analyses. They concluded that latex migrates to the coating surface more readily than does starch. They also showed that the degree of migration was a function of drying conditions and the nature of functional groups on the latex. They also used XPS line scan techniques (acquiring XPS spectrum at regular intervals along a line) and correlated variations in coating composition with mottle. This study led them to conclude that "there is a close relationship between the distribution (of coating components) in the surface direction of the coated layer and mottle."

Engstrom, Norrdahl, and Strom conducted a study on drying and its effect on binder migration and offset mottling using XPS as a tool in determining the degree of binder migration.[16] They prepared latex, starch and clay-coated light weight papers on a pilot coater and used Croda ink staining coupled with image analysis to quantify the mottle. Carbon to silicon ratios from XPS spectra were used as a measure of the pigment to binder ratio at the coating surface. By varying coating and drying conditions on their pilot coater they were able to correlate coating parameters with the degree of mottle and binder migration. This study led them to two major conclusions: (*i*) mottle increases when the evaporation rate of water from the coating decreases; and (*ii*) mottle increases when the binder content in the coating surface or the degree of binder migration decreases.

H. Fujiwara of Jujo Paper Co., Ltd. and J. Kline of Western Michigan University developed a method for the quantitative determination of coating components by XPS using curve deconvolution.[17] Their method is specific to coatings containing styrene-butadiene rubber (SBR) latex, starch, clay, and/or calcium carbonate. Using a large set of model papers of various known coating compositions, they developed a set of equations for calculating the coating composition from XPS elemental data and curve-fit for the carbon 1s region. In this method, one determines the atomic percent silicon (P_{Si}) and the atomic percent calcium (P_{Ca}) from XPS quantitative analysis and the atomic percent C–C (P_{C1}), atomic percent C–O (P_{C2}), and atomic percent C=O (P_{C3}) from curve-fits of the C1s region. These values are then inserted into the equations below in order to calculate the coating composition:

$$P_P = P_{Ca} + \left(\frac{P_{Si}}{2.37}\right) \tag{7}$$

$$Clay\ (pph) = \left(\frac{P_{Si}}{P_P}\right) \cdot 100 \tag{8}$$

$$CaCO_3\ (pph) = \left(\frac{P_{Ca}}{P_P}\right) \cdot 100 \tag{9}$$

$$Latex\ (pph) = \left[\frac{(3.591\, P_{C1} - P_{C2} - P_{C3})}{(2.004\, P_P)}\right] - 2.588 \tag{10}$$

$$Starch\ (pph) = \frac{\left[\left(\frac{P_C}{P_P}\right) - 0.619\, Lx - 0.757\right]}{0.334} \tag{11}$$

where P_P is the atomic percent total carbon and *Lx* equals the pph latex from Equation (10). All results are in pounds per hundred (pph).

The method of Kline and Fujiwara works extremely well, especially when high resolution C1s spectra can be obtained using monochromatic x-ray sources, and eliminates the need for preparing labeled compounds. Table 6 gives a comparison of experimentally determined versus actual coating composition using their method.

Carol Hemminger of Union Oil of California conducted a study of wax additives to topcoat latexes.[18] The wax additives are used to improve anti-blocking and slip properties. XPS using curve fitting techniques for the carbon 1s region was used to determine the relative concentrations of latex, wax, and clay at the coating surface. The study focussed on PVDC latex with polyethylene and carnauba wax additives. XPS was used to monitor the bloom of wax to the coating surface as a function of coating composition, wax type, and process conditions.

Table 6
XPS Analysis of Coated Papers
Comparison of XPS Results vs Known Composition

Component	Pilot Sheet XPS	Pilot Sheet Known	Commercial Sheet XPS	Commercial Sheet Known
Clay	49.9	50.0	60.2	60.0
$CaCO_3$	50.1	50.0	39.8	40.0
Latex	5.0	5.0	13.5	11.0
Starch	5.3	5.0	8.4	8.4

Results in pounds per hundred (pph) calculated using the formulas of Fujiwara and Kline[17]

While the published examples of XPS analysis to coated papers have been limited to what is summarized above, extension of XPS analysis to other coating problems should be apparent. Just as Hemminger used XPS to study wax bloom to the surface, the same technique could be used to study the bloom or migration of any coating component to the surface. This could be carried out on a representative area of a coated sheet or on defect areas. Migration effects can also be studied further. A sound technique for depth profiling coatings would enhance the prospects here. Both laser ablation and plasma etching hold promise. In principle, depth profiling of coating should be far easier than depth profiling of uncoated paper as the coating makes a better substrate having a higher inorganic concentration, flatter surface, and more homogeneous composition. It should also be easier to do further work in the area of mottle with the advent of imaging and mapping capabilities on more recent XPS instrumentation.

CELLULOSIC MATERIALS

Dorris and Gray at McGill University did a series of studies on the surface analysis of paper and wood fibers using XPS in the late 1970s. In the first study,[19] they concentrated on the surface analysis of purified cotton cellulose fibers, lignin extracted from spruce, and bleached wood pulp paper samples. Both bleached kraft pulp and bleached sulfite pulp prepared from Canadian softwoods were used in this study. The focus here was to obtain XPS reference spectra for cellulose and lignin and then to compare the ratio of cellulose to lignin on the surface of kraft and sulfite pulp. Chemical shifts corresponding to the various functional groups of cellulose and lignin were measured. Carbon 1s spectra for both cellulose and lignin are shown in Figure 8. The curve fits for the cellulose spectrum (A) shows peaks due to carbon single-bonded to oxygen (C–O) and carbon double-bonded to oxygen (C=O) in a 5:1 ratio. The small hydrocarbon peak (C–C) is not to cellulose, but rather to an impurity (probably lignin or a long-chain carboxylic acid). The curve fits for the lignin spectrum (B) shows a large hydrocarbon peak and a smaller carbon bonded to oxygen peak. The second portion of the study[20] involved XPS analysis of groundwood pulps. A series of equations for calculating the relative concentrations of cellulose and lignin of the pulp surface are presented. These

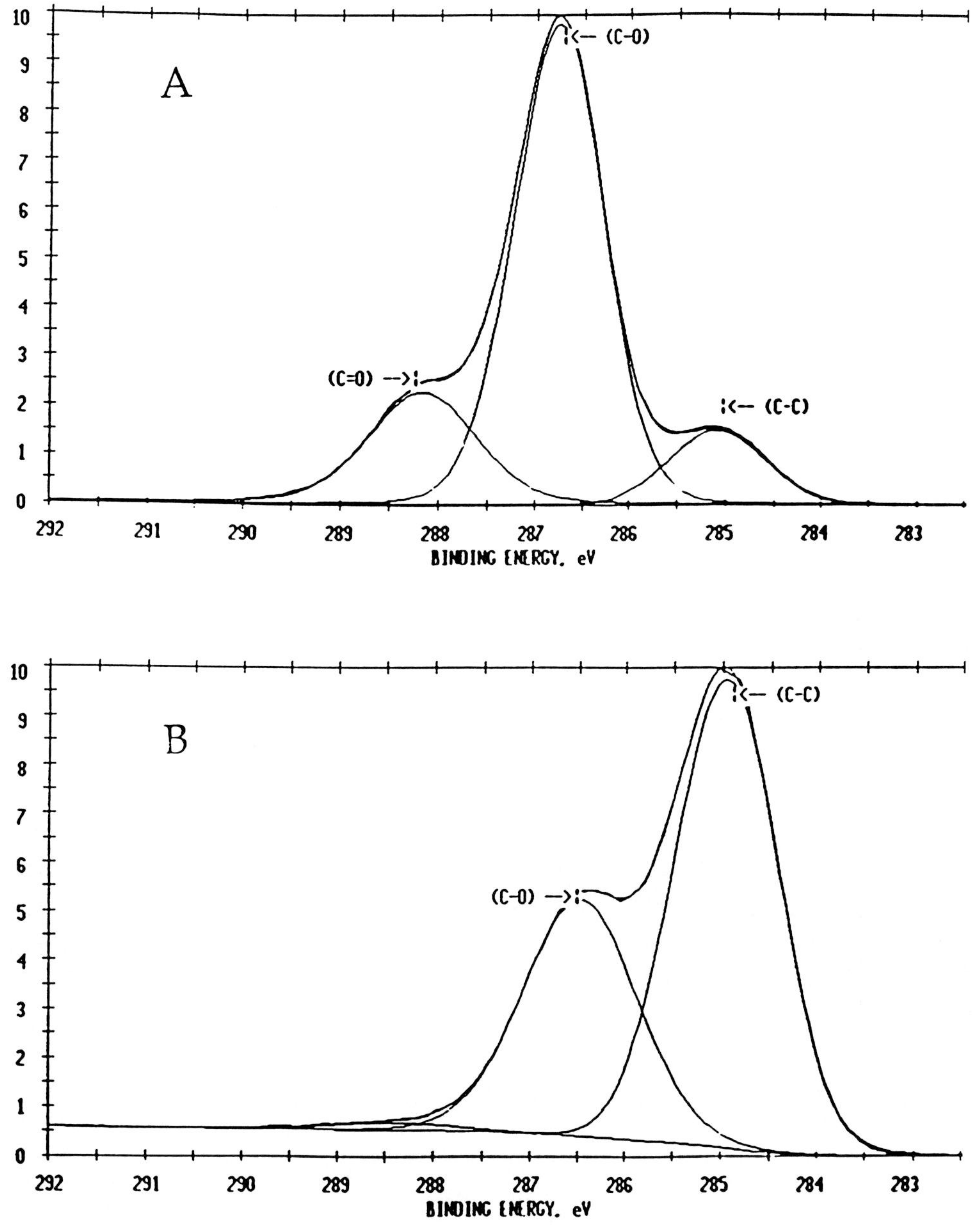

Figure 8. XPS high resolution carbon 1s spectra for cellulose (A) and lignin (B).

equations were used in conjunction with experimental XPS data to determine the lignin and cellulose concentrations on the surface of stone groundwood (SGW), thermo-mechanical pulp (TMP) and refiner groundwood (RGW). In the third and final portion of the study,[21] curve deconvolution techniques were used on the carbon 1s region to further refine the interpretation and equations developed in the first two papers.

In a more recent paper, Ahmed *et al.*,[6] of the Universite Laval, also conducted an thorough XPS study related to specifics of cellulose chemistry. In this study, it is shown that long exposure to x-rays and heat results in conversion of cellulose into an enone as shown in Figure 9. This work makes an excellent case for the need to use monochromatic x-rays and short analysis times in order to avoid this situation. The study continues to examine the XPS spectra of modified celluloses such as hydrocellulose and oxycellulose.

While both studies were limited to basic XPS spectral studies of pulp components, they nonetheless demonstrate the utility of using XPS to study the surface chemistry of pulps. These same techniques could be used in investigating the effects of changes in pulping and bleaching sequences on the surface chemistry of pulp. Information of this sort would be useful in predicting the effects of these changes on the behavior of pulps in the papermaking process. Studies of this kind have been done, although none have been published in the open literature.

PAPERMAKING ADDITIVES

XPS has been successfully used in monitoring the concentration of papermaking additives on the surface of uncoated paper and in examining the interaction of papermaking components. J.S. Brinen *et al.*[22] of American Cyanamid used XPS to monitor the concentration of a synthetic sizing agent, HABI, on paper surfaces. HABI, N,N,o-tris(octadecylcarbamoyl) hydroxylamine, is an experimental alkaline size used in alkaline papermaking. Brinen *et al.* were able to establish linear calibration curves relating the surface concentration of HABI on paper to both C1/O and N/O ratios as determined by XPS. A sample of the calibration curves obtained along with the structure of HABI are shown in Figure 10. Brinen *et al.* use this method to examine the surface concentration of HABI under different rates and levels of addition. They conclude that HABI forms "islands" on the paper surface as opposed to a continuous film and that no chemical reaction occurs between HABI and cellulose. While HABI has not become a commercially used paper sizing agent, the techniques which Brinen et al. developed could be applied to commonly used synthetic sizing agents such as alkyl ketene dimer (AKD) andallyl succinic anhydride (ASA) as well as to size press additives such as styrene maleic anhydride (SMA), styrene acrylic acid (SAA), and polyurethanes.

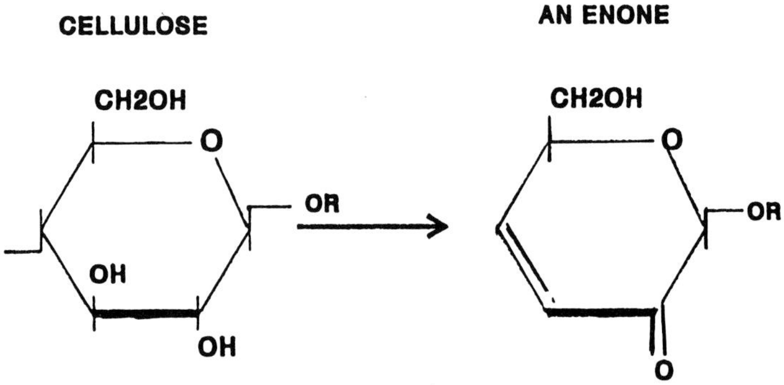

Figure 9. Ahmed *et al.*'s proposed reaction[6] for the conversion of cellulose to an enone as the result of exposure to x-rays.

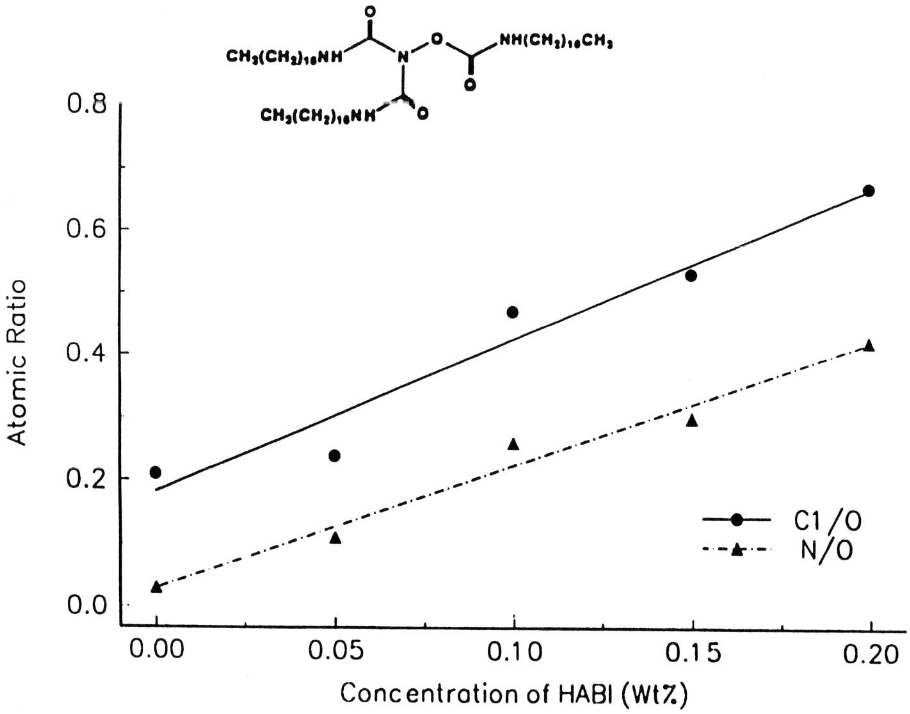

Figure 10. Calibration curves for the determination of the surface concentration of HABI size using XPS data from the carbon, nitrogen and oxygen regions. Data from Brinen *et al.*[22]

In a study conducted by the Empire State Paper Research Institute (ESPRI), Pyda *et al.* used XPS and other methods to study interactions between alkyl ketene dimer (AKD) and calcium carbonate fillers.[23] XPS was used to determine the orientation of AKD molecules on the surface of calcium carbonate. They determined that the AKD molecule was fixed in a perpendicular conformation to the calcium carbonate filler particle surface. This study points out an interesting use of XPS – the ability to determine molecular orientation of macromolecules and polymers.

Farrow *et al.* used XPS to study rosin size on the surface of acid xerographic papers.[24] Algorithms were empirically established to relate the atomic concentration of oxygen and various carbon 1s regions to surface concentrations of anhydroglucose (cellulose or starch) and hydrocarbon (rosin). Surface compositional data for a series of papers (commercial and laboratory) were related to the ability to achieve adequate toner adhesion in a xerographic type copier or printer. They concluded that there was a relationship between the surface concentration of hydrocarbon and toner adhesion.

SURFACE DEFECTS

Two general types of surface defects are observed on paper: visible (marks, discolorations, spots, etc.) and invisible (changes in coefficient of friction–slip, surface energy–wetting, etc.). XPS, used alone or with other surface characterization tools (SIMS,

FTIR, contact angle measurements, SEM, etc.) can be used to determined the location and chemical nature of these defects.

Most papers contain a surface coating, either a size press coating of starch or a pigmented coating. These coatings are thin enough to be translucent so that it is often difficult to determine whether a visible defect is on the surface of the coating or between the fiber surface and the coating. XPS can be used to determine the location of the defect. If the defect is on the surface, the XPS spectra for defect and non-defect areas will different. If the defect is beneath the coating, then the XPS spectra of the two areas will be identical. Defects of these types are usually caused by contamination on rolls, dryer cans, or fabric of the paper machine transferring to the sheet or by contaminants falling onto the sheet. Examples of contaminants include: corrosion products (metal oxides), lubricants (lithium greases, hydrocarbon oils and greases, graphite), paint chips, dried papermaking components (coatings, starch, clay, etc). XPS can easily and quickly distinguish between these general types of contaminants. Knowing the location within the sheet (on or under the coating or size press coating) and the nature of the contaminant can make it easy to eliminate the problem on the paper machine. This has been recently demonstrated by the author.[25]

Invisible surface defects can also be characterized by XPS by comparing spectra of "good" versus "bad" paper samples. One example of this type of invisible defect is changes in the coefficient of friction (COF). Coefficient of friction of paper is an important parameter in high-speed converting operations (corrugating, envelope manufacture, etc.) and high-speed printing operations where abnormal slip of the paper web or sheets can result in jamming of the equipment or poor registration. PAPRICAN[26] conducted a study in which XPS data was compared with changes in COF on newsprint samples. They found a correlation between the hydrocarbon content of the surface as determined by XPS and the measured COF of the paper. A sample of their data is shown in Figure 11.

Another example of invisible surface defects on paper is variation in surface energy. Low surface energy can contribute to low COF. Low surface energy can also contribute to poor wetting and adhesion by ink, toner, and adhesives. Surface energy considerations are becoming more critical as the printing industry shifts to water-based inks which tend to be more sensitive to surface energy considerations. Once again, XPS is a useful tool for investigating the cause of low surface energy. Studies similar to the ones done by Brinen (on sizing), Farrow (on toner adhesion), and PAPRICAN (on COF) can be done to evaluate the effect of various papermaking additives on surface energy with XPS being used to evaluate the propensity of these materials to migrate to the surface.

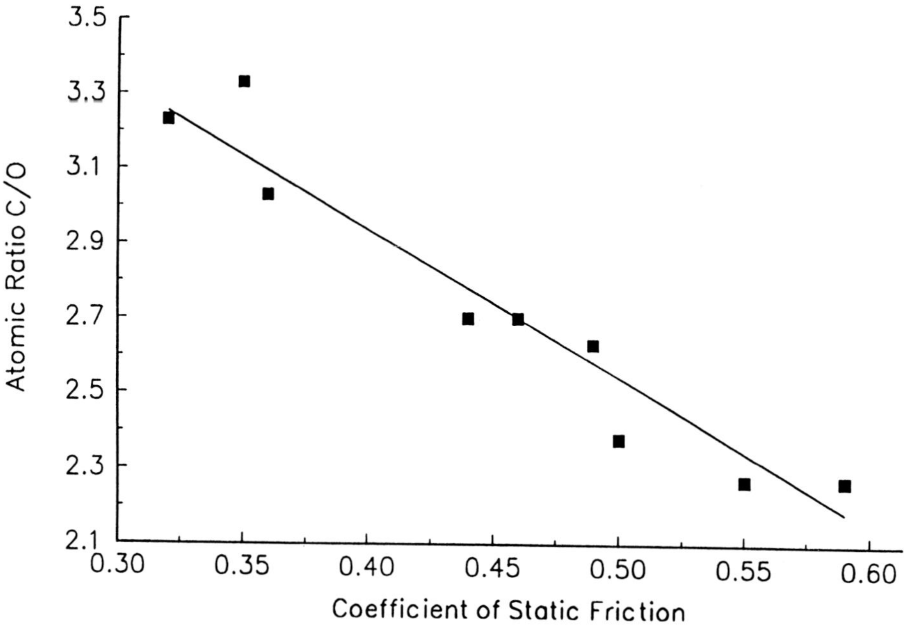

Figure 11. The effect of surface carbon content (as determined by XPS) on the static coefficient of friction. Data from Gurnagul *et al.*[35]

ADDITIONAL APPLICATIONS

The applications of XPS to studying paper surface chemistry as outlined above represents an up-to-date cross section of the open technical literature on the subject. Considerable work in this area has been done and is being done that has not yet appeared in publication; it is reasonable to expect to see an increase in publication of articles on this subject in the near future. A considerable volume of work has been published on applications of XPS in studying polymer and bio-materials surfaces. Many of the techniques discussed in these publications are readily applied to the study of paper and paper making materials.

REFERENCES

1. Siegbahn, K., Nordling, C., Fahlman, A., Nordberg, R., Hamrin, K., Hedman, J., Johansson, G., Bergmark, T., Karlson, S.E., Lindgren, I., and Lindberg, B. 1967. ESCA – Atomic, molecular and solid state structure studied by means of electron spectroscopy. *Nova Acta Regiae Societatis Sci. Upsallensis* 4(20):1–282.
2. Andrade, J. D. 1985. X-ray photoelectron spectroscopy (XPS). *In*: Surface and Interfacial Aspects of Biomedical Polymers, Vol. 1. Andrade, J. D., ed. Plenum Press, New York and London. Chap. 5.

3. Brundle, C. R., Evans, C. A. and Wilson, S. 1992. Encyclopedia of Materials Characterization. Butterworth-Heinemann, Boston.
4. Einstein, A. 1905. *Ann. Physik* 17: 132.
5. Wagner, C. D., Riggs, W. M., Davis, L. E., Moulder, J. F., Muilenberg, G. E., ed. 1979. Handbook of X-ray Photoelectron Spectroscopy. Perkin-Elmer Corporation, Eden Prairie, Chap. 2.
6. Ahmed, A., Adnot, A., Grandmaison, J. L., Kaliaguine, S. and Doucet, J. 1987. ESCA analysis of cellulosic materials. *Cellulose Chem. Technol.* 21: 483–492.
7. Wagner, C. D. and Bickman, D. M. 1989. NIST X-ray Photoelectron Spectroscopy Database, Version 1. NIST Standard Reference Database 20. National Institute of Standards and Technology (NIST), Gaithersburg, MD.
8. Beamson, G. and Briggs, D. 1992. High Resolution XPS of Organic Polymers. John Wiley & Sons, Chichester.
9. Scofield, J. H. 1976. Photoionization cross sections at 1254 and 1487 eV. *J. Electron Spectrosc.* 8:129–137.
10. Reilman, R. F., Msezane, A. and Manson, S. T. 1976. Relative intensities in photoelectron spectroscopy of atoms and molecules. *J. Electron Spectrosc.* 8: 389–394.
11 Wagner, C. D., Davis, L. E. and Riggs, W. M. 1980. Energy dependence of the electron mean free path. *Surface Interface Analysis.* 2:53–55.
12. Dolezal, H. M. 1993. Personal communication.
13. Salvati, L. 1990. XPS provides unique characterization of paper. *PHI Interface.* 12(4):10–14.
14. Arai, T., Yamasaki, T., Suzuki, K., Ogura, T. and Sakai, Y. 1988. Relationship between coating structure and print mottle. Tappi Coating Conference Proceedings pp. 187–192.
15. Arai, T. and Tomimasu, H. 1989. Measuring method of chemical distribution in coated layer by ESCA. *Japan Tappi.* 43(2):55–67.
16. Engstrom, G., Norrdahl, P. and Strom, G. 1987. Studies of the drying and its effect on binder migration and offset mottling. Tappi Coating Conference Proceedings. pp. 35–43.
17. Fujiwara, H. and Kline, J. E. 1986. ESCA Analysis of pigment coated paper. 1986 International Process and Materials Quality Evaluation Conference: Tappi Proceedings pp. 157–167.
18. Hemminger, C. S. 1985. ESCA analysis as a tool to study surfaces of treated paper. Tappi Coating Conference Proceedings pp. 55–61.
19. Dorris, G. M. and Gray, D.G. 1978. The surface analysis of paper and wood fibres by ESCA. I. *Cellulose Chem. Technol.* 12:9–23.
20. Dorris, G. M. and Gray, D.G. 1978. The surface analysis of paper and wood fibres by ESCA. II. Surface composition of mechanical pulps. *Cellulose Chem. Technol.* 12: 721–734.

21. Dorris, G. M. and Gray, D.G. 1978. The surface analysis of paper and wood fibres by ESCA. III. Interpretation of carbon (1s) peak shape. *Cellulose Chem. Technol.* 12:735–743
22. Brinen, J. S., Calbick, J. C. and Cody, R.D. 1989. Quantitative determination of alkaline sizes on paper surfaces by ESCA, Part I. HABI. *Surface Interface Analysis.* 14:245–249.
23. Pyda, M., Sidqi, M., Keller, D. S. and Luner, P. 1993. An inverse gas chromatographic study of calcium carbonate treated with alkylketene dimer. *Tappi J.* 76(4): 79–85.
24. Farrow, M. M., Miller, A. G. and Walsh, A. M. 1982. Surface chemistry of business papers: ESCA studies. *Reprographic Technology*. American Chemical Society. Chap. 23.
25. Istone, W. K. 1995. Static secondary ion mass spectrometry and x-ray photoelectron spectroscopy for the characterization of surface defects in paper products. *J. Vac. Sci. Technol. A.* 12(4):2515–2522..
26. Gurnagul, N., Wearing, J. T., Ouchi, M. D., Sparkes, D. G. and Dunlop-Jones, N. 1991. Factors affecting the coefficient of friction of paper. *PAPRICAN Pulp and Paper Reports*. PPR 871.

ADDENDA

NOTES ON SPECTRA

All XPS spectra presented in this chapter were generated by the author on the Perkin-Elmer PHI–5500 Multi-Technique System at Champion International Corporation in West Nyack, NY.

GENERAL REFERENCES

Following are some reference works that the author has found especially useful in conducting XPS analysis on paper and other polymeric materials:

Carlson, T. A. 1975. Photoelectron and Auger Spectroscopy. Plenum Press, New York and London. (An excellent work on the fundamentals on XPS).

Andrade, J. D. (ed.) 1985. Surface and Interfacial Aspects of Biomedical Polymers, Volume 1 Surface Chemistry and Physics. Plenum Press, New York and London. (Discussed surface analysis on polymers and bio-materials).

Brundle, C. R., Evans, C. A. and Wilson, S. 1992. Encyclopedia of Materials Characterization. Butterworth-Heinemann, Boston. (Describes various materials and surface analysis techniques.)

Beamson, G. and Briggs, D. 1992. High Resolution XPS of Organic Polymers. John Wiley & Sons, Chichester. (Contains reference XPS spectra of polymeric materials).

Wagner, C. D., Riggs, W. M., Davis, L. E., Moulder, J. F., Muilenberg, G. E. (eds.) 1979. Handbook of X-Ray Photoelectron Spectroscopy. Perkin-Elmer, Eden Prairie. (Reference spectra and tables for inorganic materials).

Wagner, C. D. and Bickman, D. M. 1989. NIST X-ray Photoelectron Spectroscopy Database, Version 1. NIST Standard Reference Database 20. National Institute of Standards and Technology (NIST), Gaithersburg. (Searchable database for MS-DOS computers of XPS data).

12

Principles of Electron Energy Loss Spectroscopy and its Application to the Analysis of Paper

Gianluigi Botton

University of Cambridge, Cambridge, U.K.

INTRODUCTION

In transmission electron microscopes (TEM) high energy electrons pass through a thin sample (a cross-section in the case of paper analysis) and as a result of interactions with the solid lose a fraction of their incident energy. Using an electron energy loss spectrometer (EELS) the transmitted electrons are dispersed according to their energy and their energy loss distribution retrieved. By analyzing the shape and intensity of the distribution, information is obtained about the chemical composition and electronic structure of the elements present so that their chemical state can be derived. The signal in an EELS spectrum is very high because a primary event is analyzed, as compared to x-ray and Auger electron spectroscopy where the fluorescence yield reduces the probability of an event. It is for this reason that EELS in the electron microscope has rapidly evolved since its initial development in the early 1930s as a very efficient analytical method, and it was for many years the only technique for the detection of light elements such as B, N, C and O.

The purpose of this chapter is to give a brief introduction to the technique by presenting some background information on the physical processes involved, the relevant instrumentation required, the available methods of quantification for chemical analysis, the origins of spectroscopic information and some examples of applications relevant to paper science. Since detailed reviews of the technique can now be found in the literature[1,2,3], we will concentrate on the basic background of the technique and illustrate the application of the method to problems related to paper science so that potential users can judge the effectiveness of EELS in solving particular characterization problems.

0-8493-8992-5/95/$0.00+$.50

A distinction should first be made between the surface-sensitive EELS carried out in ultra high vacuum dedicated surface analysis systems, and the related technique described in this chapter, usually carried out in transmission electron microscopes (TEM). Both methods are based on the same physical principles, but in the latter case the analysis is spatially resolved to a few nanometers and the information is provided from the thickness of the cross-section, not only from the sample surface.

BASIC PRINCIPLES

In a TEM when high energy electrons (typically 100–300 keV) pass through a thin sample (thickness ≈ 100 nm), various types of interactions occur with the atoms and the electrons of the materials. The scattering processes which result from this interaction are generally divided into two classes, elastic and inelastic scattering, defined by the magnitude of the energy lost by the incident electron. As these scattering processes are represented in different portions of the EEL spectrum, we will describe them briefly in this section.

ELASTIC SCATTERING

In elastic scattering, the incident electron trajectory is modified by the electrostatic potential of the nucleus by Coulomb interactions (also known as Rutherford scattering) and its energy is unchanged. Other processes for which the energy lost by the electron cannot be practically measured using current technology are also, in general, classified as elastic. For example, interactions with lattice vibrations (phonons) in which the energy lost is in the order of few meV are considered "quasi-elastic" and fall into this class of scattering. Mathematically, the elastic scattering can be described by a cross-section $\sigma(\theta)$ calculated from first principles that predict the probability of an electron being scattered at an angle θ (Figure 1). Related to this quantity, a mean free path λ_o can be used to represent the (mean) distance travelled in the solid between consecutive scattering events of the incident electron. With 100 keV incident electrons and for a low atomic number element such as carbon, λ_o is in the order of 200 nm, but for higher Z elements it is significantly lower (for Cu $\lambda_o \approx 20$ nm).[4] It is important to note that in solids, the angular distribution of scattered electrons cannot be solely described by $\sigma(\theta)$, which is derived for single isolated atoms as the periodicity of the lattice gives rise to electron diffraction. The angular distribution is therefore modified to include Bragg reflections. However, in amorphous solids the atomic $\sigma(\theta)$ can be considered to be approximately representative of the scattering distribution.

INELASTIC SCATTERING

Of fundamental importance to EELS are the inelastic scattering processes which result from interactions of the incident beam with electrons which are, to a different degree, bound to atoms or "shared" by atoms in the solid. Different processes can be distin-

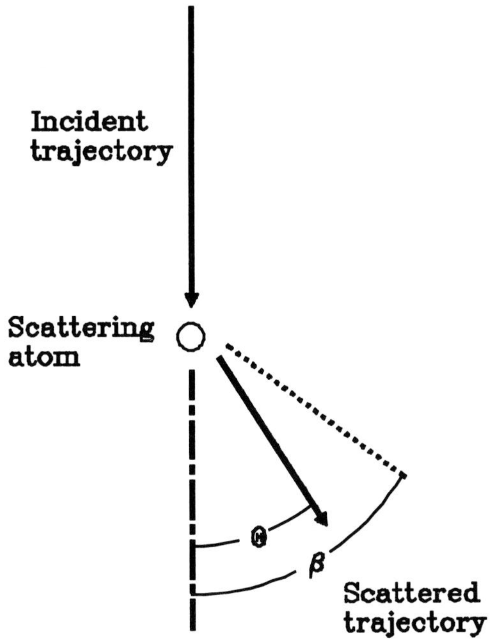

Figure 1. Electron scattering diagram showing incident and scattered trajectory of electron. θ is the scattering angle and β is the limiting collection angle of the spectrometer.

guished depending on whether one or several electrons (collectively) are involved during the interaction with the fast electron, but in all cases, the energy lost by the incident electron measured during the EELS experiment corresponds to the energy gained by one or more electrons in the solid. In single electron excitations, the fast electron causes transitions from specific occupied molecular or atomic levels into other unoccupied levels or into the continuum states. For example, ejection of an electron from atomic level (1s) in carbon will result in an absorption edge at 284 eV in the EELS spectrum (EELS spectroscopists will label this edge as a C K edge since it arises from a n=1 (1s) quantum level; similarly, L edges arise from n=2 (2s and 2p) levels, etc). Fast electrons can, however, lose a fraction of their energy from interactions with the electrons shared in the solid (the valence electrons) which react to the fast incident impulse collectively. These excitations are known as plasmons. Plasmon peaks in metals have their origin from oscillations of the electron gas which is relatively "free" to move in the lattice. In organic materials (and more generally insulators), plasmons are understood by considering oscillations of electrons in valence orbitals (*e.g.*, π and σ orbitals in C=C bonds). The position of these plasmon peaks is related to the free electron density and their width to the damping of these oscillations in the lattice. These various inelastic processes are observed in different portions of the EEL spectrum as described in Figure 2. Mathematically, the energy-loss spectrum, which is dominated by low loss events in the plasmon region, is related to the imaginary part of the dielectric function of the material $\epsilon(q,E)$ by[5]

$$Spectrum \propto Im\left(\frac{1}{\epsilon(q,E)}\right) \quad (1)$$

Each interaction process is also described by an angular and energy-dependent cross-section contributing to ϵ where q (the momentum transfer which is related to the scattering angle θ) and E (the energy loss) represent the angular and the energy dependence of the scattering process respectively. The spectrum is therefore representative of the electronic properties of the material. From a low loss spectrum it is possible via Kramers–Kronig analysis to retrieve the imaginary and real parts of the dielectric function.[5] For each specific process "i" (single electrons from different energy levels, for plasmon excitation etc.) and for the total inelastic scattering, a mean free path (λ_i or λ_T respectively) can be described as has been done for the elastic scattering case. For C, the total inelastic mean free path (involving the probability of an electron losing some energy) is 80 nm at 100 keV. We will see later how this parameter can easily provide the sample thickness through spectrum analysis.

THE ENERGY LOSS SPECTRUM

In a schematic diagram of the different portions of a typical spectrum it is possible to represent the energy range of the elastic and various inelastic processes described above with the respective molecular or atomic transitions (Figure 2). It is important to note the large dynamic intensity range present in the spectrum, as the probability of ionization for a core level is much smaller than that of the excitation of a plasmon.

INSTRUMENTATION

SPECTROMETER

In order to record the energy loss spectrum in a TEM, the electrons transmitted through the sample are dispersed according to their energy by a spectrometer. The intensity at a particular energy loss is recorded using either serial (*i.e.*, channel after channel) or parallel (*i.e.*, all channels acquired simultaneously) detection. Several types of spectrometers exist, but this chapter will concentrate on the most common post-column magnetic prisms which can be fitted to existing TEMs.

After interacting with the specimen, electrons travel through the magnetic lenses of the microscope and are projected onto the fluorescent viewing screen where images are usually observed. The prism is added to the system after the viewing chamber, and the magnetic field of the prism bends the trajectory of the electrons entering the spectrometer with a radius related to their velocity /energy. Electrons having suffered an energy loss will be focused at different positions (ΔX) in the dispersion plane with respect to electrons that have not suffered any energy loss (Figure 3). In the serial detection approach, spectra can be recorded by electrostatically or magnetically shifting the

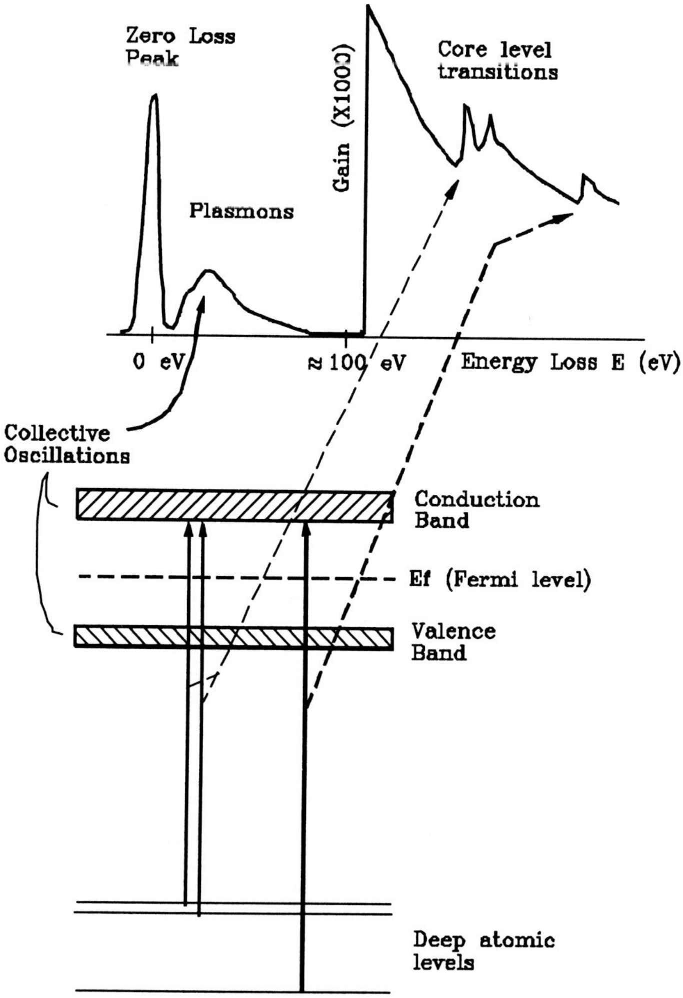

Figure 2. Energy loss spectrum and associated excitations in an energy diagram.

dispersed spectrum in front of an energy selecting slit which is followed by a scintillator and a photomultiplier. The intensity at that energy is recorded in one channel of memory of the multichannel analyzer (MCA). With such a system, it is possible to deal with the large dynamic range of spectra by switching the counting mode from voltage-to-frequency conversion (for the high intensities in the low loss region of the spectrum) to single electron counting (in the core loss energy region). Furthermore, gain changes are available in the spectrometer electronics to increase sensitivity. Acquisition times can be changed for any particular channel. The spectrum acquisition time is determined by the dwell time at each channel and the number of acquired channels. For core edges (*e.g.*, C K) and typical microscope conditions, the recording time is in the order of a minute.

Recently, parallel detection spectrometers have been developed[6] which record the energy distribution in parallel (all energies, 1024 channels, at the same time). The

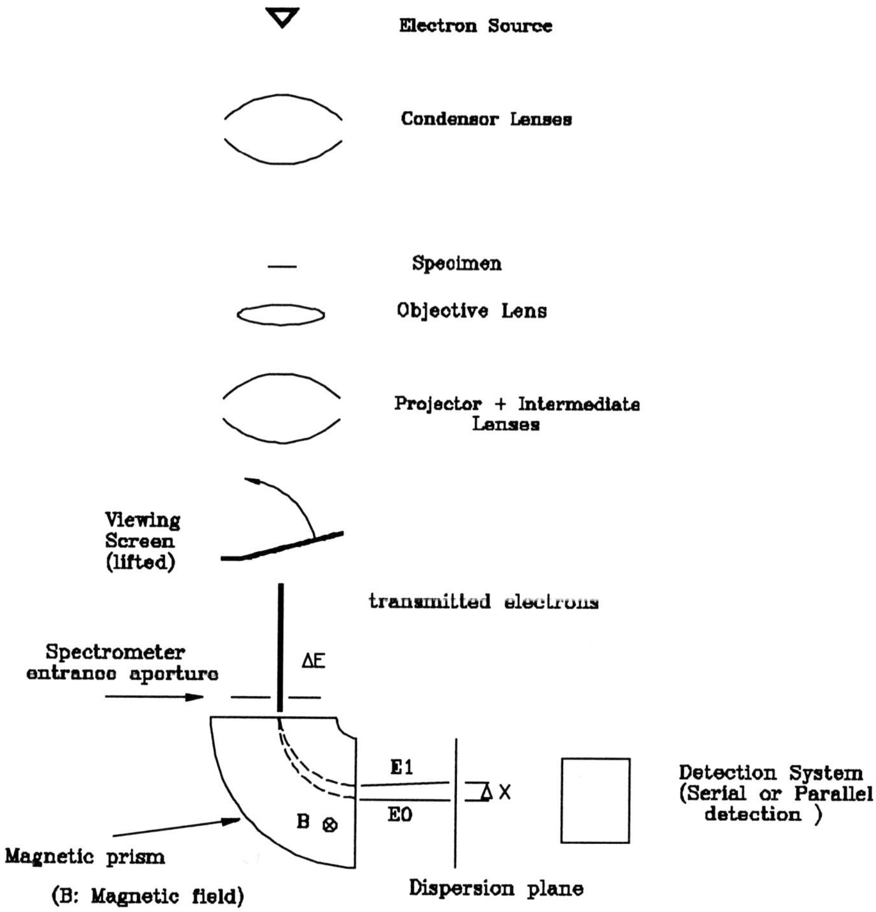

Figure 3. Schematic diagram of the TEM and EELS spectrometer system.

dispersed electrons illuminate a scintillator, the emitted light is recorded by a charge coupled device (CCD) detector, and the intensity is successively read and displayed on the MCA. The spectrum acquisition time is determined for a desired energy window by the dwell time of each channel (all channels are exposed simultaneously). Spectra can, therefore, be obtained in times as short as 50 to 100 msec. The dynamic range is a limitation, however, given that the detector saturates at 16K counts and it is not possible to change the gain at any one particular channel. If several edges at significantly different energies are of interest, several spectra of different acquisition times have to be acquired in order to optimize the counting statistics. These conditions have to be used to record core loss and low loss spectra, and an effective gain change is calculated in the overlapping regions of the two spectra.

The significant reduction in acquisition times with parallel detectors decreases radiation damage to the sample caused by extensive exposure and improves the spatial resolution as sample drift is decreased. Because of the reduced instabilities of the energy source and electronics during the shorter acquisition time and the parallel recording, energy drifts in the spectra are eliminated and the resolution improved. An exhaustive discussion of the optimized operation and evaluation of the only commercially available parallel electron energy loss spectrometer (PEELS model 666 by Gatan Inc.) has been presented by Egerton *et al.*[7] Other types of commercially available spectrometers (*e.g.*, the Castaing–Henry and the Castaing–Senoussi (also known as Ω) filters) are used in dedicated microscopes (the EM902 and EM912 models manufactured by Zeiss Inc.) as they are integrated into the column. As the prism is in the column, it is possible with such instruments to obtain energy filtered images and thus to map the chemical distribution of elements.[8] Similar systems but with post column filters allowing chemical mapping have been recently been developed.[9] Mapping with parallel detectors has been presented by Hunt and Williams[10] and Botton and L'Esperance.[11]

ENERGY RESOLUTION

The energy resolution measured by the full-width-at-half-maximum of the zero loss peak is in part determined by the energy spread of the electron source, the instabilities of the microscope–spectrometer system and the spectrometer resolution. In conventional TEMs equipped with thermionic sources (*e.g.*, LaB_6) the resolution can be as low as 0.9 eV if care is taken to eliminate all sources of stray magnetic fields which interfere with the operation of the system.[7] Generally, values of 1.0 to 2.5 eV are obtained depending on the saturation of the filament, extraction voltage and voltage of the microscope. With cold field emission guns (FEGs) available in dedicated scanning transmission microscopes (STEMs) and in the recently developed FEG–TEMs resolution can be as low as 0.35 eV, which is close to the energy spread of the electron source. High energy resolution is required for the fine structure analysis of the edges but should not be of great concern for routine chemical analysis.

MICROSCOPE–SPECTROMETER COUPLING

As presented in Figure 3, the spectrometer is situated after the microscope's projection chamber. In these conditions, either an image or a diffraction pattern is present on the viewing screen depending on the selection made by the operator. For these two modes, the area from which the spectra are obtained and the collection conditions (the limiting angle of acceptance of the spectrometer) are determined differently. In the image mode, spectra are collected from the area on the viewing screen (magnified by the projector lenses) equal to the diameter of the spectrometer entrance aperture. The collection angle is determined by the size of the objective aperture that may be used to observe the image. Because of chromatic aberrations, however, the area analyzed is generally larger and is influenced by the chromatic aberration coefficient and the energy loss.[12] In the diffraction mode, which is generally preferred to avoid the aberration effects mentioned, the analyzed area is selected either by focusing the probe onto the area of interest or by using a small selected area aperture; the collection angle is determined by the portion of the diffraction pattern (dependent on the camera length) entering the spectrometer aperture.

In modern analytical TEMs, probe sizes of 5 nm can be obtained with enough current (0.05 mA) to acquire spectra with sufficient signal for low energy edges (*e.g.*, Al L) at short acquisition times (in the order of 100 msec).[11] In the recently developed FEG–TEMs (cold, thermally assisted or Schottky sources) or dedicated STEMs, probe sizes of 1 nm with 1 nA current can be achieved. For the same signal (for example, 1 nA) these instruments allow an improvement in resolution of about 30 times.[11]

With the energy distribution is also associated an angular distribution of scattering as indicated in Equation (1). At energies just above the ionization threshold, the distribution is forward peaked while at energies considerably higher than the edge (or plasmon peak) the distribution is peaked at a non-zero angle (see reference 12, page 44). Given the different behaviors of the background and the edge angular distributions, optimization of the collection angles to increase the signal to noise ratio should be carried out with reference to the full-width-at-half-maximum of the angular distribution θ_E which is proportional to the energy loss E. Typical collection angles used are in the order of 10 mrad. A more detailed analysis of the optimization of the signal is beyond the scope of this chapter and interested readers are referred to Egerton[1,12] for more details.

SAMPLE PREPARATION AND THICKNESS

The relatively small mean free path for inelastic scattering imposes stringent conditions on the samples to be analyzed, as the visibility of the edges with respect to the background is impaired and quantitative chemical analysis affected if these conditions are not respected.[13] Generally, sample thickness (t) relative to the mean free path λ for total inelastic scattering should be guided by $t/\lambda \leq 0.5$ to avoid any difficulties caused by multiple scattering events. The value of t/λ can be evaluated from a low-loss spectrum by[1]

$$\frac{t}{\lambda} = \ln\left(\frac{I_t}{I_0}\right) \tag{2}$$

where I_t and I_0 are the intensities of the entire spectrum (including the zero-loss peak) and the zero-loss peak, respectively, as shown in Figure 4. Similarly, if the mean free path is known, the sample thickness can be estimated using Equation (2).

Ultramicrotomy is a suitable sample preparation technique whereby wood pulp fibers or paper are embedded in epoxy resin and thin sections are cut using a diamond or a glass knife.[14] Embedding of paper in vacuum using a low viscosity medium has proven to

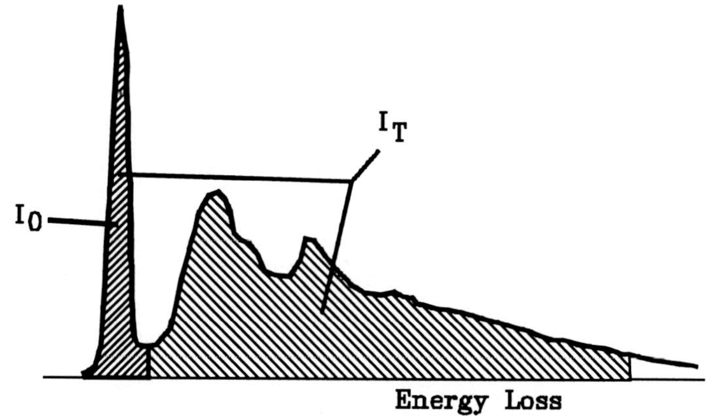

Figure 4. Low energy loss spectrum. I_o and I_t are the intensities used in the calculation of the relative thickness t/λ.

be very efficient as discussed by de Silveira.[15] Using this method, sections of about 50 nm thickness were obtained over large areas (several microns) showing cross-sections of fibers and the coating layer, if present (Figure 5). Sections are usually supported on a colloidal carbon film held on a copper microscope grid. Other techniques for sample preparation are also available[16,17] but are not well suited for paper analysis. If powders are analyzed, a dispersion of the powder in methanol can be dropped on a carbon film supported on a copper grid. The use of electrons of higher accelerating voltages (300–400 keV) can alleviate the difficulty in preparing thin sections as λ increases with increasing voltage. Instruments capable of these accelerating voltages are, however, considerably more expensive and samples often become sensitive to knock-on beam damage as is the case for some oxides and sensitive organic and inorganic polymers.

CHEMICAL ANALYSIS

The signal of a particular ionization edge "a" (S_a) is related to the number of atoms of the element "a" (N_a) irradiated by the electron beam, the ionization cross-section (σ) and the incident number of electrons (I):

$$S_a = N_a \sigma\,(\Delta,B)\,I \tag{3}$$

where $\sigma(\Delta,\beta)$ is a function of experimental parameters, as will be seen below. By extracting the signal of an edge from the preceding background, it is possible to determine the absolute number of atoms contributing to the signal. The cross-section $\sigma(\Delta,\beta)$ is dependent on the collection angle β and the width Δ of the integration window of the signal, and can be calculated using computer programs based on hydrogenic models.[1] Short computer codes exist to calculate cross-sections of K and L edges and can easily be run on personal computers. Generally, commercial spectrum acquisi-

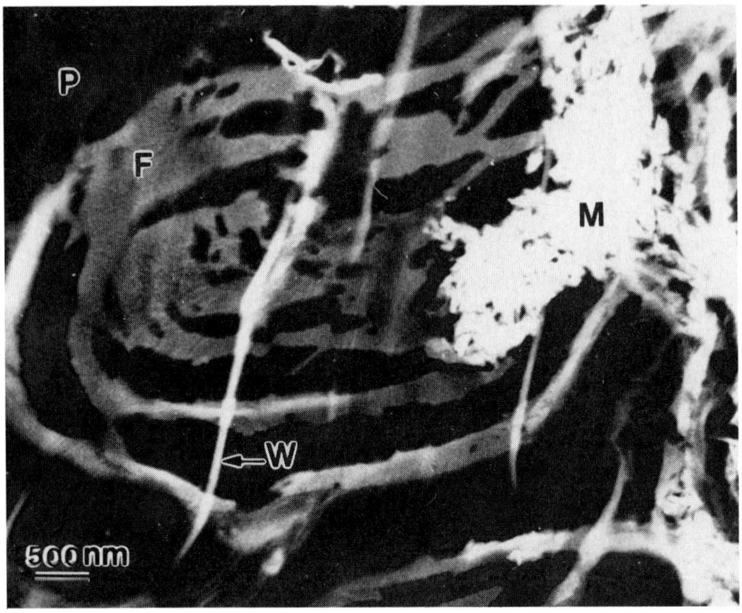

Figure 5. (S)TEM (dark field) image of a wood pulp fiber cross-section (F) embedded in a polymer (P) with mineral pigments (M). White lines (W) correspond to wrinkles and folds of the colloidal carbon film which is being supported by the Cu grid.

tion and processing packages include built-in functions which can be evaluated according to the experimental parameters.[18]

The accuracy of σ calculations is about 5% for K edges and 10% for L edges. More sophisticated models for σ are the Hartree–Slater methods which require significantly more computations.[19,20] These methods are more accurate but not readily available to the general user. If the absolute number of atoms is not required, it is often experimentally easier to calculate the relative concentration. The signal of two edges "a" and "b" is used:

$$\frac{C_a}{C_b} = \frac{S_a \sigma_b (\Delta, \mathrm{B})}{S_b \sigma_a (\Delta, \mathrm{B})} \tag{4}$$

In order to measure S_i, the background preceding the edge is extrapolated and subtracted (Figure 6). The signal is then integrated over an energy window Δ. In the energy range typically used for chemical analysis ($E \geq 100$ eV), the background follows a power law of the type AE^{-R}, where A and R are coefficients and E is the energy loss that can be fitted using linear least squares of the data (transformed on a logarithmic scale) over a window Γ. Edges of different type (K or L) and integration interval Δ can be used provided the appropriate $\sigma(\Delta, \beta)$ is calculated. In the case of absolute quantification, I is

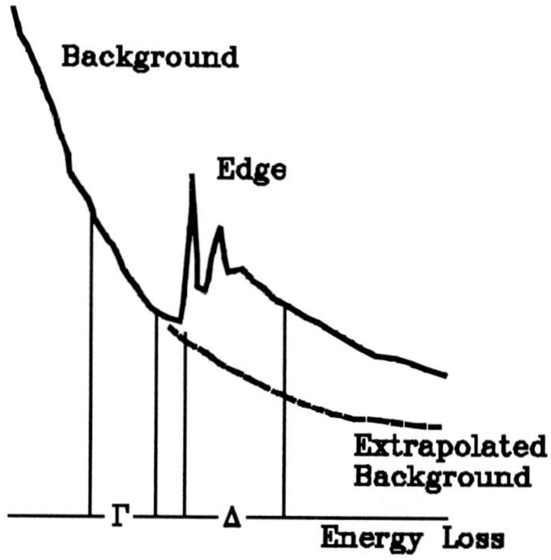

Figure 6. Background extrapolation and signal extraction. S_i is the edge signal. Γ is the fitting window and Δ is the signal integration window.

usually measured from the low-loss spectrum (see Figure 4) and gain changes have to be considered.

Approaches other than least square fitting of the background preceding the edge exist and allow optimal estimation of the parameters A and R.[21] Although these methods can be easily implemented on a personal computer, they are not generally available on commercial systems. Statistical analysis of data is performed using programs found in the literature[22] to calculate errors, variances and detection limits. Statistics have also been implemented in quantitative EELS imaging.[11] For situations where there are very low concentrations of elements of interest and weak signals consequently occur, other methods are better suited to take advantage of parallel detectors in enhancing the edges visibility and retrieving the signal.

A recently developed method is the first difference (FD) or second difference (SD) technique[23] with which a detection limit of 30 ppm of calcium in a carbon matrix has been reached. Second difference spectra are obtained by acquiring three spectra shifted in energy (one with respect to the other) with a voltage scan module. If the first spectrum (*I1*) is acquired with an initial offset E eV (as is usually the case), the second and the third spectra (*I2* and *I3*) will be recorded with an offset E+ΔE and E-ΔE respectively. Spectra are then linearly combined so that the second difference spectrum is SD=2(I1)-I2-I3. The value of ΔE (generally 3–5 eV) is optimized according to the width of the fine structure details.[23] First difference (FD) spectra are obtained by subtracting two spectra (again shifted).

This method of measurement reduces the effect of channel-to-channel gain variations in spectra and therefore increases the visibility of the edges from the background noise. Quantification is carried out by multiple least square fitting of the SD or FD spectra with reference to SD spectra. A detailed description of the procedure can be found in Leapman.[24] The difficulty in the analysis of these spectra is that only the first few eV at the edge threshold are used in the quantification; the spectra are, therefore, very sensitive to the chemical environment. The same element in different compounds (*e.g.*, the Si L edge in Si and SiO_2, the C K in amorphous carbon and graphite) can exhibit startlingly different near-edge structures that can be used to quantify the amounts of various polytypes of the same element (*e.g.*, graphite, diamond and amorphous carbon) in a multiphased small volume probed by the electron beam.[25] This points to the importance of quantifying spectra with standards of similar nature.

Another method that significantly reduces channel-to-channel variations is the gain averaging method[26] in which spectra are recorded at different energy offsets. In this technique, spectra are averaged together and the resultant spectra exhibit the usual topology as normal spectra. A large number of spectra can be combined with dramatic improvement in the signal-to-noise ratio. Statistical analysis of this method has been presented by Schattschneider and Jonas.[27]

For the quantification of spectra, the user is not restricted to calculated cross-sections since standards can also be used. A "K-factor" approach, as used in energy dispersive x-ray spectrometry in TEM, can be employed[28] if great accuracy (limited by statistical precision and not by the errors in the parameter estimation and the calculations) is required in the concentration determination. Often this is the only viable approach for the analysis of M, N and O type edges[29,30] for which calculated cross-sections are very inaccurate or non existent.

DETECTION LIMITS

It is important to review the sensitivity of the techniques as well as to distinguish between the "minimum detectable fraction" (MDF) and the "minimum detectable mass" (MDM) notions that are often discussed in the literature. In the MDF measurement the lowest concentration of an element uniformly dispersed in a matrix is determined, whereas in the MDM experiment the smallest number of atoms in a cluster in a matrix is measured. The experimental requirements for these measurements are different because spatial resolution is not a stringent parameter for the MDF. Low MDM is achieved in FEG–(S)TEMs in which small probes with intense currents are available, as discussed. In such instruments it has been possible to detect single atoms of heavy elements on a carbon matrix.[31,32] Low MDF can be achieved with thermionic and thermally-assisted or Schottky field emission sources in modern TEMs as large currents are available regardless of the beam size.

Leapman and Hunt[33] compared the detection limits for energy dispersive x-ray spectrometry (EDS) and EELS. Measurements carried out on well-characterized standards (*e.g.*, Leapman and Newbury[34]) determined that: *i*) for transition elements (filling 3d and 4f shells) having sharp L and M edges it is possible to detect concentrations from 10 ppm (atomic) to 100 ppm; *ii*) for elements with K edges (and L edges for low atomic number Z) and for transition metal elements with L edges (filling d shell) it is possible to detect concentrations from 100 ppm to 1000 ppm; and *iii*) for the other elements detectable concentrations >1000 ppm (atomic).[35] A clear advantage of the EELS technique over EDS is in the analysis of low atomic number elements ($Z \leq 8$) for which the fluorescence yield is very low. For these elements, x-rays can be measured only with ultrathin or windowless EDS detectors. Furthermore, there is significant absorption of these soft x-rays in the sample, in the detector window and in the ice which is often present on the detector. Given these sensitivities and limitations it is considered that, for routine chemical analysis, the two techniques are considered to be complementary.

NEAR EDGE STRUCTURES

The fine variation of intensity in the detail of an edge in the first 10–20 eV from the threshold is called the "near edge structure." As stated earlier, the edges in an EELS spectrum represent transitions of electrons from occupied levels to the first unoccupied levels and the unbound states. Because of solid state effects due to bonding to other atoms, the energy and the distribution of these electronic states are modified and depend strongly on the nature of the chemical bond and the crystallographic environment. The difference between the chemical effects as observed in photoelectron spectroscopy (XPS) and EELS (which is similar to x-ray absorption spectroscopy, XAS) is that the former technique is only sensitive to core level shifts whereas, for the latter technique, the effect is a combination of core and final state shifts and the sensitivity to the density distribution of the unoccupied states. Although providing more information, the interpretation of EELS spectra is more difficult. To this simplified picture other effects contribute to the modification of the fine structure as multi-electron and excitonic effects alter the detailed structure. Although the detailed modelling of these effects is starting to show some success,[36–38] the "fingerprinting" approach is often used to identify or distinguish: *i*) the nature of the environment (*e.g.*, the O K edges in FeO, α-Fe_2O_3, γ-Fe_2O_3, Fe_3O_4 are different[39]); *ii*) the coordination and symmetry of the atomic site (Si L edge in the SiO_4 unit in different minerals); and *iii*) the bonding character (*e.g.*, Al K in the metal and in an oxide are different). These effects have been recently reviewed by Brydson *et al.*[40] Interesting examples in organic chemistry are variations in the fine structure on the C K edge of the nucleic acids[41] and various other polymeric molecules.[42]

APPLICATIONS TO PAPER FIBERS

Previous applications of EELS to problems in the pulp and paper industry have not been documented in the literature. However, as different portions of the EELS spectrum can be used to characterize a material and to provide information about the chemical composition and electronic structure, the technique should be useful for fundamental investigations or for the study of contaminants. For example, a comparison of the low-loss spectra of an embedding polymer and paper fiber indicates a small shift of the plasmon energy to lower values for the paper fiber (Figure 7); further analysis of the different dielectric functions of these two materials can be carried out using Kramers–Kronig analysis of the spectra. This method will not be discussed here, although it is useful for some studies. Differences in the near edge structure of the C edge in these two materials can also be seen (Figure 8). In the polymer, the π^* peak (representing transitions from the 1s level to the π^* anti-bonding orbital) is narrower when compared with that of the fiber. Although small compared to changes reported for various structural forms of carbon (*e.g.*, Fink *et al.*[42]), this difference indicates a richer variation in types of organic bonds in the paper fibers, as the width of the peak arises from the contribution of various separate peaks with similar close energies. This agrees with XPS spectra[43], but in our analysis this information is spatially resolved to within 50 nm.

Chemical analysis of fibers reveals the presence of a Ca L edge at 346 eV (Figure 9) concentrated mostly in the center of the fiber wall. This finding indicates that during the pulping process most of the calcium present is extracted from the inner layer surrounding the lumen and the outer layer of the fiber wall but its residual concentration remains

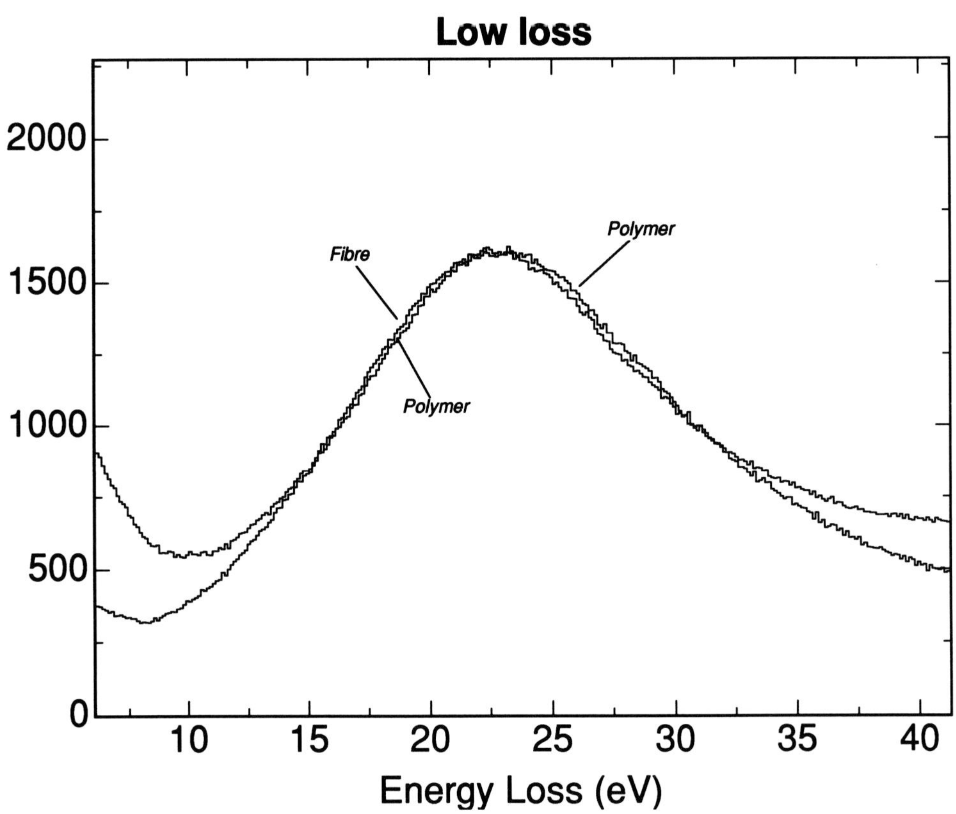

Figure 7. Low loss spectra of the polymer and the fiber. Differences at energies above 35 eV are caused by relative thickness (t/λ) changes between the two analyzed areas. Y-axis indicates the number of counts.

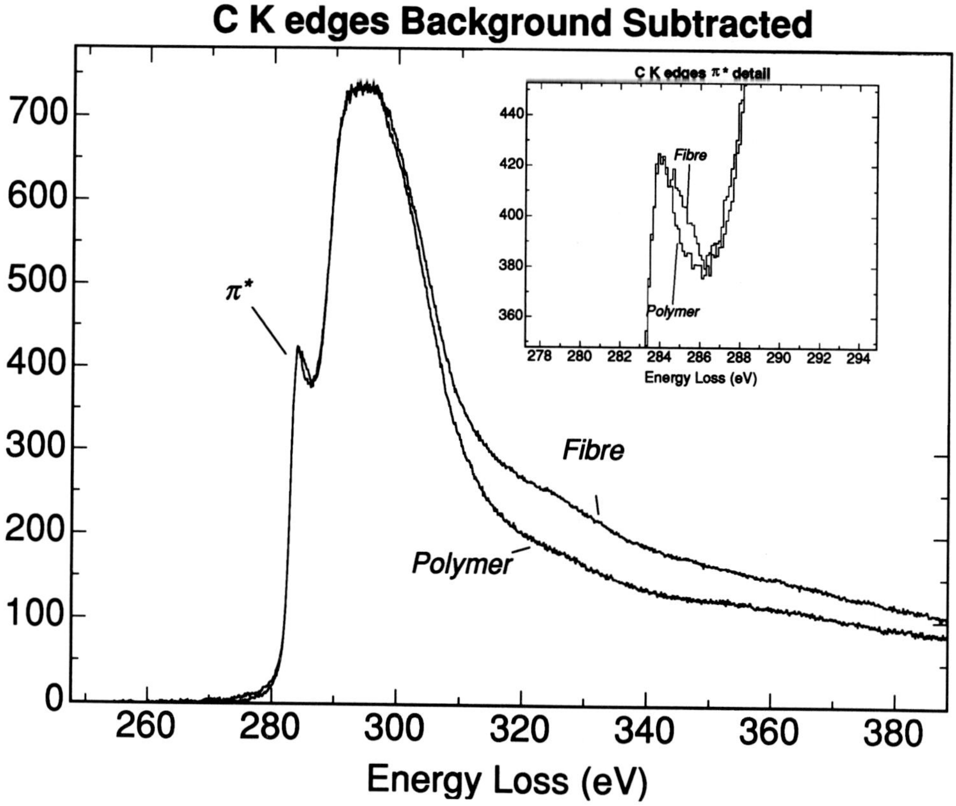

Figure 8. Carbon K edges of the embedding polymer and the wood pulp fiber. In the inset, detail of the π^* peak. Differences in the shape of the edges at energies above 300 eV are caused by relative thickness (t/λ) changes. Y-axis indicates the number of counts.

higher in the center of the wall. In the microstructural characterization of paper by TEM, heterogeneities of the structure can be easily visualized at high magnifications. For example, very fine particles dispersed in the lumen have been observed (Figure 10a). Given the high spatial resolution of EELS, it is possible to analyze these particles. Spectra from this specimen reveal that the Fe L and O K edges are present (Figure 10b and 10c). The fine structure of these edges can be used to distinguish the exact type ofoxide if comparison with other standards (*e.g.*, FeO, α-Fe_2O_3, γ-Fe_2O_3, Fe_3O_4) is carried out. With reference to the work of Colliex *et al.*,[39] our spectra indicate that the oxide is most likely FeO.

SUMMARY

This chapter presented the basic principles of the EELS technique and reviewed the instrumentation and the procedures for the quantification of spectra. The technique provides information about the composition and chemical structure of the fiber with very

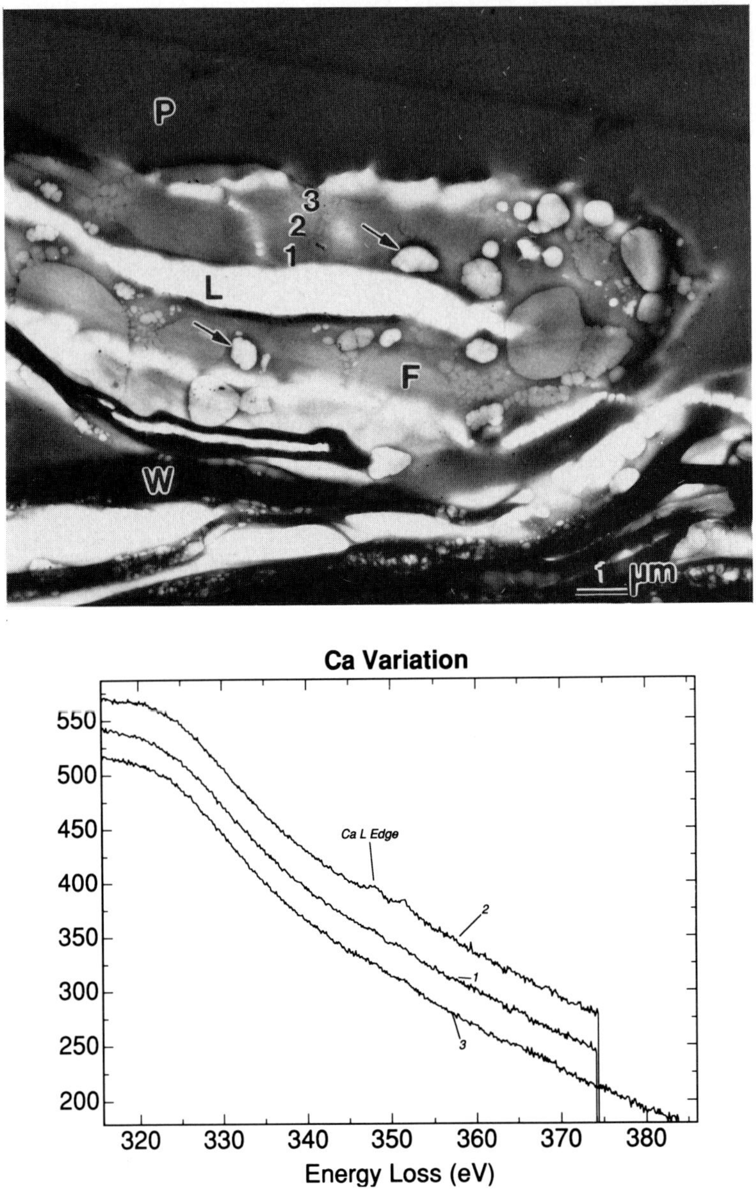

Figure 9. TEM (bright field) image of a fiber cross-section and EELS spectra collected at different positions of the fiber wall. Letters indicate: embedding polymer (P), fiber wall (F), lumen (L) and carbon film wrinkles and/or folds (W). Stains indicated by arrows show damage to the colloidal film due to the wet cross-section during preparation. The EELS spectra show variations in Ca L edge collected from positions 1,2 and 3. Spectra were normalized to the C signal. Y-axis indicates the number of counts.

high spatial resolution, and can be used to detect minor constituents dispersed in the fiber matrix or to analyze very small areas. Information can be determined not only about the elements present but also about their chemical state. EELS has not been previously documented in pulp and paper investigations, but it may be useful for laboratory studies of basic principles and pulping mechanisms or for analysis of contaminants.

(a)

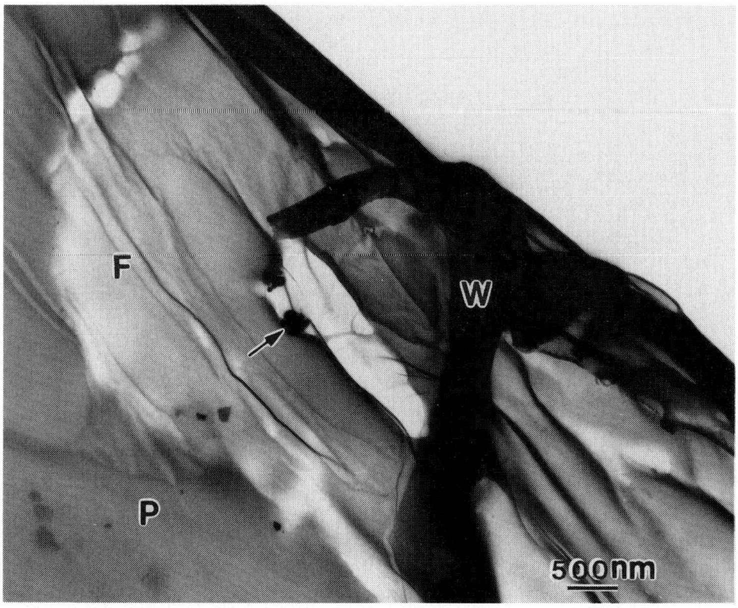

(b)

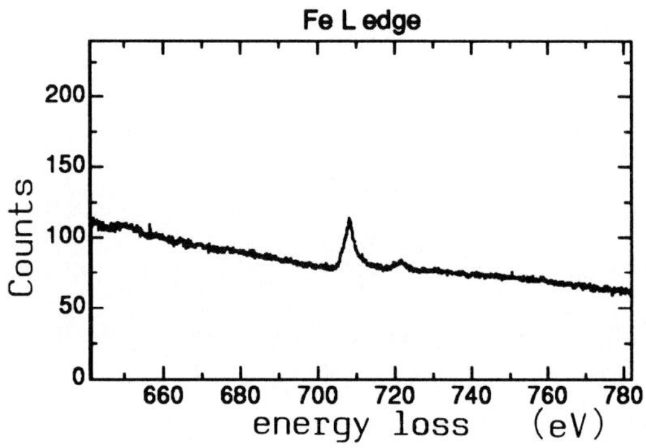

(c)

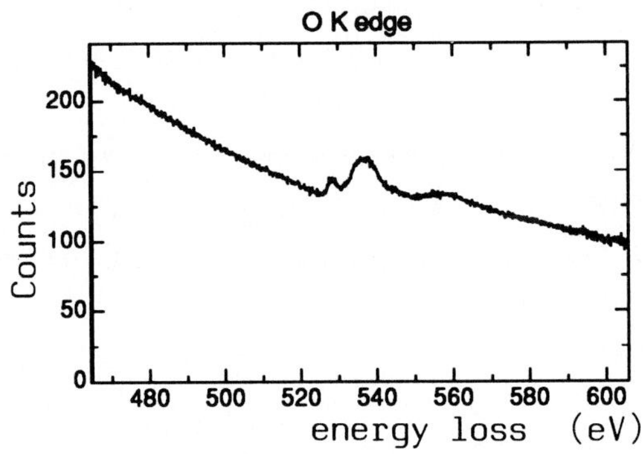

Figure 10. (a) TEM (bright field) image of a fiber with particles in the lumen. Arrow indicates the analyzed particles. Labels used are the same as in Figure 9 spectra showing: (b) Fe L edge and (c) O K edge from the particles.

REFERENCES

1. Egerton, R.F. Electron energy loss in the electron microscope. Plenum Press, NY. 1986.
2. Colliex, C. Electron energy loss spectroscopy in the electron microscope. *In*: Advances in Optical and Electron Microscopy, ed. V.E. Cosslett and R. Barer, Academic Press, London, Vol. 9:65–177. 1984.
3. Disko, M.M., Ahn, C.C. and Fultz, B. 1992. Transmission Electron Energy Loss Spectrometry in Materials Science. The Minerals, Metals & Materials Society, Warrandale, PA.
4. Newbury, D.E. 1986. Electron beam-specimen interactions in the analytical electron microscope. *In*: Principles of Analytical Electron Microscopy. Joy, D.C., Romig, A.D., Jr. and Goldstein, J.I., eds. Plenum Press, NY, pp.1–27.
5. Raether, H. 1980. Excitations of Plasmons and Intraband Transitions by Electrons. Springer Tracts in Modern Physics, 88, Springer-Verlag, Berlin.
6. Krivanek, O.L., Ahn, C.C. and Keeney, R.B. 1987. Parallel detection electron spectrometer using quadropole lenses. *Ultramicroscopy* 22:103–115.
7. Egerton, R.F., Yang, Y.Y. and Chen, S.C. 1993. Characterization and use of the Gatan 666 parallel near-edge electron energy loss spectrometer. *Ultramicroscopy* 48:239–250.
8. Ottensmeyer, F.P. 1984. Electron spectroscopic imaging: parallel energy filtering and microanalysis in the fixed beam electron microscope. *J. of Ultrastructure Research* 88(2):121–134.
9. Krivanek, O.L., Gubbens, A.J., Dellby, N. and Meyer, C.E. 1992. Design and first applications of a post-column imaging filter. *Microscopy, Microanalysis, Microstructures* 3(2/3):187–199.
10. Hunt, J.A. and Williams, D.B. 1991. Electron energy loss spectrum imaging. *Ultramicroscopy* 38:47–73.
11. Botton, G. and L'Esperance, G. 1994. Development, quantitative performance and application of a parallel electron energy loss spectrometer imaging system. *J. Microsc.* 173(1):9–25.
12. Egerton, R.F. 1992. Experimental techniques and instrumentation. *In*: Transmission Electron Energy Loss Spectrometry in Materials Science. Disko, M.M., Ahn, C.C., and Fultz, B., eds. The Minerals, Metals & Materials Society, Warrandale, PA, pp. 29–46.
13. Zaluzec, N.J. 1983. The influence of specimen thickness in quantitative electron energy loss spectroscopy. *In*: Proc. 41st Ann Meet. of the Electron Microscopy Society of America. Bailey, G.W., ed. pp. 388–389.
14. Reid, N. 1975. Ultramicrotomy, *In*: Practical Methods in Electron Microscopy, Vol. 3. Glauert, A.M., ed. Elsevier, Amsterdam, The Netherlands.

15. de Silveira, G. 1991. An improved method for the preparation of fibre cross sections. Pulp and Paper Report (PPR) No. 862. Pulp and Paper Research Institute of Canada, Montreal, Quebec, Canada.
16. Goodhew, P.J. 1973. Thin foil preparation for electron microscopy. *In:* Practical Methods in Electron Microscopy, Vol. 1. Glauert, A.M., ed. Elsevier, Amsterdam, The Netherlands.
17. Goodhew, P.J. Specimen preparation in materials science. 1985. *In*: Practical Methods in Electron Microscopy, Vol. 11. Glauert, A.M., ed. Elsevier, Amsterdam, The Netherlands.
18. Kundman, M., Chabert, X., Truong, K. and Krivanek, O.L. 1990. ELP Software for Macintosh II Computer. Gatan Inc., Pleasanton, California.
19. Leapman, R.D., Rez, P. and Mayers, D.F. 1980. K, L and M shell generalized oscillator strengths and ionization cross-sections for fast electron collisions. *J. Chem. Phys.* 72(2):1232–1243.
20. Rez, P. 1982. Cross-sections for energy loss spectrometry. *Ultramicroscopy* 9:283–288.
21. Unser, M., Ellis, J.R., Pun, T. and Eden, M. 1987. Optimal background estimation in EELS. *J. Microsc.* 145(3):245–256.
22. Trebbia, P. 1988. Unbiased method for signal estimation in electron energy loss spectroscopy, concentration measurements and detection limits in quantitative microanalysis: methods and programs. *Ultramicroscopy* 24:399–408.
23. Shuman, H. and Somlyo, A.P. 1987. Electron energy loss analysis of near-trace element concentrations of calcium. *Ultramicroscopy* 21:23–2.
24. Leapman, R.D. 1992. EELS quantification analysis. *In*: Transmission Electron Energy Loss Spectrometry in Materials Science. Disko, M.M., Ahn. C.C., and Fultz, B., eds. The Minerals, Metals & Materials Society, Warrandale, PA, pp. 47–83.
25. Hunt, J.A. and Williams, D.B. 1992. The current state of spectrum imaging. *In*: Proc. 50th Ann. Meet. of the Electron Microscopy Society of America. Bailey, G.W., Bentley, J., and Small, J., eds. pp.1200–1201.
26. Boothroyd, C.B., Sato, K and Yamada, K. 1990. The detection of 0.5 at% boron in Ni_3Al using parallel electron energy loss spectroscopy. *In*: Proc. XII Int Congress for Electron Microscopy. Peachey, L.D. and Williams, D.B., eds. Vol.2:80–81.
27. Schattschneider, P. and Jonas, P. 1993. Iterative reduction of gain variations in parallel electron energy loss spectrometry. *Ultramicroscopy* 49:179–188.
28. Malis, T.F. and Titchmarsh, J.M. 1985. K-factor approach to EELS. *Inst. Phys Conf. Ser.* 78:181–184.
29. Egerton, R.F. 1993. Oscillator-strength parametrization of inner-shell cross-sections. *Ultramicroscopy* 50:13–28.
30. Hofer, F. 1991. Determination of inner-shell cross-sections for EELS quantification. *Microscopy, Microanalysis, Microstructures* 2(2/3):215–230.

31. Colliex, C. 1992. Spatially-resolved electron energy loss spectroscopy (SREELS). *In*: Transmission Electron Energy Loss Spectrometry in Materials Science. Disko, M.M., Ahn, C.C., and Fultz, B., eds. The Minerals, Metals & Materials Society, Warrandale, PA, pp. 85–106.
32. Krivanek, O.L., Mory, C., Tence, M. and Colliex, C. 1991. EELS quantification near the single atom detection level. *Microscopy, Microanalysis, Microstructures* 2(2/3): 257–267.
33. Leapman, R.D. and Hunt, J.A. 1991. Comparison of detection limits for EELS and EDXS. *Microscopy, Microanalysis, Microstructures* 2(2/3):231–244.
34. Leapman, R.D. and Newbury, D.E. 1992. Trace analysis of transition elements and rare earths by parallel EELS. *In*: Proc. 50th Ann. Meet. of the Electron Microscopy Society of America. Bailey, G.W., Bentley, J., and Small, J., eds. pp.1250–1251.
35. Leapman, R.D. 1993. Philips-Gatan Workshop on EEL Spectroscopy and Imaging. Eindhoven, The Netherlands.
36. Sawatzky, G.A. 1991. Theoretical description of near-edge EELS and XAS spectra. *Microscopy, Microanalysis, Microstructures* 2(2/3):153–158.
37. Rez, P., Weng, X. and Hong Ma. 1991. The interpretation of near-edge structure. *Microscopy, Microanalysis, Microstructures* 2(2/3):143–151.
38. Rez, P. 1992. Energy loss fine structures. *In*: Transmission Electron Energy Loss Spectrometry in Materials Science. Disko, M.M., Ahn, C.C., and Fultz, B., eds. The Minerals, Metals & Materials Society, Warrandale, PA, pp. 107–129.
39. Colliex, C., Manoubi, T. and Ortiz, C. 1991. Electron energy loss spectroscopy near-edge structure in iron-oxygen systems. *Phys. Rev.*, B 44, 20:11402–11411.
40. Brydson, R., Sauer, H. and Engel, W. 1992. Electron energy loss near-edge structure as an analytical tool. *In*: Transmission Electron Energy Loss Spectrometry in Materials Science. Disko, M.M., Ahn, C.C., and Fultz, B., eds. The Minerals, Metals & Materials Society, Warrandale, PA, pp. 131–154.
41. Isaacson, M. 1972. Interactions of 25 keV electrons with nucleic acids, bases, adenine, thyamine and uracil. II Inner shell excitations and inelastic cross-sections. *J. Chem. Phys.* 56(5):1813.
42. Fink, J., Muller-Heinzerlung, T., Pfluger, J., Bubenzer, A., Koidl, P. and Crecelius, G. 1983. Structure and bonding of hydrocarbon plasma generated carbon films: an electron energy loss study. *Solid State Comm.* 47(9):687–691.
43. Gurnagul, N., Ouchi, M.D., Dunlop-Jones, N., Sparkes, D.G. and Wearing, J.T. 1993. Factors affecting the coefficient of friction of paper. *J. Appl. Polymer Sci.* 46: 805–824.

13

Surface Analysis Using Millimeter-Wave Resonant Instruments

Jon S. Martens

Conductus, Inc., Sunnyvale, California, U.S.A.

INTRODUCTION

There is always a need for additional and improved techniques of non-invasive characterization of materials and surfaces in the paper industry. Millimeter-wave analysis of dielectrics and conducting surfaces is one such class of techniques offering reasonable sensitivity and resolution with more noise/vibration immunity than some optical techniques. A large number of 'electromagnetic' techniques have been employed for paper and wood characterization. These include: *i*) extraction of lumped electrical parameters from circuit behavior;[1] *ii*) the measurement of transmission of an electromagnetic wave through the sample,[2] and *iii*) the measurement of parameters from resonant techniques. At least one version of the last approach, based on a quasi-optical resonator, can provide sensitivity and spatial resolution that will be of interest to the paper community. In particular, the analysis of formation and water content in paper and the detection of corrosion (on equipment) by a resonant technique will be discussed in this chapter.

RESONANT TECHNIQUES

The use of cavity resonant techniques for materials analysis has been known for many years.[3] A cavity is formed by some largely conducting structure in which an electromagnetic wave propagates and is reflected off the various surfaces such that at some frequencies, the various reflected waves add in phase. Energy is then stored in the system and quantities such as resonant frequency and the quality of the resonance (Q, related to the linewidth and defined by a ratio of stored to dissipated energy within the system) can be measured fairly easily. Since the response depends critically on the relative phasing of the various reflected waves, the details of the material in the wave path becomes critical. If the wave passes through a dielectric material, its electric path length changes consider-

0-8493-8992-5/95/$0.00+$.50

ably since the phase velocity of the wave is modified. This alters the frequency at which the system waves will add in phase, hence changing the resonant frequency. This allows the detecting of changes in density or thickness of a dielectric sample such as paper. Changes in loss in the wave path also dramatically affect measurement results. A material with any loss (a lossy dielectric, finitely conducting cavity walls, etc.) decreases the stored energy-to-power lost ratio hence decreasing Q. This is how water content or a corrosive change in the surface of a conductor can be detected.

CONFOCAL RESONATORS

The confocal resonator is one example of a mm-wave resonant structure[4,5] that has been investigated extensively for use in detecting corrosion.[6] It is quasi-optical in that two conducting surfaces form the resonant structure and the rest of the system is free space. The conductors can be both planar, or one or both may be spherical in shape. In terms of field confinement and practical implementation, it is convenient to use one spherical mirror and one planar conductor (see Figure 1). This structure has some advantages over closed cavities. Samples can be laterally scanned more easily, the system is fundamentally scalable to higher frequencies (intricate machining is not required) where there are often resolution and sensitivity advantages, and a single mirror system can be used over a fairly wide range of frequencies.

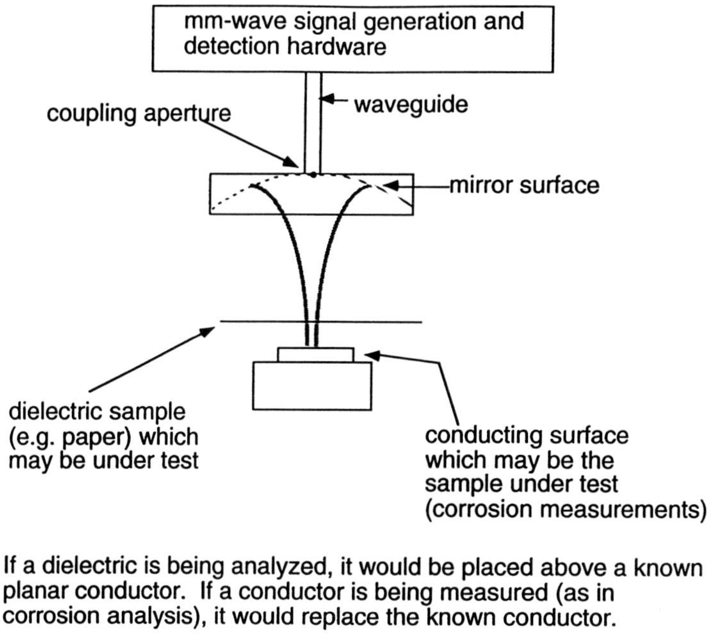

Figure 1. The structure of a confocal resonator with appropriate sample placement locations. A dielectric sample need not be in contact with a conductor but it should be in the beam path (outline shown).

The field structure in such a resonator is well-known.[4] Since the sample under test will be near the planar reflector or will be the planar reflector, the fields near that surface are of the most interest. The mode is designated transverse electromagnetic (TEM) which indicates that the electric and magnetic field vectors are linearly polarized and orthogonal to each other. This has some implications for grain detection since the polarization can be rotated and the measured loss of a grain-oriented material will increase dramatically if the electric field is parallel to the grain. The spatial variation of the electric field amplitude is given by:[4]

$$E_x = E_0 \sqrt{\frac{2}{1+\xi^2}}\, e^{\left(\frac{-kr^2}{b(1+\xi^2)}\right)} \sin\left(k\left[\frac{b}{2}(1+\xi) + \frac{\xi r^2}{b(1+\xi^2)}\right] - \frac{\pi}{2} + \phi\right) \qquad (1)$$

where $\xi=2z/b$ ($z=0$ on the planar conductor surface), $\phi=\tan^{-1}((1-x)/(1+x))$, b=the radius of curvature of the mirror, k=the wave number and r is the radial distance from the center axis. This equation is presented mainly to illustrate where the field energy is concentrated and, hence, what portions of the sample will contribute most heavily to the measurement. At or near the surface of the planar conductor, the spatial dependence is approximately

$$E_x(r,\phi) = E_1 e^{\left(\frac{-kr^2}{b}\right)} \qquad (2)$$

where r is the radial distance from the center. For common operating frequencies, the $1/e$ distance (distance from the center at which the field amplitude is reduced to about 36% of its maximum value) is a few millimeters. Hence the measurement is a spot analysis on the scale of larger test objects. Images can be generated by rastering the measurement across the sample. To observe detail smaller than the spot size, deconvolution methods can be employed.

In a typical measurement, the measured quantity is reflected energy (measured by a mm-wave reflectometer). The source is swept in frequency over a narrow range and at resonance (when the mirror-reflector spacing is precisely a multiple of half wavelengths), a large amount of power is absorbed by the cavity. During this period, the reflected signal provides detailed information about the reflectors and what materials are between them. The analysis of this reflected energy *vs.* frequency data is used to gather material data on the samples of interest.

The measurements to be described here can be divided into two categories: the sample under test is a planar reflector (as in corrosion analysis and metals studies) and the sample under test is a dielectric (*e.g.*, paper). In the first case, there are typically no other materials present besides the sample (plus any corrosion products on top) and the mirror. The electrical properties to be extracted here are called surface impedance (real and imaginary parts, $Z_s = R_s + j X_s$). Since the properties of the mirror and calibration samples are known *a priori*, the sample surface impedance properties are extracted from the following simple formulae:

$$R_{s,sample} = \frac{\omega_0 \mu_0 b}{4Q_0} - R_{s,\ mirror} \tag{3}$$

$$X_{s,sample} = X_{s,cal} - \frac{\mu_0 b}{2}\left(\omega_{0,sample} - \omega_{0,cal}\right) \tag{4}$$

In a perfect metal, the real and imaginary parts are equal, near textbook values and stable in time. If a metal corrodes, all of these change. The easiest to detect is an absolute change in the real part (surface resistance). Clean copper may have a surface resistance of 70 mΩ at 94 GHz but after a day's exposure to a high humidity environment, it may increase to 85 mΩ. More severe corrosion, that affecting structural integrity, produces even greater changes.

The analysis is slightly more complicated in the case of dielectric samples (Figures 2 and 3). Here the sample is placed between two reflectors. (The analysis to follow assumes that the dielectric is near the planar conductor but it is fairly trivial to modify the analysis for other cases as long as the sample is closer to the planar reflector than the mirror). When a dielectric is inserted in the beam path, two things happen: the effective surface reactance changes because the electrical beam path has been lengthened (dielectric constant >1) and the effective surface resistance increases because of loss in the dielectric. Since the reflector properties are known, the complex permittivity of the dielectric can be extracted from these simple analyses. For the purposes of paper, the imaginary part of permittivity tracks most strongly with water content. The effective real part of permittivity tracks with density and thickness. This allows the extraction of formation data by looking at variations in ϵ' as a function of position:

$$\varepsilon' \approx 1 + \frac{2\ \Delta f_0}{X f_0} \tag{5}$$

$$\varepsilon'' \approx \frac{1}{X}\left(\frac{1}{Q_{old}} - \frac{1}{Q_{new}}\right) \tag{6}$$

$$X = e^{-2} Im \left[\frac{e^{j2kd}}{kb - j2kd} \left(1 - \frac{1}{kb - j2kd} \right) \right] + \frac{(1 - e^{-2}) \sin[2kd]}{bk} - \frac{2d}{b} \quad (7)$$

where b is the radius of curvature of the mirror, d is the average sample thickness and $\Delta f_0 = f_{0,new} - f_{0,old}$. The *old* and *new* refer to *before* and *after* sample insertion, respectively. If the paper is placed arbitrarily, the position of the paper relative to the reflectors must be known but the dependency of the results on this parameter is weak.

ANALYSIS TECHNIQUES

ANALYSIS OF PAPER

Paper clearly falls into the category of dielectric analysis. Two properties are normally of most interest: water content and formation. We will use the latter term here to mean uniformity of electrical thickness. Since density and thickness variations are rarely correlated in precisely inverse ways, this translates into uniformity of density and thickness (more commonly, they are positively correlated[7]). Moisture content is the easiest to analyze since it tracks fairly closely with the imaginary part of permittivity. This is easy to see because of the high absorptivity of water in all of the microwave spectrum. As has been discussed by many authors, water will be by far the dominant loss mechanism. The imaginary part of permittivity (ϵ'') does tend to scale quite directly with water content as can be seen from test on pine samples illustrated in Figures 2 and 3. Although this is not a fair comparison for paper studies, these particular wood samples were relatively thin and low in moisture content so that the parameter ranges being studied are not unreasonable.

Similar results have been seen with a standard 20 lb. bond paper that started out dry (kept in a desiccator for several weeks) and was then exposed to a 40 °C, 50 % relative humidity environment for an hour. The imaginary part of permittivity for both is shown in Figure 4.

The extraction of formation is somewhat more complicated because of the possible coupling between density and thickness effects. Assuming the density and thickness variations are not exactly coupled such that $\epsilon^{0.5}d$ is constant, variations in either will appear as a change in ϵ'. In general, for a given sheet, density increases tend to accompany thickness increases which is fortunate from a data extraction perspective. By correlating with gross variations in ϵ'' it is possible to separate out the variations due to the two effects. For generic uniformity analysis, a plot of ϵ' *vs* position is quite useful. By itself, it does not separate effects of density and thickness but it may be enough for some

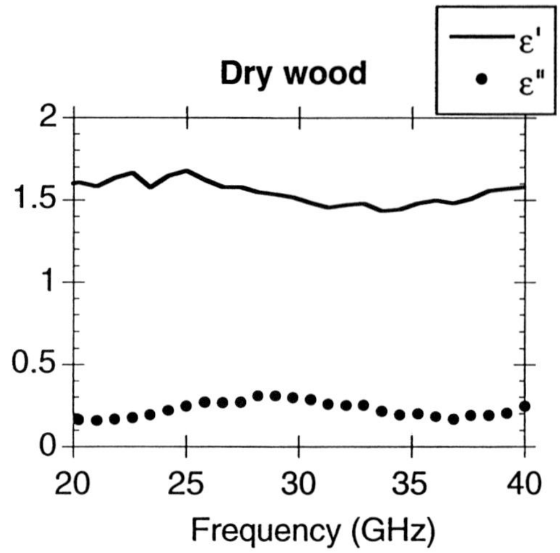

Figure 2. Real and imaginary parts of permittivity (normalized to e_0) for a dried wood sample (southern pine).

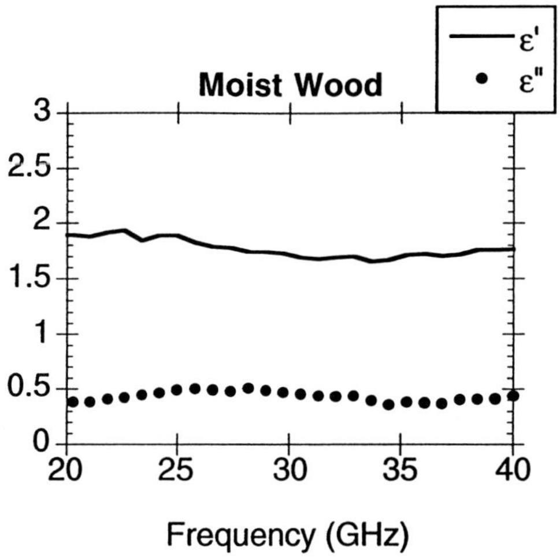

Figure 3. Real and imaginary parts of permittivity normalized to e_0 of a moist wood sample (southern pine physically immersed in water for a few seconds and patted dry).

analyses. The plotted disturbance index scales inversely as effective electrical thickness. The results for a variety of papers, ranging from inexpensive dot-matrix printer paper to high grade bond paper are shown below (Figure 5). The differences in gradient are quite obvious.

The above data represent a scan across samples with a spatial step size of 0.3 millimeters. As the energy distribution on the paper is somewhat larger than this, there is actually more variation present than is indicated. This is particularly true if the spatial period of variations is small. Deconvolution techniques can be used to extract a clearer picture of the variations involved (Figure 6). This follows from the simple observation

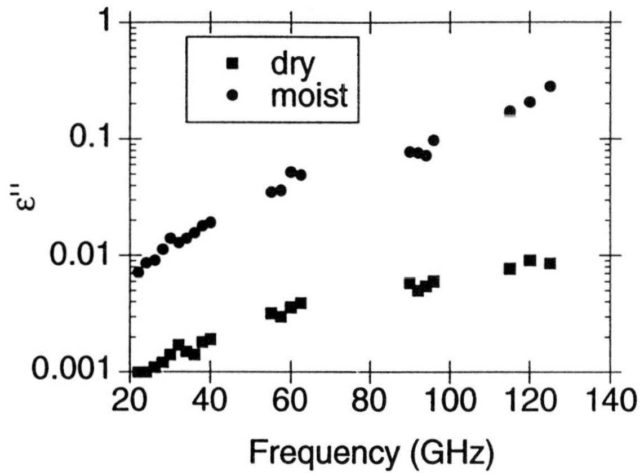

Figure 4. Imaginary part of permittivity *vs* frequency for a dry paper sample and one exposed to a humid atmosphere as described in the text.

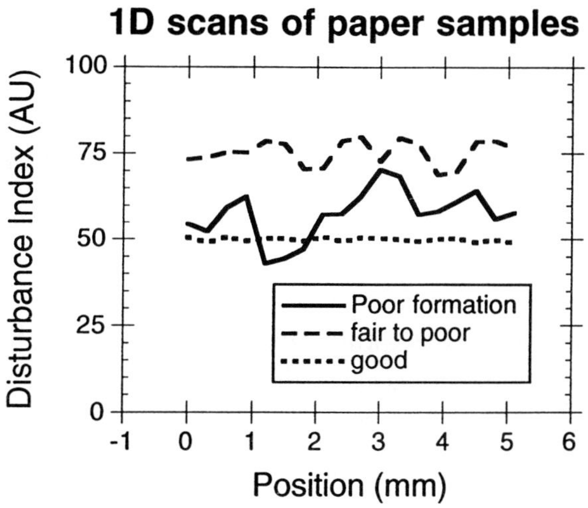

Figure 5. One dimensional scans of three paper samples of varying homogeneity. The disturbance index scales inversely as effective electrical thickness.

that a measured quantity K(x,y) (closely related to resonant frequency in the case of the above figure) is actually described by

$$K_{measured}(x,y) = K_{actual}(x,y) \otimes \left(Ae^{\frac{-kr^2}{b}} \right) \tag{8}$$

where A is a constant and k and b were defined earlier (the ⊗ operator stands for deconvolution). The second function represents the field distribution at the sample. The deconvolution process removes the filtering performed by the measurement. This, of course, places stringent requirements on the noise in the actual data since any removal of filtering will magnify any scatter in the raw data.

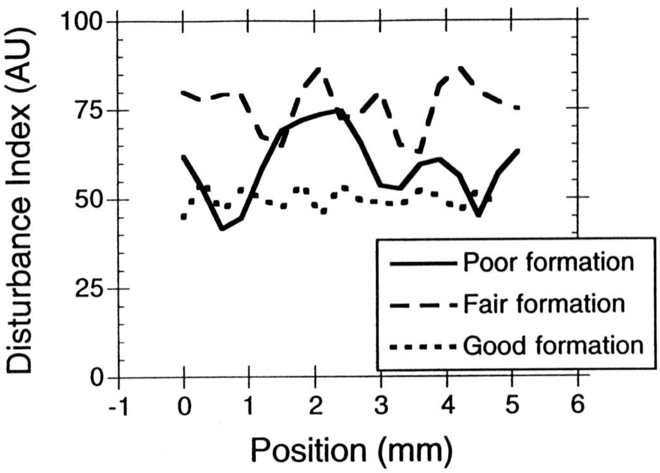

Figure 6. One dimensional scans of three paper samples from Figure 5 but with deconvolution routines applied to more clearly reveal the variations.

An important attribute of the system is the ability to test on-line. Since the resonator makes no contact with the sample, this is at least a plausible implementation. One can imagine the structure shown in Figure 1 with the resonator being placed above the paper stream and a planar reflector below the paper stream. The paper will not be directly on the reflector as in the previous analysis but this is only a minor complication if the paper is closer to the planar reflector than to the mirror. The issue, then, is whether the instrument can do the measurement with meaningful results at the paper velocities that would be typical (2500–4500 fpm). Since the scale of variations that are interesting are only a few mm, the paper should move less than 1 mm during the time of measurement. This translates to a maximum allowable measurement time of about 40 μs. The physics of the resonator and the measurement apparatus force a minimum measurement time as well. There are two main items placing such a lower limit on measurement time. The first, a fundamental one, arises because of energy storage in the resonator. There is a finite time required to change the energy distribution which is given by

$$\tau_{cavity\ response} = \frac{2Q_0}{\omega_0} = \frac{Q_0}{\pi f_0} \tag{9}$$

For typical confocal parameters this number is below 1 ms, hence it is not a major concern (worst case: Q=50000 and f_0 = 30 GHz leading to $t_{cavity} \approx$ 0.5 μs). The second hurdle is the practicality of data acquisition. The mm-wave detection circuitry can respond on the order of a few ms, as can the amplifiers. Rather than feeding the data directly to a computer, a long storage buffer may be used with measurements of some duty cycle to allow the data to be analyzed. Thus the paper stream would probably not be monitored constantly, but one out of every two to four sections (each about one mm long) could be analyzed.

CORROSION ANALYSIS

Corrosion can be looked at as the addition of a perturbative material. The sample starts as a lossy conductor and then a dielectric or poor conducting layer is added. By use of a reference, it is easy to obtain information about this layer from changes in Q and resonant frequency. A surface can be periodically or continuously probed to determine changes in surface electromagnetic properties. These changes tend to be noticeable long before massive structural change has occurred except in very thin materials.[6]

This is an example of the extraction of vertical structure information from the analysis of surface impedance as a function of frequency. The depth that a wave can penetrate into a given lossy or conducting material (skin depth) decreases with increasing frequency (scaling as $f^{-0.5}$ for metals). By measuring surface impedance versus frequency, a picture of electromagnetic behavior versus depth can be obtained. The shape of the field distribution must be deconvolved from the data to extract the true material structure. This can be most easily done if a stratified system is assumed and if it is treated as a series of transmission lines.

For many corrosion systems, the simple two layer model shown in Figure 7 will suffice. As an example, layer one could be Fe and layer two, FeO_x. The analysis of these structures is described in detail elsewhere.[6] The model is basically a series of two transmission lines terminated in the conducting layer one. We can directly measure the reflected energy from this system of transmission lines as a function of frequency. The first transmission line (free space between the sample and the mirror) is well-known but the second transmission line (the layer two) has unknown impedance, propagation characteristics and length. From the necessity of physicality and some knowledge of the layer two material, it is possible to reduce the number of variables to be determined.

Assume layer one to be a reasonable conductor of conductivity s_1 and that layer two is a poor dielectric of complex permittivity $\epsilon' - j\epsilon''$ and thickness t_2. The measured admittance $Y_{s,eff}$ (comes directly from the resonator measurements) can then be modeled by Equation (10):[6]

$$Y_{s,eff} = \left(\frac{1}{\sqrt{2\,\omega\,\mu_0\,\sigma_1}\left[\frac{1}{2\,\sigma_1} + \omega\,\mu_0\,t_2^2 + 2\,t_2\sqrt{\frac{\omega\,\mu_0}{2\sigma_1}}\right]} \right) \div$$

$$\left(\left[1 + 2\omega t_2 \varepsilon'' \sqrt{\frac{\omega\mu_0}{2\sigma_1}} + \omega^2\mu_0 t_2^2(\varepsilon' + \varepsilon'')\right] + j\left[-1 + 2\omega t_2\varepsilon'\sqrt{\frac{\omega\mu_0}{2\sigma_1}} + \omega^2\mu_0 t_2^2(\varepsilon' - \varepsilon'') - \omega\mu_0 t_2\sqrt{\frac{2\sigma_1}{\omega\mu_0}}\right] \right)$$

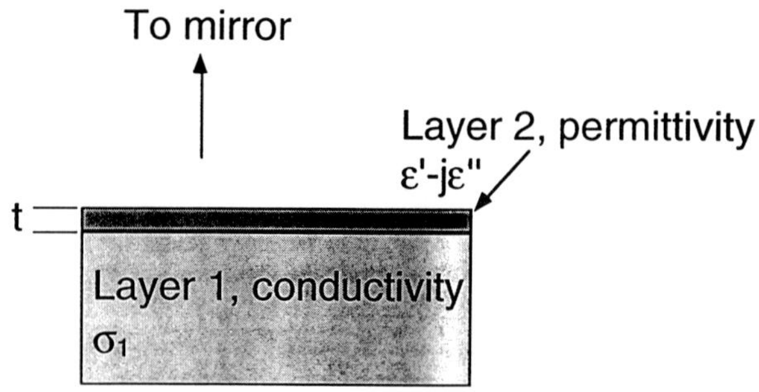

Figure 7. Two layer structure adequate for most corrosion studies. The measurement of the surface impedance of this sample *vs* frequency enables one to deduce the parameters of layer 2 (the corrosion product).

The real and imaginary parts can be measured ($Y_{s,eff} = 1/Z_{s,eff}$) and the only unknowns are t_2, ϵ' and ϵ". We can assume ϵ' to be frequency independent[3] over the frequency ranges used. The frequency dependence of ϵ" can be treated as a fitting parameter p as in $\epsilon'' = K\omega^p$ (for a conduction dominated loss, p=-1).[3] The real and imaginary parts can be fit to the measured data over the resonances measured with (at most) four fitting parameters to find the unknown physical and material parameters. Because the number of data points can exceed the number of fitting parameters by up to an order of magnitude and the fit parameters are very independent, the fitting is generally not difficult. When the fit is complete, the corrosion layer thickness and electrical properties have been determined.

Among corrosion products examined are the sulfidation products on copper and the oxidation of aluminum and solder. Of particular interest here, however, are oxidation of stainless and non-stainless steels. In one experiment, two grades of stainless steel were examined. Both started out with approximately the same surface resistance. Both were then exposed to a 40 °C, 80% relative humidity environment for an extended period of time with the surface impedance being periodically measured. Even from the raw surface resistance data shown in Figure 8, it is clear that one sample changed much more rapidly than the other. Until the end of the first week, however, no difference was noted with an optical microscope. The layer thickness, calculated using the above formulation shows the difference even more clearly (see Figure 9). The main point of this experiment is that significant corrosion can be detected quite automatically before visible damage has occurred.

SUMMARY

Resonant measurement techniques offer sensitivity that is difficult to match compared to transmission methods. This is exemplified by the extraction of corrosion information and the detail available on extracting the real part of permittivity. While these approaches certainly do not have the lateral resolution of optical techniques, the sensitivity to

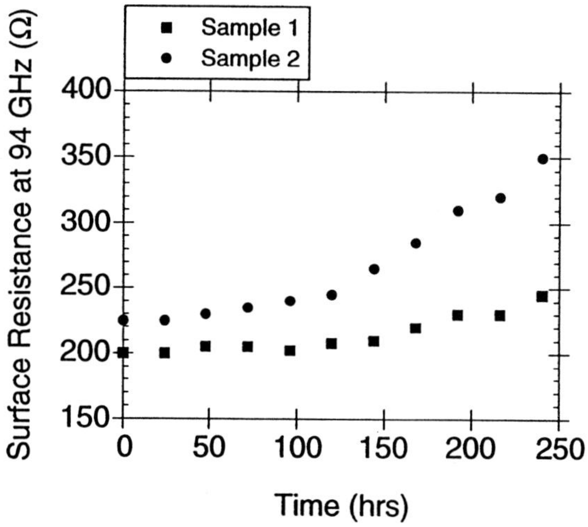

Figure 8. Surface resistance of two stainless steel samples of varying grades. No changes on either samples were detectable with an optical microscope until about 170 hours. The time refers to length of exposure to a humid atmosphere.

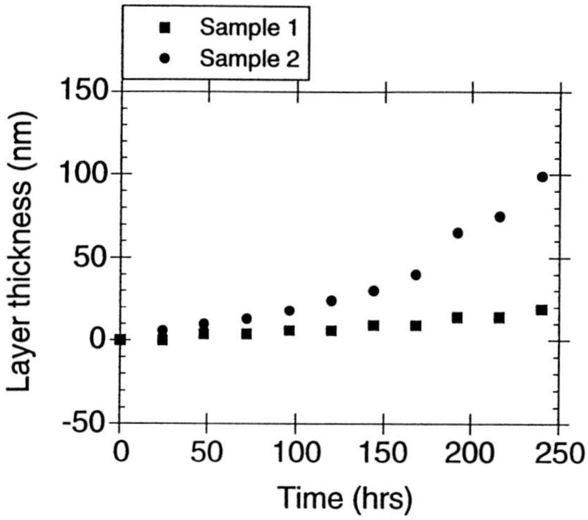

Figure 9. Calculated thickness of corrosion layer based on resonant measurements for the two stainless samples described above.

vibration and spurious reflectances from the sample are less of a problem. The non-invasiveness is certainly an issue when one compares this technique to electrochemical and radiological approaches. Although it has not yet been demonstrated, the approach, when used with a quasi-optical resonator like that described here, seems amenable to on-line measurements.

ACKNOWLEDGEMENTS

The author would like to acknowledge the many fruitful discussions with and assistance from Terry Conners of the Mississippi Forest Products Laboratory on the subject of this technique.

REFERENCES

1. Bergmanis, K.A., Klotinish, E.E. and Matis, I.G. 1972. Device for measuring permittivity of materials. US Patent No. 3,694,742.
2. Kinanen, I. and Duncker, J. 1978. Method for classifying and measuring timbers. US Patent No. 4,123,702.
3. Harrington, R. F. 1961. Time-Harmonic Electromagnetic Fields. McGraw-Hill, New York. Chapters 2 and 7.
4. Boyd, G.D. and Gordon, J.P. 1961. Confocal multimode resonator for millimeter through optical wavelength masers. *Bell Sys. Tech. Jour.* 40:489–508.
5. Martens, J.S., Hietala, V.M., Ginley, D.S., Zipperian, T.E. and Hohenwarter, G.K.G. 1991. Confocal resonators for measuring the surface resistance of high temperature superconducting films. *Appl. Phys. Lett.* 58:2543–2545.
6. Martens, J.S., Ginley, D.S. and Sorensen, N.R. 1992. A novel millimeter-wave corrosion detection method. *J. Electrochem. Soc.* 139:2886–2890.
7. Skaar, C. 1988. Wood–Water Relations. Chapter 6. Springer-Verlag, New York.

14

Atomic Force Microscopy

Shaune J. Hanley and Derek G. Gray

PAPRICAN and McGill University, Montreal, Quebec, Canada

INTRODUCTION

The atomic force microscope (AFM) was first introduced by Binnig, Quate and Gerber in 1986.[1] Its success has led to the development of a family of related techniques based on the same principle, grouped under the name of scanning force microscopy (SFM). The AFM itself was developed from the scanning tunneling microscope.[2,3]

The AFM can be used to image the surface topography of conducting and nonconducting samples at scales ranging from the macroscopic down to the atomic.[1,4–8] Imaging may be performed in air, in a vacuum[9] or in a fluid environment,[10] and even at low temperatures.[11,12] This has allowed the study of a wide range of materials and processes that include biological molecules[10,13,14] and real-time observation of biological processes such as blood clotting[10] and cell dynamics,[15,16] electrochemical processes *in situ*[17] and surface properties of polymers.[18–20] The AFM may also be used to measure surface forces[21,22] and nanomechanical properties.[23]

Other scanning force microscopes have further broadened the experimental possibilities. Mate *et al.*[24] introduced the lateral force microscope (LFM), opening the field of nanotribology to SFM.[25] By varying the tip material, forces such as magnetic[26] and electrostatic[27] forces may also be mapped. A number of reviews of SFM have appeared.[5,8,28–34]

The wide applicability of these techniques and the availability of commercial instruments[35] suggest that they may be of value in examining the surface properties of paper and of other materials of interest in pulp and paper science. With this in mind, we here describe the basic AFM method and some current modifications, and mention a few applications to cellulose and wood.

0-8493-8992-5/95/$0.00+$.50

OPERATING PRINCIPLES

The basis of the force microscope is the measurement of force between a sharp tip and the sample. This force is measured as an elastic deflection of the cantilever which supports the tip.[31] A sensor is used to measure the deflection. For a cantilever beam with a given spring constant, k, the force on the tip may be calculated from Hooke's Law, $F = -kz$, where z is the displacement of the cantilever.[33] Forces of the order of 10^{-13} to 10^{-4} N can be measured with this system and a lateral resolution of the order of Ångströms can be achieved.[31]

The force between the tip and sample may be measured using either non-modulated or modulated techniques. In the first case, it is simply the displacement of the cantilever on contacting the surface which is used. The displacement signal of the cantilever may be used either directly as the z data, or a feedback loop may be used to keep the cantilever at constant deflection (hence force) by moving the piezo in the z direction. The digitized signal from the feedback loop is used as the z data. The two modes are termed "deflection mode" and "height mode", respectively. The height mode has a larger z range but is limited to slower scan rates.

In modulated techniques[36,37] the cantilever is induced to oscillate at or near its resonant frequency. As the tip is brought close to the surface there is an interaction between the tip and the surface which causes a change in the resonant frequency.[38,39] The sample may then be imaged with a feedback loop set to keep the resonance frequency fixed, thereby mapping the surface at a constant value on the force gradient.[38,39]

A representative force–distance curve is shown in Figure 1. The tip starts at rest with zero deflection at point I. Consider a sample surface moving towards the tip. As the sample approaches the tip, the cantilever is deflected towards the sample due to the attractive force between the tip and surface. Just before point II, the tip contacts the surface. Point II is the maximum attractive force. At zero deflection, point III, the attractive force equals the repulsive force. The sample then continues to travel a set distance, deflecting the cantilever to some point IV, at which the direction of sample movement is reversed. At point IV the sample starts to be retracted from the tip. Point V is the maximum adhesive force between tip and surface and the tip then returns to the rest position.[40]

The scanning force microscope can image the sample at any point along the force curve. The first atomic force microscope[1] was configured to work with the tip in contact with the sample in the repulsive regime of the force curve (Figure 1). Later, noncontact methods[36,37] were introduced using a modulated cantilever to sense the force gradient in the attractive region of the force curve (before point II, Figure 1). The first method is currently most widely used, however, as it offers the highest spatial resolution.

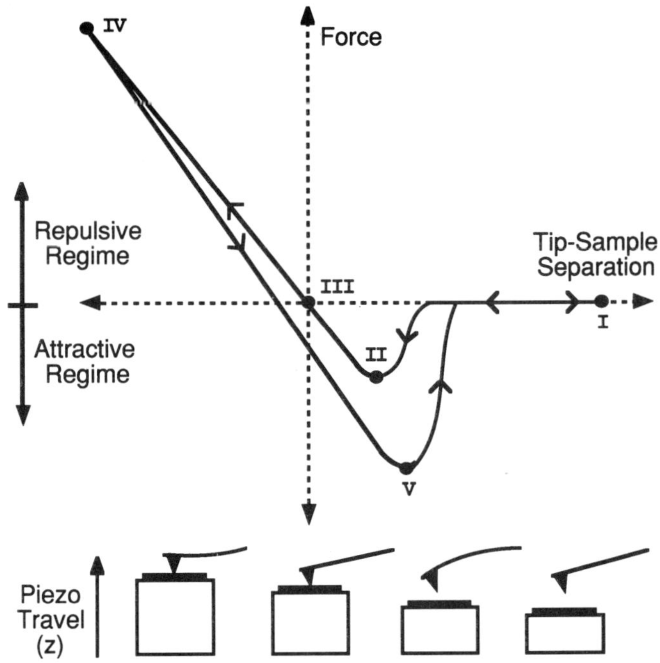

Figure 1. Representative force–distance curve and schematic of tip and piezo movement. Point I, rest position; Point II, maximum attractive force; Point III, zero cantilever deflection and zero load applied to the sample; Point IV, maximum load applied to the sample; Point V, maximum adhesive force.

The force–distance curves obtained have been used to measure van der Waals, double-layer, solvation, electrostatic and magnetic forces between surfaces[21,22,40,41] Colloidal particles have been attached to the tip for the direct measurement of colloidal interactions.[42–44] Ducker *et al.*[45] have used the AFM to measure the forces between silica particles and air bubbles. The forces between toner particles and carriers used in xerography, and surfaces of varying composition have been measured by Ott *et al.*[46] Hoh *et al.*[47] have measured quantized adhesion in water between a standard silicon nitride tip and glass. The force measurement may also be used to measure nanomechanical properties.[23,48,49]

CONTACT FORCE MICROSCOPY

Figure 2 shows a typical commercial configuration for a contact AFM. The cantilever deflection is sensed with a laser beam that is reflected off the back of the cantilever onto a position–sensitive photodiode.[9,50] The deflection is measured as the difference between the upper (A+B) and lower (C+D) halves of the photodiode. This method is commonly used as it is both reliable, simple, and sensitive enough for atomic resolution. Other methods to monitor the cantilever deflection include STM,[1] optical interferometry[36,37] and capacitance.[51]

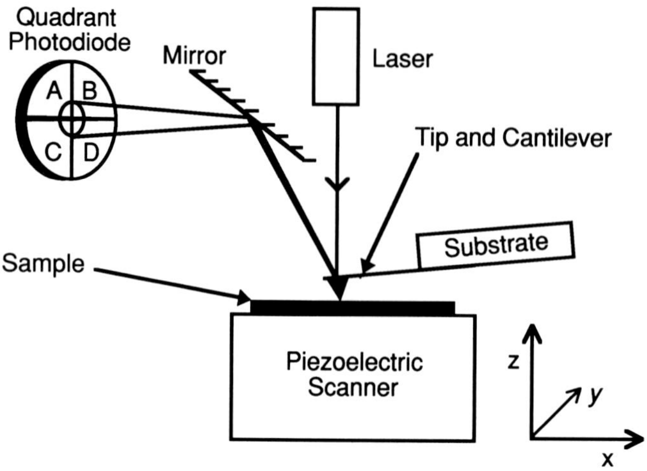

Figure 2. Schematic of a scanning force microscope, with an optical deflection system to measure the cantilever displacement. The system shown here may be used simultaneously as an atomic force microscope (AFM) and as a lateral force microscope (LFM).

The imaging takes place in the repulsive region of the force–distance curve. Soft cantilevers are used with spring constants of 0.01 to 1 Nm^{-1}. In order to minimize the damage to the sample and/or the tip during scanning, the force can be reduced from the initial contact force (point II in Figure 1) by retracting the sample away from the tip while keeping the tip in contact with the surface. The force between tip and sample is typically 10^{-7} to 10^{-8} N in air.[31] These forces may be lowered by an order of magnitude by imaging in either a liquid or a vacuum to reduce the attractive forces felt by the tip, primarily by eliminating the capillary forces due to the contamination layer.[31,38] An example of a liquid cell used in AFM is shown in Figure 3.

LATERAL FORCE MICROSCOPY

Mate *et al.*[24] used a modified atomic force microscope to measure the atomic-scale friction force between a tungsten tip and graphite. The lateral deflection of the tip was monitored with a optical interferometer, allowing atomic-scale lateral friction force images to be collected on graphite[24] and mica.[52] Lateral or frictional force microscopy[24] (LFM) is usually performed simultaneously with atomic force microscopy.[53,54] Since the surface friction is specific to a given material, the combined AFM–LFM (Figure 2) has the advantage not only of imaging topography, but also can distinguish different materials[55,56] as shown in Figure 4. This combination of topographical data with lateral frictional force data should provide a powerful tool for tribological studies at the atomic scale. To date, mica,[52] graphite[24] and the surface of Langmuir–Blodgett films[55,56] have been investigated by LFM.

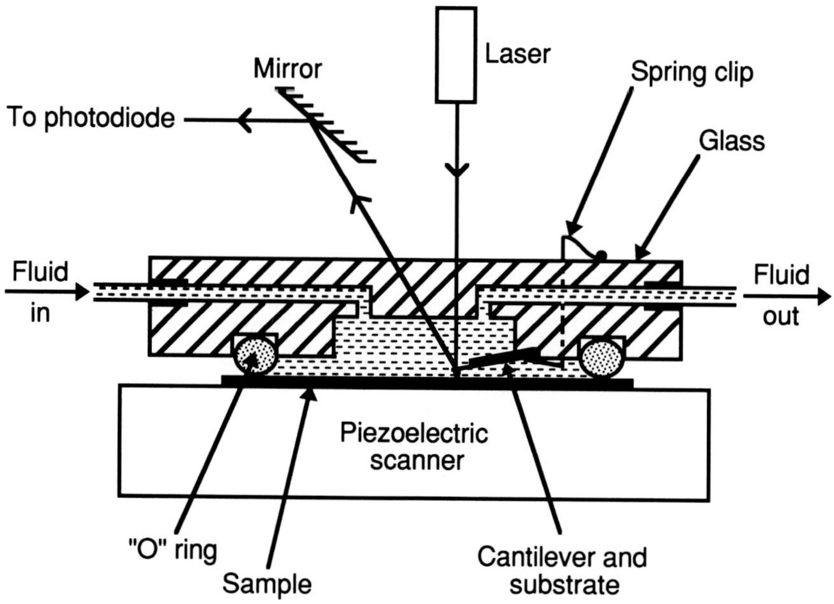

Figure 3. Schematic of AFM liquid cell.[84]

NON-CONTACT FORCE MICROSCOPY

This method[36,37] works in the attractive region of the force curve (before point II in Figure 1). A stiff cantilever with a spring constant, k, of 20–50 Nm^{-1} is used. A piezo electric ceramic modulates the cantilever at or near resonance frequency, ω_R, in the range of 200–400 Hz. This induces a low amplitude oscillation of the order of 0.2–1.0 nm in the cantilever tip. As the sample is brought close to the tip, the surface interacts with the tip and changes the effective spring constant of the cantilever, $k_{eff} = k_c + \delta F/\delta z$, where k_c is the spring constant of the cantilever and $\delta F/\delta z$ is the force gradient acting on the tip due to the surface interactions. The resonance frequency is given by $(\omega_R)^2 = \{(k_c + \delta F/\delta z)/m\}$, where m is the effective mass of the cantilever.[39] The sample may be imaged with a feedback loop set to keep the resonance fixed, hence profiling at a constant value on the force gradient.[36] The first modulated techniques[36,37] used a cantilever at a constant frequency and monitored the change in amplitude with lock-in amplifier techniques. Albrecht *et al.*[39] later introduced a method with an FM detector. The signal from the detector is used to keep the cantilever at its resonant frequency and the amplitude is also maintained at a constant level.

This method is ideal for imaging delicate samples, as any tip–surface contact is only inadvertent, and by comparison to contact force microscopy the force resolution is higher and susceptibility to thermal drift is lower.[33] However, the tip sample separation governs the spatial resolution.[57] When imaging in air, this separation must be of the order of 5–10 nm to avoid the tip getting caught in the contamination layer, thus limiting the resolution. Furthermore, artifacts due to imaging of the contamination layer rather than the actual sample surface may occur.

FORCE MODULATION MODE

Force modulation[58,59] is normally performed simultaneously with contact force microscopy. An image of the surface topography and a measure of the sample surface compliance are measured simultaneously. This method, like lateral force microscopy (LFM), can distinguish different materials. Figure 4 shows a comparison of AFM, LFM and force modulation images. In the topographic image, the surface planes appear uniform, but in both the LFM and force modulation images the contaminants on the surface are apparent.

Pethica and Oliver[60] suggested a method in which an oscillating force is applied to the tip and the resulting phase shift in the displacement is observed. Maivald *et al.*[58] have performed force modulation during contact imaging using a feedback loop to keep the force constant (height mode). The compliance measurements were performed by moving the piezo 25 nm in the z direction and measuring the cantilever deflection. This measurement was then repeated across the sample. The ratio between the input movement and the cantilever deflection was used to create a force modulation image. Radmacher *et al.*[59] have used a similar system to simultaneously investigate both the viscous and the elastic properties of the surface during contact imaging by measuring the amplitude and phase shift with lock-in techniques. Anselmetti *et al.*[61] used a combined STM/AFM to collect force modulation measurements.

TAPPING MODE

The AFM tapping mode[62] is a variation of the modulated non-contact force microscopy described above, the difference being that there is a brief contact between the microscope tip and the sample during each oscillation of the cantilever. A stiff cantilever (k = 20 to 50 N/m) oscillates either at or near its resonance frequency, 200–400 kHz, with an

Figure 4. Three images taken on the same area of a grooved glass sample showing topographic data (AFM), lateral force data (LFM) and force modulation data. While the topography image (left) allows the measurement of the groove height, both the lateral force data (center) and the force modulation (right) data show regions of different materials on the surface on the right hand side of the sample. Illustration courtesy of Topometrix.

oscillation amplitude of 20 to 100 nm. A larger amplitude than that used for non-contact imaging gives the cantilever sufficient energy to escape the contamination layer. The intermittent tip-surface contact modifies the amplitude. A feedback system is used to image the surface at a set modified oscillation amplitude. The force exerted on the surface is of the order of 0.1 to 10 nN and the lateral shear force is eliminated. The lateral resolution is limited by the tip radius and not by the height of the tip above the sample as in contact methods.[63,64]

ARTIFACTS

The main source of artifacts in SFM is tip-sample convolution.[65–69] An image is actually a combination of the tip and sample shapes. As long as the tip is sharper than the features which it is imaging, the true sample profile is collected. The first commercially available tips were square pyramidal in shape with an aspect ratio of 1:1 and a tip radius of approximately 100 nm. Recently, a number of methods have been used to manufacture sharp, high aspect ratio tips.[71–73] Figure 5 illustrates how problems arise when imaging features sharper than the tip.

Artifacts can also arise from tip asymmetry, which may result either during tip fabrication[70] or from tip contamination by the sample.

The collection of a true three-dimensional image also relies on the piezoelectric scanner(s) moving the sample under the tip in a linear and reproducible way. However, piezoelectric materials do not behave in an ideal manner due to nonlinearity, hysteresis, creep and cross-coupling between either X or Y scanner movement and the Z component.[74,75] These nonlinearities require correction either by compensation in the instrument software, or by hardware techniques which measure the actual scanner motion.[76–78]

The feedback loop, if not properly optimized, can lead to image artifacts. If the feedback gain is too high, the system can start to oscillate, and if the gains are set too low the sensor output is no longer constant.

The sample itself may cause artifacts in the image. Soft samples may be damaged or deformed by the tip. Lower scanning forces may be used to reduce these effects. Other techniques, such as tapping force microscopy[62] avoid linear shear forces during scanning and hence reduce sample damage.[64] Loose particles on the sample surface and tip-sample adhesion also create problems.

Other effects which may cause artifacts during image collection are mechanical vibration, thermal instability and electronic noise.

Once an image has been collected, a number of techniques may be used to enhance the data. Sample tilt and high frequency noise may be removed. Images may be filtered by fast Fourier transform (FFT) techniques to eliminate unwanted frequencies. Care is

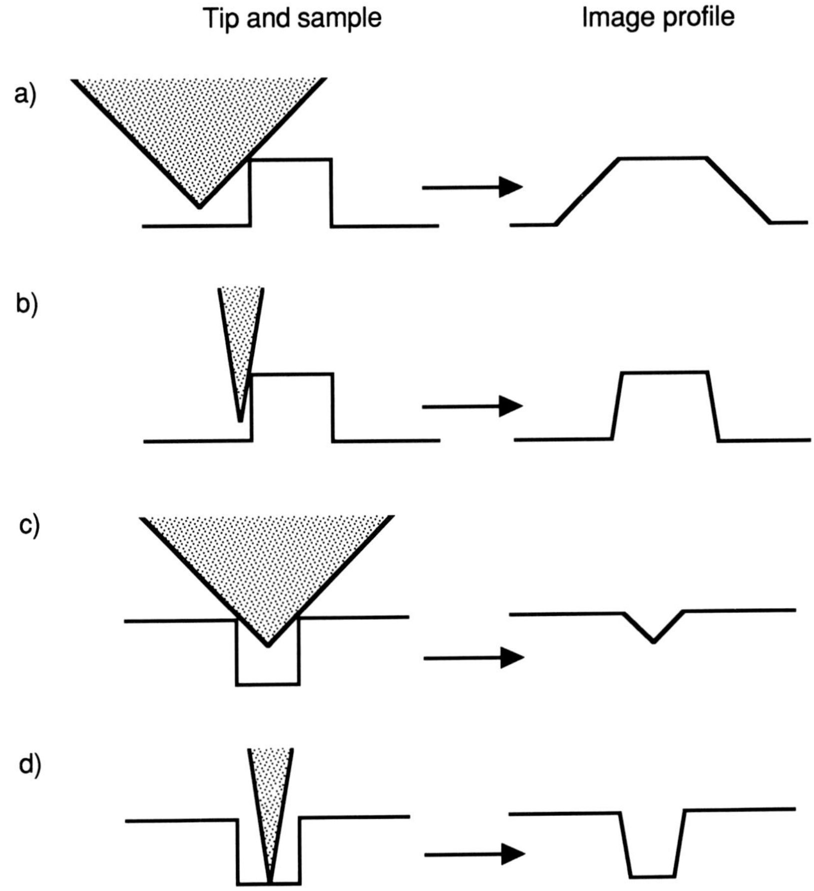

Figure 5. Illustration of the effect that tip geometry can have on the image profile. (a) The lateral distortion created in the image by a low aspect ratio tip. A sharper tip is required to measure a truer profile. (b) Note that the true sample height is measured in both cases. When a sample has deep trenches, a low aspect ratio tip may not be able to touch the bottom (c), resulting in an inaccurate depth measurement. A high aspect ratio tip (d) has to be used in such instances.

needed in applying these techniques to avoid image distortion[74,75] (or even image creation).

PULP AND PAPER AND RELATED FIELDS

The application of scanning force microscopy to the characterization of surfaces of interest to paper science and technology is at an early stage. The useful properties of paper result from the mechanical and surface properties of cellulose and the interactions of cellulose with water, so an obvious place to start is to see if AFM can generate useful images of cellulose surfaces. As with any new microscopic technique, it is best to look at a well-characterized surface with a high degree of order, and also to check results against a second technique. For this reason, highly crystalline cellulose microfibrils isolated from a marine algae (*Ventricaria*[79]) were selected as a sample; by comparing AFM and transmission electron microscope images of the same microfibrils,[66] the complementarity of the two techniques was demonstrated. Furthermore, it was possible to observe periodic structure at the molecular level with the atomic force microscope (Figure 6a). Line profiles taken along the long axis of the microfibril (the cellulose chain direction), show reproducible periodicities of 1.07 and 0.53 nm, (Figure 6b). The unit cell of *Ventricaria* cellulose has a fiber repeat distance of 1.038 nm, containing two glucose units.[80] The periodicities of 1.07 and 0.53 nm observed with the AFM are therefore likely to correspond to the fiber repeat and glucose unit length, respectively. Similar crystallographic data have been obtained from AFM scans of synthetic polymers.[18–20,81–83] The observed registry perpendicular to the cellulose chain direction only extends over a small distance. The absence of long-range order in this direction may suggest that at the outer surface of the crystallites the cellulose chains are parallel to the fiber axis but show poor lateral registry. AFM results at molecular resolution are only to be expected for uncontaminated, highly ordered, uniform surfaces with a regularly repeating structure. It would be unrealistic to expect, for example, that useful molecular information might be obtained from regular AFM scans of lignin surfaces. Nevertheless, information about the crystallinity of cellulose at the surface of pulp fibers might be relevant to some end uses.

Perhaps of more direct relevance is the observation that AFM can give unexpectedly detailed topographical information on wood and fiber surfaces at lower resolutions. The surface of microtomed sections of black spruce (*Picea mariana*) wood has been studied with the atomic force microscope.[84] Wood samples were embedded in a low viscosity Spurr resin[85] and 90-nm thick sections were prepared on a Reicht-Jung Ultracut E microtome equipped with a diamond knife. Since it is the topography of the section surface which is imaged, the samples were simply mounted on a clean flat surface such as mica.

The surface of microtomed sections is not flat and featureless.[86,87] An image of the surface of a transverse section through a bordered pit is shown in Figure 7. The essential features of the pit, torus and the primary and secondary wall of adjacent tracheids are clearly visible. The structure of the cell wall is more clearly seen in the composite images of a radial section of the tangential wall as shown in Figure 8. The features in the cell wall are revealed by the AFM as topographical changes on the surface of the section. The surface relief apparent in these figures reflects the response of the heterogeneous mechanical structure of the wood to the forces exerted on it during the embedding and sectioning processes. While the embedding material is homogeneous and relatively easy to cut, the fiber wall is heterogeneous, and the knife edge will encounter different resistances in different parts of the cell wall. For example, microfibrils orthogonal to the direction of motion of the knife might be expected to be more resistant than those aligned parallel to the direction of the cut. Furthermore, the force exerted by the wedge-shaped knife must include a component normal to the surface of the section,

(a)

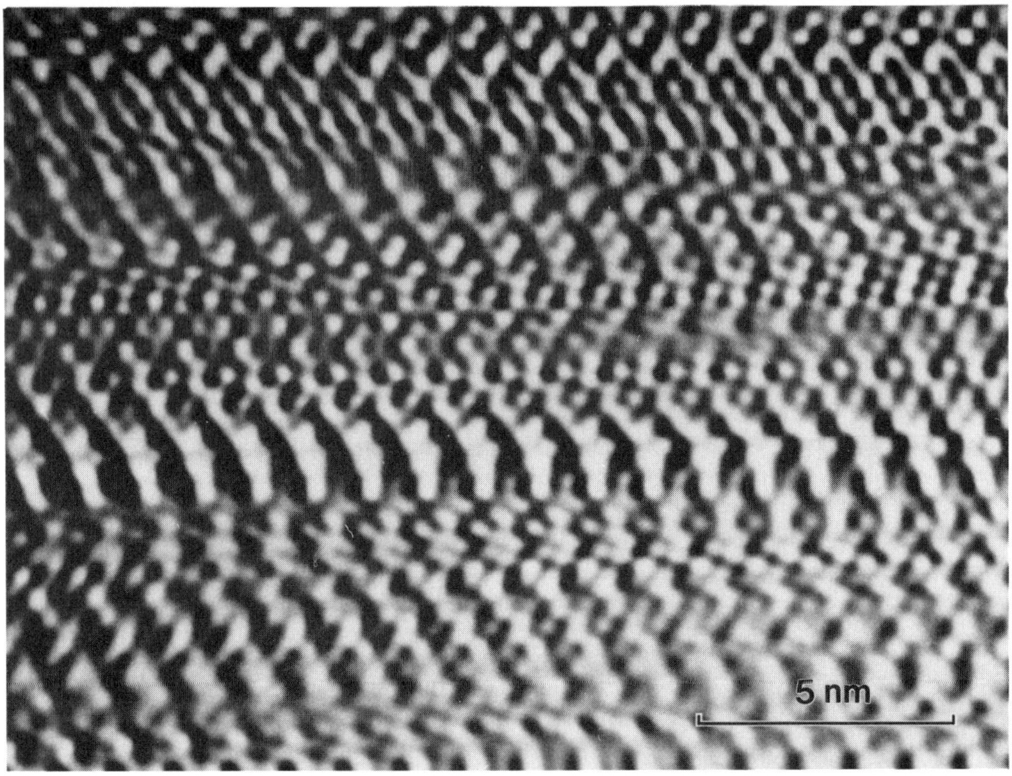

(b)

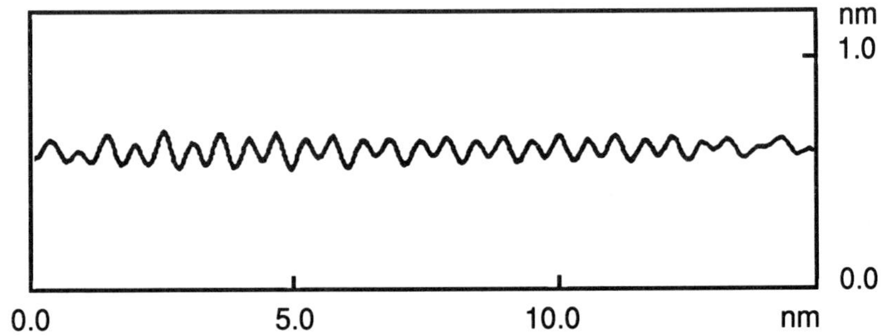

Figure 6. (a) Image of the surface of *Ventricaria* cellulose microfibril on mica. The fiber axis, and hence the cellulose chain direction, is from left to right. The image has been filtered with a two-dimensional fast Fourier transform technique. (b) A representative line profile taken from the image shown above.[66]

resulting in deformations that generate variations in surface topography of the section. After microtoming, the sections are floated away on water in a reservoir behind the diamond knife blade. While on the water surface, the sections swell visibly. Subsequent drying causes the sections to shrink again. This swelling and shrinkage again must depend on the structure and orientation of elements in the section; different responses may lead to variations in thickness or other deformations that are observable by AFM.

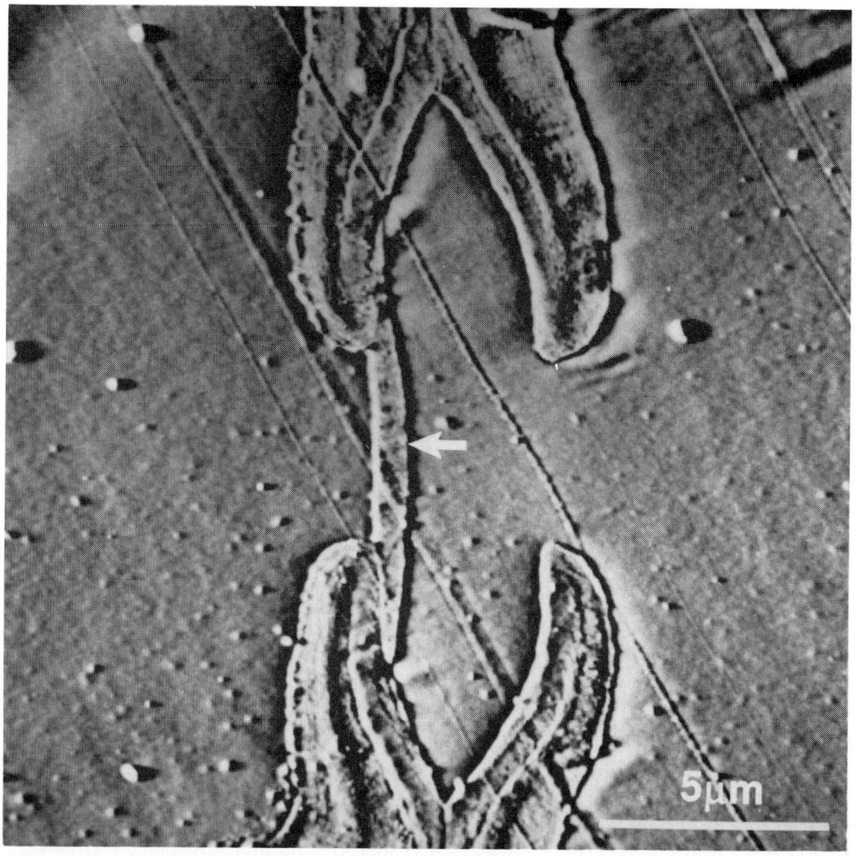

Figure 7. The surface of a transverse section of black spruce (*Picea mariana*), showing a bordered pit. The torus is indicated by the arrow. The middle lamella, primary walls and secondary walls may be distinguished. The knife direction is indicated by the diagonal lines seen in the image.[84]

In Figure 8 the middle lamella (M) is visible in the center of the image, bordered on either side by a layer composed of the primary wall and outer secondary wall (P+S1). The middle and inner secondary wall, S2 and S3 respectively, may be clearly distinguished. Furthermore, the transition regions from the S1 to S2 and S2 to S3 layers[88,89] may also be distinguished as a gradual change in surface texture in these regions. The microtome direction is indicated by the grooves left by defects in the diamond knife in the surface of the resin filled lumen (L) at the extreme left-hand side of the figure. It is also apparent that the texture of the adjacent walls is different on opposite sides of the compound middle lamella. The reasons for this are discussed elsewhere.[84]

Individual cellulose microfibrils that have been isolated from the cell wall can be imaged with the AFM. However, of greater interest in pulping and papermaking is the question of whether the dimensions and orientations of individual microfibrils can be detected within the cell wall using the AFM. Looking at Figure 8, there is a strong temptation to assume that the lines apparent in the S2 layers correspond to individual

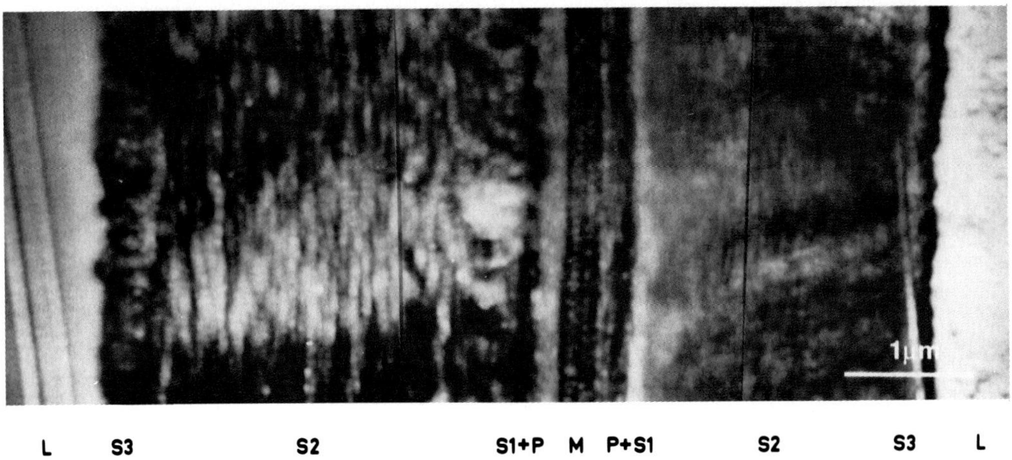

Figure 8. A composite of three AFM images taken successively across adjacent fiber walls of black spruce. The section has been taken radially through the tangential wall. The middle lamella (M), primary wall (P), secondary wall (S1, S2 and S3) and resin filled lumen (L) may be distinguished clearly. The knife direction is indicated by the marks in the left lumen.[84]

microfibrils. However, the spacing of these lines varies from 30 nm to 200 nm, which is far too large for individual microfibrils in the wood cell wall. As discussed previously in the section on artifacts, the shape of the AFM tip has been shown to distort lateral dimensions in this size range. We have, however, scanned with tips of different end diameter and seen no significant change in the line spacings.[90] This may indicate that the layer widths are approximately correct, and that we are actually seeing bundles of microfibrils. Radial sections of radial cell walls showed regions without any microfibrillar texture, presumably due to lignin in the middle lamella. The sections also intersected bordered pits and showed the microfibrils oriented radially along the edge of the pit.

The AFM is not limited to imaging the thin sections which may be affected by compression and swelling during sectioning and collection. The section block face may also be imaged[90] as shown in Figure 9. The layer widths measured within the fiber wall did not change significantly, compared to those measured on a thin section. These block faces have also been imaged under water in a liquid cell and the swelling of the walls can be imaged.[90] As one might expect, the middle lamella did not swell as much as the secondary wall (Figure 10).

While the surfaces of highly crystalline microfibrils of cellulose and microtomed sections of wood have proved to be amenable to AFM imaging, it is the surfaces of pulp fibers which are of fundamental importance to the pulp and paper industry. The AFM is normally limited to 5–10 μm in the z range, and this can limit the samples that may be examined, so sample choice and preparation is very important. Fibers can be mounted by simply drying down a very dilute aqueous suspension onto a glass coverslip.

The major difference found between unbeaten and beaten kraft pulp fibers was that, in the latter case, extensive fibrillar material was observed on the glass surface beside the

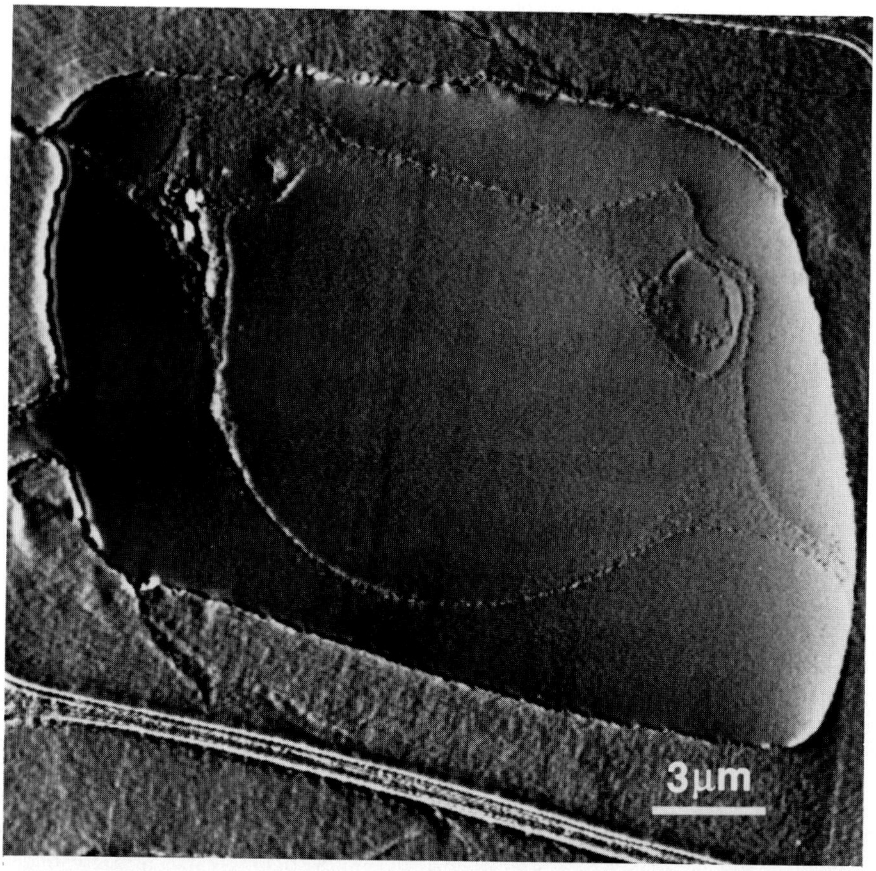

Figure 9. AFM image of the surface (transverse section) of a section block face of embedded black spruce wood. The middle lamella, primary walls, secondary walls and the cytoplasm may be distinguished.

fiber as shown in Figure 11.[84] However, no significant difference was noted between images of the fiber surfaces themselves. The surfaces of the fibers appeared to consist of rounded elements, with the dimensions of the order of 100 nm. This may indicate lignin-coated microfibrils. Once again, the possible distortion of the lateral dimensions due to tip shape may play a part.

Figure 12 shows a network of criss-crossed microfibrils over a more ordered layer on the surface of an unbeaten pulp fiber, in accord with the recognized structure of the outer layers of the wood cell.[88,89]

One strength of the AFM technique is that samples may be imaged while immersed in a fluid.[10] Sample selection and preparation is very important because of sample height limitations and the need to keep the sample fixed to the substrate during imaging. Also, swollen samples may be softer and more easily damaged by the tip. Figure 13 shows a ribbon of a cell wall from TMP which was imaged in deionized water.[84] In subsequent work, the surfaces of kraft pulp fibers and paper surfaces have been imaged under water

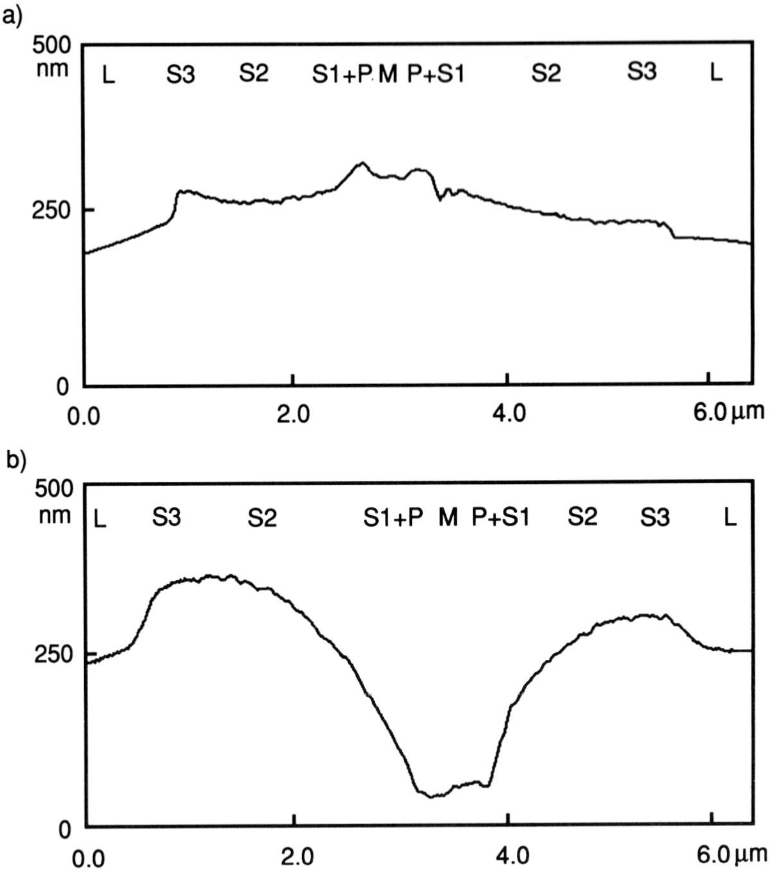

Figure 10. Line profiles of the face of a section block face of embedded black spruce wood, in air (a) and in deionized water (b). Adjacent fiber walls have been microtomed radially through the tangential wall. Comparing the profiles, it may be clearly seen that the secondary walls swell in water more than the middle lamella.

to the microfibrillar level. Swelling of individual fibers and the paper surface are readily observable upon immersion in water.

A final question is whether AFM can be applied to commercial paper surfaces. As mentioned previously, the restriction here is the z range limit of 5–10 μm. Figure 14 shows the surface of a kraft handsheet made from unbeaten, unbleached black spruce. Individual fibers in the paper web may be distinguished, thus allowing study of individual fiber surfaces. We have also used AFM to study the effects of different calendering conditions on the surface of the newsprint and to study ink thickness variations on coated papers.[90]

Obviously, further work is necessary to fully exploit the remarkable ability of AFM to image the surface topography of materials of interest to the pulp and paper industry. Research is progressing at two levels. The ability of AFM to image surfaces at atomic resolution is being used to examine highly crystalline polysaccharide samples such as

Valonia cellulose and β-chitin. Satisfactory images have already been obtained; they will be interpreted by comparison with data from TEM, crystallography and molecular modelling. Wood fiber surfaces and cross-sections are being studied at low magnification, with the aim of determining surface morphology, secondary wall structure, fibrillation of fiber surfaces by mechanical treatment and disruption of fiber-fiber bonds. Images of Langmuir-Blodgett polymer micelle films are also being obtained in cooperation with Professors Lennox and Eisenberg. Useful applications of other scanning force microscopies in this field are also confidently expected in the future.

ACKNOWLEDGEMENTS

We thank G. Honeyman (Digital Instruments) and E. Robinson (Topometrix) for information and the Government of Canada for support through the NCE Mechanical Pulps Network. S.H. thanks PAPRICAN and DuPont Canada for fellowships.

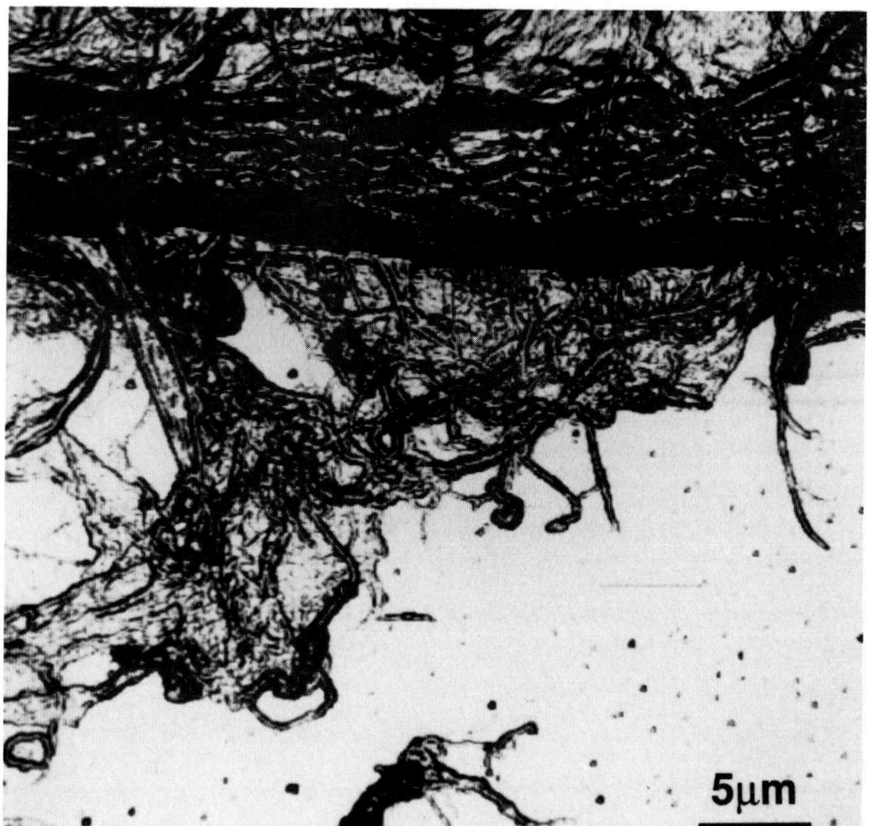

Figure 11. The edge of an unbleached beaten black spruce kraft pulp fiber. The fibrillar material seen at the edge of the fiber was not observed for the unbeaten fibers.[84]

Figure 6, from reference 66, is reproduced by permission of the publishers, Butterworth Heinemann Ltd. Figures 3, 7, 8, 11, and 13 are reproduced with permission from *Holzforschung* (reference 84).

Figure 12. The surface of an unbleached beaten black spruce kraft pulp fiber. The fiber axis is from left to right.

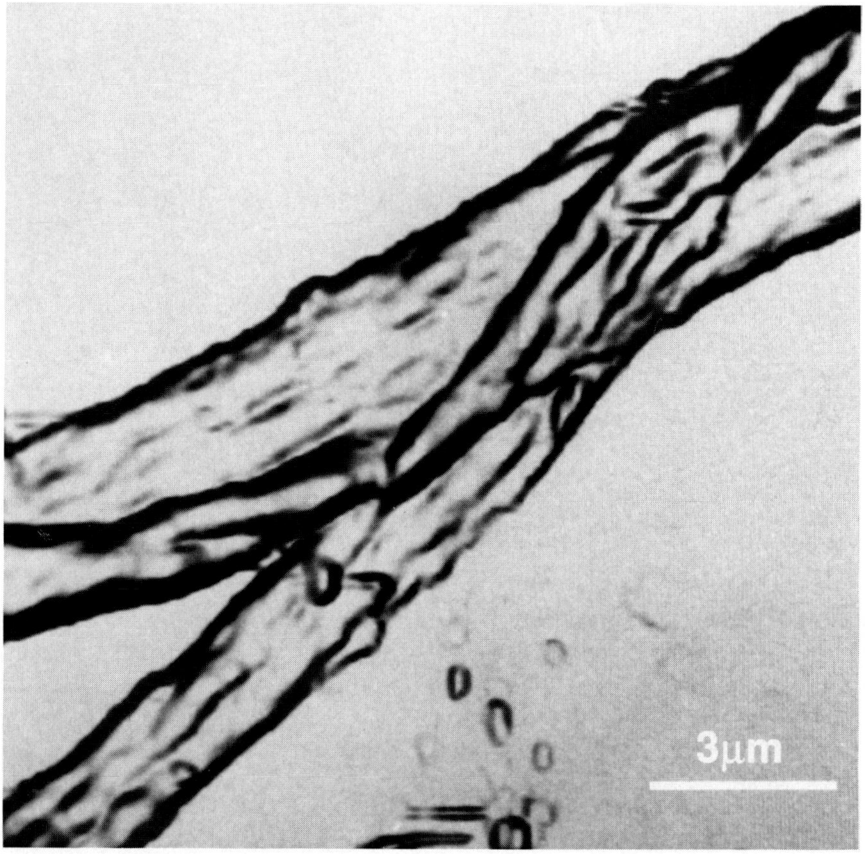

Figure 13. A ribbon of fiber wall from a black spruce thermomechanical pulp. The image was collected in a liquid cell filled with deionized water.[84]

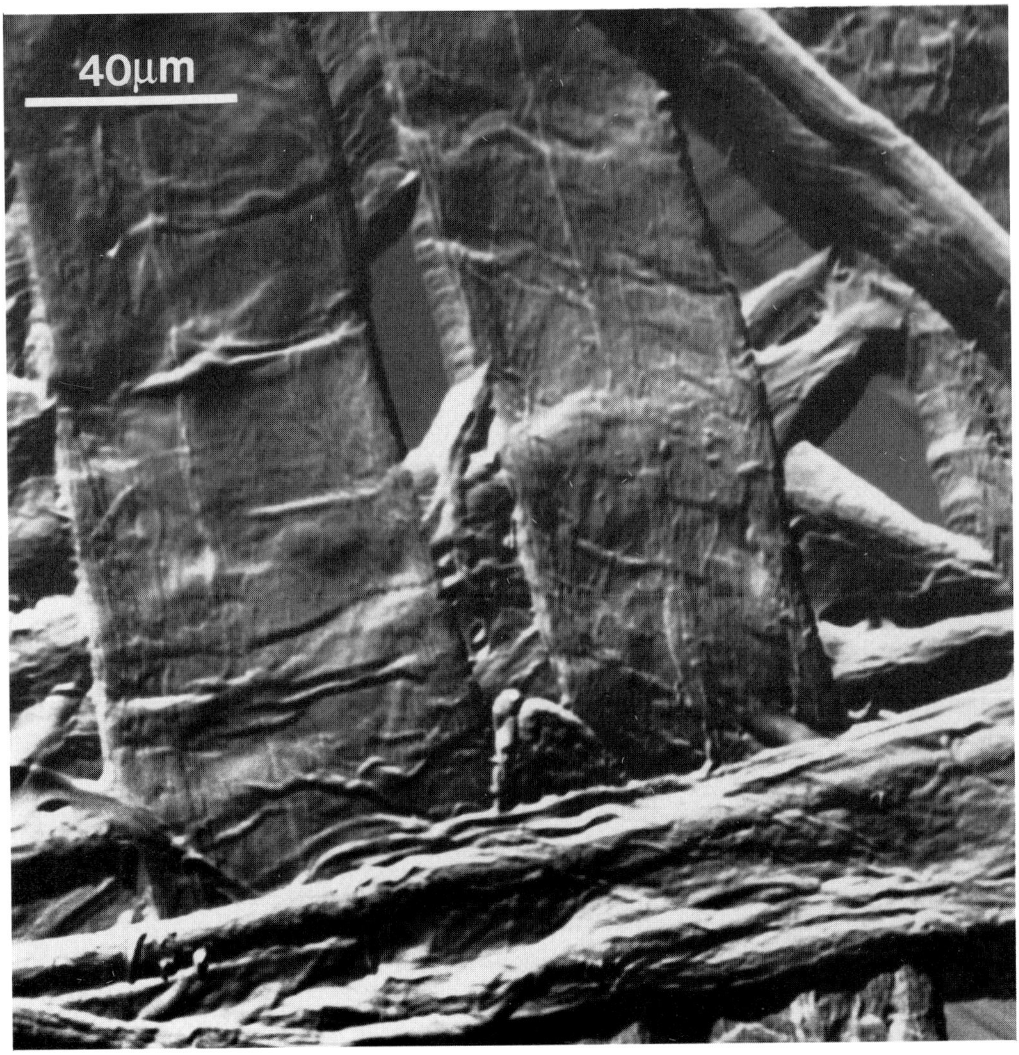

Figure 14. AFM image of the surface of a unbeaten, unbleached black spruce kraft handsheet.

REFERENCES

1. Binnig, G., Quate, C. F., and Gerber, C. 1986. Atomic force microscope. *Physical Review Letters* 56(9):930–933.
2. Binnig, G., Rohrer, H., Gerber, C., and Weibel, E. 1982. Surface studies by scanning tunneling microscopy. *Physical Review Letters* 49(1):57–61.
3. Bonnell, D. A., ed. 1993. Scanning Tunneling Microscopy and Spectroscopy: Theory, Techniques, and Applications. VCH Publishers, Inc., New York..
4. Albrecht, T. R. and Quate, C. F. 1987. Atomic resolution imaging of a nonconductor by atomic force microscopy. *J. of Applied Physics* 62(7):2599–2602.
5. Gould, S. A. C., Drake, B., Prater, C. B., Weisenhorn, A. L., Manne, S., Hansma, H. G., Hansma, P. K., Massie, J., Longmire, M., Elings, V., Northern, B. D., Mukergee, B., Peterson, C. M., Stoeckenius, W., Albrecht, T. R., and Quate, C. F. 1990. From atoms to integrated circuit chips, blood cells, and bacteria with the atomic force microscope. *J. of Vacuum Science and Technology* A8(1): 369–373.
6. Snyder, S. R. and White, H. S. 1992. Scanning tunneling microscopy, atomic force microscopy, and related techniques. *Analytical Chemistry* 64:116R–134R.
7. McGuire, G. E., Ray, M. A., Simko, S. J., Perkins, F. K., Brandow, S. L., Dobisz, E. A., Nemanich, R. J., Chourasia, A. R., and Chopra, D. R. 1993. Surface characterization. *Analytical Chemistry* 65:311R–333R.
8. Ultramicroscopy 1992. 42–44.
9. Meyer, G. and Amer, N. M. 1988. Novel optical approach to atomic force microscopy. *Applied Physics Letters* 53(12): 1045–1047.
10. Drake, B., Prater, C. B., Weisenhorn, A. L., Gould, S. A. C., Albrecht, T. R., Quate, C. F., Cannell, D. S., Hansma, H. G., and Hansma, P. K. 1989. Imaging crystals, polymers, and processes in water with the atomic force microscope. *Science* 243: 1586–1589.
11. Kirk, M. D., Albrecht, T. R., and Quate, C. F. 1988. Low-temperature atomic force microscopy. *Review of Scientific Instruments* 59(6):833–835.
12. Prater, C. B., Wilson, M. R., Garnaes, J., Massie, J., Elings, V. B., and Hansma, P. K. 1991. Atomic force microscopy of biological samples at low temperature. *J. of Vacuum Science and Technology* B9(2) (Mar/Apr):989–991.
13. Gould, S. A. C., Marti, O., Drake, B., Hellemans, L., Bracker, C. E., Hansma, P. K., Keder, N. L., Eddy, M. M., and Stucky, G. D. 1988. Molecular resolution images of amino acid crystals with the atomic force microscope. *Nature* 332:332–334.
14. Weisenhorn, A. L., Egger, M., Ohnesorge, F., Gould, S. A. C., Heyn, S.-P., Hansma, H. G., Sinsheimer, R. L., Gaub, H. G., and Hansma, P. K. 1991. Molecular-resolution images of Langmuir-Blodgett films and DNA by atomic force microscopy. *Langmuir* 7:8–12.

15. Haberle, W., Horber, H., Ohnesorge, F., Smith, D. P. E., and Binnig, G. 1992. *In situ* investigations of single living cells infected by viruses. *Ultramicroscopy* 42–44: 1160–1167.
16. Henderson, E., Haydon, P. G., and Sakaguchi, D. S. 1992. Actin filament dynamics in living glial cells imaged by atomic force microscopy. *Science* 257:944–946.
17. Manne, S., Hansma, P. K., Massie, J., Elings, V. B., and Gewirth, A. A. 1991. Atomic-resolution electrochemistry with the atomic force microscope: Copper deposition on gold. *Science* 251:183–186.
18. Albrecht, T. R., Dovek, M. M., Lang, C. A., Grütter, P., Quate, C. F., Kuan, S. W. J., Frank, C. W., and Pease, R. F. W. 1988. Imaging and modification of polymers by scanning tunneling and atomic force microscopy. *J. of Applied Physics* 64(3):1178–1184.
19. Maganov, S. N., Qvarnström, K., Elings, V., and Cantow, H.-J. 1991. Atomic force microscopy on polymers and polymer related compounds. *Polymer Bulletin* 25(6): 689–694.
20. Hansma, H., Motamedi, F., Smith, P., and Hansma, P. 1992. Molecular resolution of thin, highly oriented poly(tetrafluoroethylene) films with the atomic force microscope. *Polymer Communications* 33(3):647–649.
21. Weisenhorn, A. L., Hansma, P. K., Albrecht, T. R., and Quate, C. F. 1989. Forces in atomic force microscopy in air and water. *Applied Physics Letters* 54(26):2651–2653.
22. Mate, C. M., Lorenz, M. R., and Novotny, V. J. 1989. Atomic force microscopy of polymeric liquid films. *J. of Chemical Physics* 90(12):7550–7555.
23. Burnham, N. A. and Colton, R. J. 1989. Measuring the nanomechanical properties and surface forces of materials using atomic force microscopy. *J. of Vacuum Science and Technology* A7(4):2906–2913.
24. Mate, C. M., McClelland, G. M., Erlandsson, R., and Chiang, S. 1987. Atomic-scale friction of a tungsten tip on a graphite surface. *Physical Review Letters* 59(17):1942–1945.
25. Overney, M. and Meyer, E. 1993. Tribological investigations using friction force microscopy. *MRS Bulletin* May:26–34.
26. Martin, Y. and Wickramasinghe, H. K. 1987. Magnetic imaging by "force microscopy" with 1000 Å resolution. *Applied Physics Letters* 50(20):1455–1457.
27. Stern, J. E., Terris, B. D., Mamin, H. J., and Rugar, D. 1988. Deposition and imaging of localized charge on insulator surfaces using a force microscope. *Applied Physics Letters* 53(26):2717–2719.
28. Wickramasinghe, H. K. 1989. Scanned-probe microscopes. *Scientific American* 261(4):98–105.
29. Rugar, D. and Hansma, P. 1990. Atomic force microscopy. *Physics Today* 43 (October):23–30.

30. Engel, A. 1991. Biological applications of scanning probe microscopes. *Annu. Rev. of Biophysics and Biophysical Chem.* 20:79–108.
31. Meyer, E. and Heinzelmann, H. 1992. Scanning force microscopy (SFM). *In*: Scanning Tunneling Microscopy II. Weisendanger, R., and Güntherodt, H.-J., eds. Springer-Verlag, Berlin. pp. 99–149.
32. Radmacher, M., Tillmann, R. W., Fritz, M., and Gaub, H. E. 1992. From molecules to cells – Imaging soft samples with the AFM. *Science* 257:1900–1905.
33. Burnham, N. A. and Colton, R. J. 1993. Force microscopy. *In*: Scanning Tunneling Microscopy and Spectroscopy. Bonnell, D.A., ed. VCH Publishers, Inc., New York. pp. 191–249.
34. Sarid, D. 1991. Scanning Force Microscopy: With Applications to Electric, Magnetic and Atomic Forces. Oxford University Press, New York.
35. For example: Digital Instruments, 520E. Cortona Drive, Santa Barbara, CA 93103, USA; Topometrix, 5403 Betsy Ross Drive, Santa Clara, CA 95054-1162, USA; Park Scientific Instruments, Sunnyvale, CA, USA.
36. Martin, Y., Williams, C. C., and Wickramasinghe, H. K. 1987. Atomic force microscope–force mapping and profiling on a sub-100 Å scale. *J. of Applied Physics* 61(10):4723–4729.
37. McClelland, G. M., Erlandsson, R., and Chiang, S. 1987. Atomic force microscopy: General principles and a new method. Review of progress in quantitative non-destructive evaluation. Thompson, D.O., and Chimenti, D.E., eds. Vol. 6B., Plenum, New York pp. 1307–1314.
38. Hues, S. M., Colton, R. J., Meyer, E., and Güntherodt, H.-J. 1993. Scanning probe microscopy of thin films. *MRS Bulletin* January:41–49.
39. Albrecht, T. R., Grütter, P., Horne, D., and Rugar, D. 1991. Frequency modulation detection using high-Q cantilevers for enhanced force microscope sensitivity. *J. of Applied Physics* 69(2):668–673.
40. Burnham, N. A., Colton, R. J., and Pollock, H. M. 1991. Interpretation issues in force microscopy. *J. of Vacuum Science and Technology* A9(4):2548–2556.
41. Weisenhorn, A. L., Maivald, P., Butt, H.-J., Hansma, P. K. 1992. Measuring adhesion, attraction, and repulsion between surfaces in liquids with the atomic force microscope. *Physical Review* B45(19):11226–11232.
42. Ducker, W., Senden, T. J., and Pashley, R. M. 1991. Direct measurement of colloidal forces using the atomic force microscope. *Nature* 353:239–241.
43. Li, Y. Q., Tao, N. J., Pan, J., Garcia, A. A., and Lindsay, S. M. 1993. Direct measurement of interaction forces between colloidal particles using the scanning force microscope. *Langmuir* 9:637–641.
44. Rutland, M. W. and Snedin, T. J. 1993. Adsorption of the poly(oxyethylene) nonionic surfactant C12E5 to silica: A study using atomic force microscopy. *Langmuir* 9:412–418.

45. Ducker, W. A., Xu, Z., and Israelachvili, J. N. Measurement of forces between particles and bubbles. Presented at Colloid and Surface Science Symposium, University of Toronto, June 20–23, 1993.
46. Ott, M. L., Mizes, H. A., and Nash, R. J. Adhesion at polymer interfaces measured with the atomic force microscopy. Presented at Colloid and Surface Science Symposium, University of Toronto, June 20–23, 1993.
47. Hoh, J. H., Cleveland, J. P., Prater, C. B., Revel, J.-P., and Hansma, P. K. 1992. Quantized adhesion detected with the atomic force microscope. *J. of the American Chemical Society* 114:4917–4918.
48. Salmeron, M., Folch, A., Neubauer, G., Tomitori, M., and Ogletree, D. F. 1992. Nanometer scale mechanical properties of Au(111) thin films. *Langmuir* 8: 2832–2842.
49. Cohen, S. R. 1992. An evaluation of the use of the atomic force microscope for the studies in nanomechanics. *Ultramicroscopy* 42–44:66–72.
50. Alexander, S., Hellemans, L., Marti, O., Schneir, J., Elings, V., and Hansma, P. K. 1989. An atomic-resolution atomic force-microscope implemented using an optical lever. *J. of Applied Physics* 65(1):164–167.
51. Neubauer, G., Cohen, S. R., McClelland, G. M., Horne, D., and Mate, C. M. 1990. Force microscopy with a bidirectional capacitance sensor. *Review of Scientific Instruments* 61(9):2296–2308.
52. Erlandsson, R., Hadziioannou, G., Mate, C. M., McClelland, G. M., and Chiang, S. 1988. Atomic scale friction between the muscovite mica cleavage plane and a tungsten tip. *J. of Chemical Physics* 89(8):5190–5193.
53. Marti, O., Colchero, J., and Mlynek, J. 1990. Combined scanning force and friction microscopy of mica. *Nanotechnology* 1:141–144.
54. Meyer, G. and Amer, N. M. 1990. Simultaneous measurement of lateral and normal forces with an optical-beam-deflection atomic force microscope. *Applied Physics Letters* 57(20):2089–2091.
55. Overney, R., Meyer, E., Frommer, J., Brodereck, D., Lüthi, R., Howard, L., Güntherodt, H.-J., Fujihira, M., Takano, H., and Gotoh, Y. 1992. Friction measurements on phase separated thin films with the atomic force microscope. *Nature* 359:133–134.
56. Meyer, E., Overney, R., Brodbeck, D., Howard, L., Lüthi, R., Frommer, J., and Güntherodt, H.-J. 1992. Friction and wear of Langmuir-Blodgett films observed by friction force microscopy. *Physical Review Letters* 69:1777–1780.
57. Moiseev, Y. N., Mostepanenko, V. M., Panov, V. I., and Sokolov, I. Y. 1988. Force dependences for the definition of the atomic force microscopy spactial resolution. *Physics Letters* A 132(6,7):354–358.

58. Maivald, P., Butt, H. J., Gould, S. A. C., Prater, C. B., Drake, B., Gurley, J. A., Elings, V. B., and Hansma, P. K. 1991. Using force modulation to image surface elasticities with the atomic force microscope. *Nanotechnology* 2:103–106.
59. Radmacher, M., Tillman, R. W., and Gaub, H. E. 1993. Imaging viscoelasticity by force modulation with the atomic force microscope. *Biophysical J.* 64:735–742.
60. Pethica, J. B. and Oliver, W. C. 1987. Tip surface interactions in STM and AFM. *Physica Scripta* T19:61–66.
61. Anselmetti, D., Gerber, C., Michel, B., Güntherodt, H.-J., and Rohrer, H. 1992. Compact, combined scanning tunneling/force microscope. *Review of Scientific Instruments* 63(5):3003–3006.
62. Trademark of Digital Instruments
63. Manual for the Multimode atomic force microscope (MMAFM), 1993. Digital Instruments, 520E. Montecito Street, Santa Barbara, CA 93103, USA.
64. Zhong, Q., Inniss, D., K.Kjoller, and Elings, V. B. 1993. Fractured polymer/silica fiber surface studied by tapping mode atomic force microscopy. *Surface Science* 290(1–2):L688–L692.
65. Patil, R., Kim, S.-J., Renecker, D. H., Weisenhorn, A. L. 1990. Atomic force microscopy of dendritic crystals of polyethylene. *Polymer Communications* 31: 455–457.
66. Hanley, S. J., Giasson, J., Revol, J.-F., and Gray, D. G. 1992. Atomic force microscopy of cellulose microfibrils; comparison with transmission electron microscopy. *Polymer* 33(21):4639–4642.
67. Allen, M. J., Hud, N. V., Balooch, M., Tench, R. J., Siekhaus, W. J., and Balhorn, R. 1992. Tip-radius-induced artifacts in AFM images of protamine-complexed DNA fibers. *Ultramicroscopy* 42–44:1095–1100.
68. Zenhausern, F., Adrian, M., Emch, R., Taborelli, M., Jobin, M., and Desconts, P. 1992. Scanning force microscopy and cryo-electron microscopy of tobacco mosaic virus as a test specimen. *Ultramicroscopy* 42–44:1168–1172.
69. Eppell, S. J., Zypman, F. R., and Marchant, R. E. 1993. Probing the resolution limits and tip interactions of atomic force microscopy in the study of globular proteins. *Langmuir* 9:2281–2288.
70. Grütter, P., Zimmermann-Edling, W., and Brodbeek, D. 1992. Tip artifacts of microfabricated force sensors for atomic force microscopy. *Appied. Physics Letters* 60(22):2741–2743.
71. Albrecht, T. R., Akamine, S., Carver, T. E., and Quate, C. F. 1990. Microfabrication of cantilever styli for the atomic force microscope. *J. Vacuum Science and Technology* A 8(4):3386–3399.
72. Russell, P., and Ximen, H. 1992. Microfabrication of AFM tips using focused ion and electron beam techniques. *Ultramicroscopy* 42–44:1526–1532.
73. Cassidy, R. 1993. Pick the right probe for your needs. *R & D Magazine* pp. 57–58.

74. Artifacts in SPM. Topometrix technical report. Topometrix Corporation, 5403 Betsy Ross Drive, Santa Clara, CA 95054-1162. pp. 1–20.
75. How to buy a scanning probe microscope. Park Scientific Instruments, Sunnyvale, CA. pp. 15–40.
76. Vieira, S. 1986. The behaviour and calibration of some piezoelectric ceramics used in the STM. *IBM J. Res. Develop.* 30(5):553–556.
77. Griffith, J. E., Miller, G. L., and Green, C. A. 1990. A scanning tunneling microscope with a capacitance-based position monitor. *J. of Vacuum Science and Technology* B8(6): 2023–2027.
78. Barrett, R. C., and Quate, C. F. 1991. Optical scan-correction system applied to atomic force microscopy. *Review of Scientific Instruments* 62(6):1393–1399.
79. Olsen, J. L. and West, J. A. 1988. Ventricaria (Siphonocladales-Cladophorales Complex, Chlorophyta), a new genus for Valonia Ventricosa. *Phycologia* 27(1): 103–108.
80. Gardner, K. H. and Blackwell, J. 1974. The structure of native cellulose. *Biopolymers* 13:1975–2001.
81. Lotz, B., Wittmann, J.-C., Stocker, W., Maganov, S. N., and Cantow, H.-J. 1991. Atomic force microscopy on epitaxially crystallized isotactic polypropylene. *Polymer Bulletin* 26:209–214.
82. Snévity, D., and Vansco, G. J. 1992. Selective visualization of atoms in extended-chain crystals of oriented poly(oxymethylene) by atomic force microsocpy. *Macromolecules* 25:3320–3322.
83. Snévity, D., Guillet, J. E., and Vansco, G. J. 1993. Atomic force microscopy of polymer crystals: 4. Imaging of oriented isotactic polypropylene with molecular resolution. *Polymer* 34(2):429–431.
84. Hanley, S. J. and Gray, D. G. 1994. Atomic force microscope images of black spruce wood sections and pulp fibres. *Holzforschung* 48(1):29-34.
85. J.B. EM Services, Inc.
86. Bouligand, Y. 1972. Twisted fibrous arrangements in biological materials and cholesteric mesophases. *Tissue and Cell* 4(2):189–217.
87. Giraud-Guille, M.-M. 1986. Direct visualization of microtomy artifacts in twisted fibrous extracellular matrices. *Tissue and Cell* 18(4):603–620.
88. Preston, R. D. 1974. Flowering plants; secondary walls. *In*: the Physical Biology of Plant Cell Walls. Chapman and Hall, London. pp. 276–326.
89. Fujita, M. and Harada, H. 1991. Ultrastructure and formation of wood cell wall. *In*: Wood and Cellulosic Chemistry. Hon, D.N.-S., and Shiraishi, N., Eds., Marcel Dekker Inc., New York. pp. 3–57.
90. Hanley, S.J. and Gray, D.G., unpublished work.

15

Using the Photon Tunneling Microscope to View Paper Surfaces

Timothy B. Arnold,* Terrance E. Conners,** and Glenn L. Dyer*

*Dyer Energy Systems, Inc., Tyngsboro, Massachusetts, U.S.A.
**Mississippi Forest Products Laboratory, Mississippi State University, U.S.A.

INTRODUCTION

Light passing through one medium to another with a lower index of refraction (*i.e.*, $\eta_1 > \eta_2$) is totally reflected at the medium interface if the angle of incidence is greater than some critical angle[1]

$$\textit{Critical Angle} = \sin^{-1}\left(\frac{\eta_2}{\eta_1}\right) \tag{1}$$

This can occur in a microscope, for example, when light passes successively through a glass cover slip and an air gap above a specimen; total internal reflection occurs within the microscope objective for light that impinges upon the specimen at greater than the critical angle. If we assume that the index of refraction of a glass cover slip is about 1.523 relative to air (at $\lambda = 0.589$ micrometers, sodium light),[2] then the critical angle would be about 41°; as the index of refraction is dependent upon the wavelength of light used, the critical angle will also vary with the wavelength of light used. Light that is reflected from the interface between the two media at angles greater than the critical angle will also pass a narrow gap in the rarer of the two media (η_2) to enter a third medium if the refractive index of this medium (η_3) is greater than that of the second medium. This phenomenon, termed frustrated total internal reflection, is sometimes described as "photon tunneling" by analogy to quantum mechanics predictions of wave-like α-particle passage through a potential barrier (Figure 1). A change in the distance between the aforementioned interface and the specimen surface modulates the amount of photon tunneling that occurs

0-8493-8992-5/95/$0.00+$.50

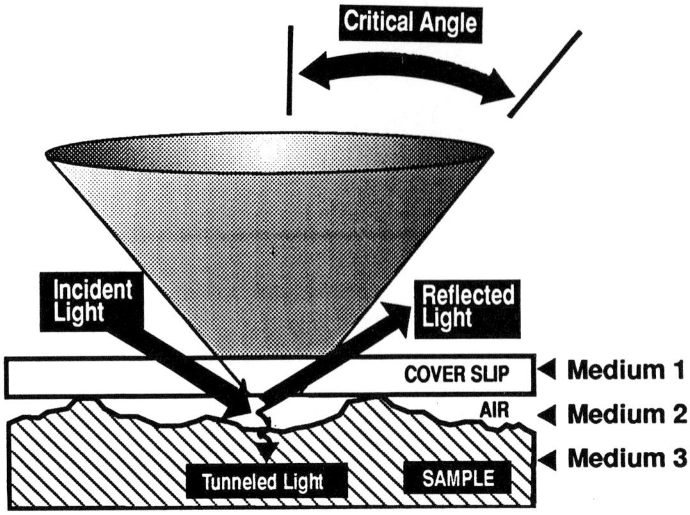

Figure 1. Total internal reflection (TIR) occurs when light strikes the specimen beyond the critical angle. Frustrated TIR (photon tunneling) occurs when light strikes a specimen at some angle greater than the critical angle *and* when the index of refraction of the sample (medium three) is greater than that of medium two.

across the second medium (and consequently, the amount of light that is reflected back up through a microscope objective to an observer). Photon tunneling, therefore, creates a gray-scale map of the surface topography of the specimen under observation. For many materials, distinctions of surface height are limited only by the observer's ability to visually or electronically discriminate gray levels in the resulting image. However, changes in optical properties (*e.g.*, fibers with different dielectric characteristics) will result in the observation of Z-dimensional changes even in perfectly flat surfaces.

Photon tunneling microscopy predates scanning tunneling microscopy (STM) by nearly 20 years in the form of the frustrated total internal reflection microscope.[3] There are two forms of photon tunneling microscopy at present: *i*) the photon tunneling microscope (PTM) and *ii*) the photon scanning tunneling microscope (PSTM).[4,5] The current PTM was developed as a profilometer and surface characterization tool by John Guerra at Polaroid Corporation beginning in 1985. In the PTM the light tunnels across an air gap to a dielectric sample from a flat reference surface called a transducer. The gray-scale pattern of the entire field is displayed at a rate of 30 video frames/second. In PSTM, the (transparent) sample is the source of the light beyond the critical angle. A probe picks up light that tunnels across the air gap, and the probe is scanned over the surface of the transparent sample (at a constant height in one configuration) in a manner similar to that of the scanning tunneling microscope[6] and the atomic force microscope.[7] Each microscope is only useful with appropriate samples. As the PSTM requires a transparent sample it cannot be used to examine the surfaces of paper. The PTM, however, requires only that a sample be a dielectric with uniform optical properties; it can thus be used on a much wider range of materials.

USE OF THE PHOTON TUNNELING MICROSCOPE

THE MICROSCOPE CONFIGURATION

Photon tunneling microscopy uses a reflecting optical microscope along with total internal reflection at the transducer interface to create images. The overall instrument is depicted in Figure 2. The optical microscope shown is a good quality metallurgical microscope; the light source is inside the microscope itself so that the sample is illuminated from above by light that has come down the axis of the instrument. A high-numerical aperture objective is required to collect the reflected light from the specimen. Reflected light from the specimen returns up the instrument, and the sample can then be viewed either directly through the eyepieces or on the monitor connected to the video camera mounted atop the microscope. The video signal can also be converted from a gray scale image to a 3D image composed of multiple line-scans on an XYZ oscilloscope. Processing allows perspective changes of the 3D display, and in either case the results can be recorded by the Polaroid camera and the VCR at the same time as they are displayed on the monitor. Use of a frame grabber for single line-scans permits complete statistical roughness evaluations. A transducer is included in the instrument to permit built-in calibration at four steps with 50 nm intervals.

The process by which the photon tunneling image is produced is shown in Figure 3. The first requirement for viewing this image is that the sample must be in good optical contact with the transducer. The objective lens must have a numerical aperture (N.A.) greater than 1.0, and the micrographs accompanying this work were taken using an objective lens with an N.A. of 1.25. Light passes down the axis of the optical microscope and enters the glass or flexible plastic transducer via the immersion oil between the objective lens and the transducer. The incident light of greatest interest lies between the horizontal plane and the critical angle, *i.e.*, the angle at which the light suffers total internal reflection at the glass (transducer) to air (between the transducer and the sample) interface. If there is no sample present, all the photons are reflected at the interface and return back up the optical microscope so that a bright field is seen. However, if a sample is brought up from below until it touches the transducer at the high spots, all the photons tunnel across to the sample and none return to the optical microscope. These lost photons mean that the field will appear black at these contact points. Similarly, parts of the sample that are too far from the transducer appear white because the gap is too large for tunneling, and those that are in between appear as varying shades of gray corresponding to intermediate degrees of tunneling. The wavelength of the light used should affect the working depth (the focal plane is the bottom side of the transducer), as the tunneling range is about one-half of the illumination wavelength. With the light used in this work the tunneling range was approximately 350 nm, although with the use of infrared light the imageable depth can probably be extended to 1 or 2 μm.

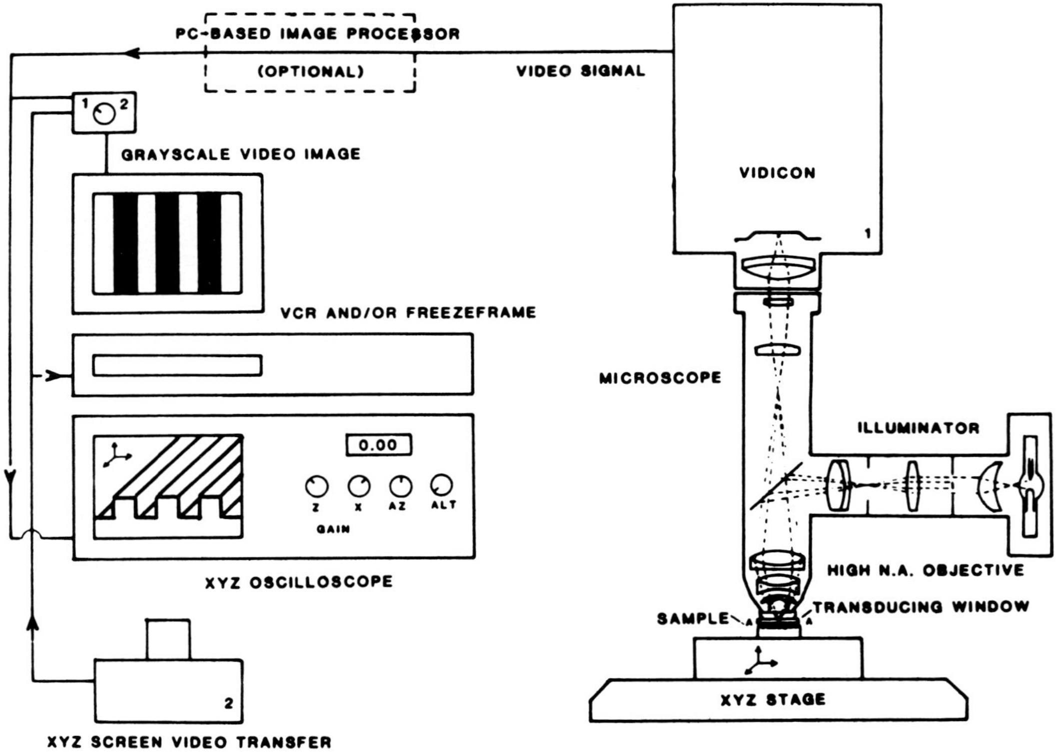

Figure 2. A schematic view of the photon tunneling microscope used in this work, including the electronics and the 3D oscilloscope display.

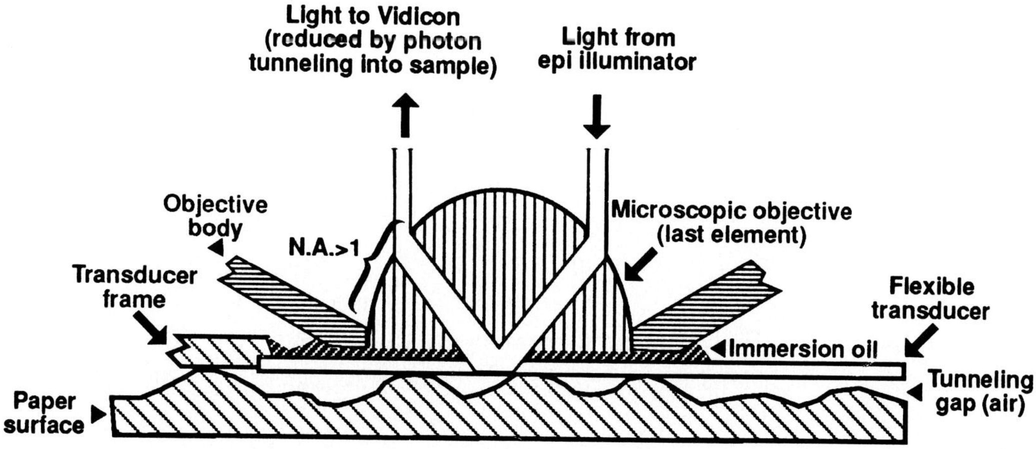

Figure 3. Diagram depicting how the paper sample is examined in the photon tunneling microscope.

SPECIMEN PREPARATION

Photon tunneling microscopy requires little in the way of specimen preparation. The main requirement for specimens to be examined is that they should be good dielectric materials with uniform optical properties. As a practical matter, this indicates that PTM might not be useful either for metal samples or biological materials, particularly wet ones. Metal samples can be observed with the use of replicas as used in transmission electron microscopy. The surfaces of biological specimens can be replicated with (for example) cellulose acetate, but this technique precludes observations of processes-in-progress. Wet specimens also may present difficulties because the presence of water in the interface can prevent tunneling from occurring (although with an objective of numerical aperture greater than 1.36 it is possible to tunnel through water). To the best of our knowledge, examinations of paper samples with intermediate levels of moisture have not been reported with this technique.

EXAMINATION OF PAPER SPECIMENS

In the case of viewing paper, as with other samples, one would expect to see the parts of the fibers that are within 350 nm of the surface. For some papers (*e.g.*, rougher papers) this means that the entire surface may not be imaged, but with smooth papers it should be possible to observe fine details both of paper structure and of inks or thermofused toners. Although we have examined a number of different types of papers, including currency papers, newsprint and others, the samples depicted in this report were chosen to illustrate the potential usefulness of this observation technique. These samples include both coated and uncoated papers, printed papers, and a photocopy paper observed during wetting conditions. We found that some samples, especially very glossy papers, had a "dished" appearance in the microscope; this is an artifact related to the surface reflectivity and should not be construed as representing the physical appearance of the paper surface.

PAPER SAMPLES

Fax paper: The first sample, a thermal fax paper, is shown in Figure 4. This paper is shown in the direct-viewing mode, and portions of raised letters can be seen on the right hand side of the image. “Direct viewing” refers to an image acquired using the optical microscope 10X objective; as this is not an oil immersion objective, the viewing takes place with an air gap between the objective and the transducer and photon tunneling does not take place. The apparent topography in this image is due to 3D software processing and is not related to the actual topography, measurable only with PTM.

Glossy coated paper: The second sample is a piece of glossy coated paper from a Playboy magazine. Two images are shown here: the first (Figure 5) is an oblique view showing how individual traces can be used to depict the surface topography, and the second (Figure 6) shows selected traces from that same image that could be used for roughness calculations. The features are about 50–100 nm in height.

Laser-jet printing: Figure 7 shows three laser jet dots on a 25 by 20 μm section of a photocopy paper. Each dot is about 110 nm high and the larger dots are about 10 μm across. (The vertical and horizontal scales are dissimilar).

Figure 4. Low magnification (10X objective) view of thermal fax paper showing part of a raised loop on the right. Direct viewing, not PTM.

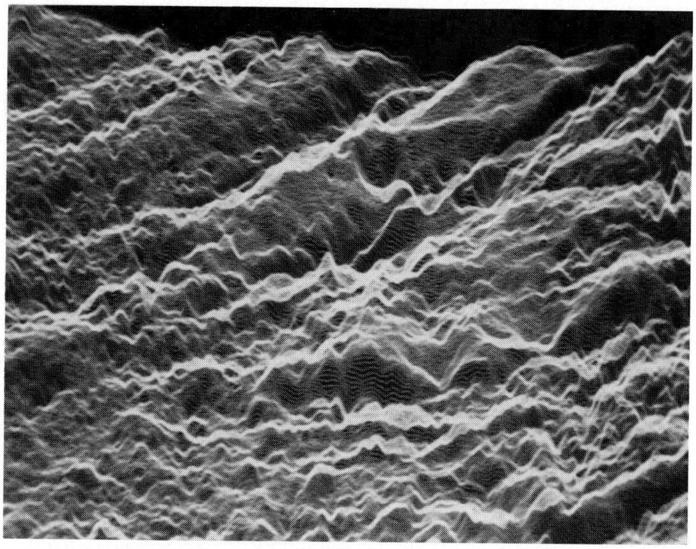

Figure 5. Glossy coated paper, PTM viewing mode. Note how the individual traces depict the topography in this oblique view. Field of view 95 μm x 75 μm.

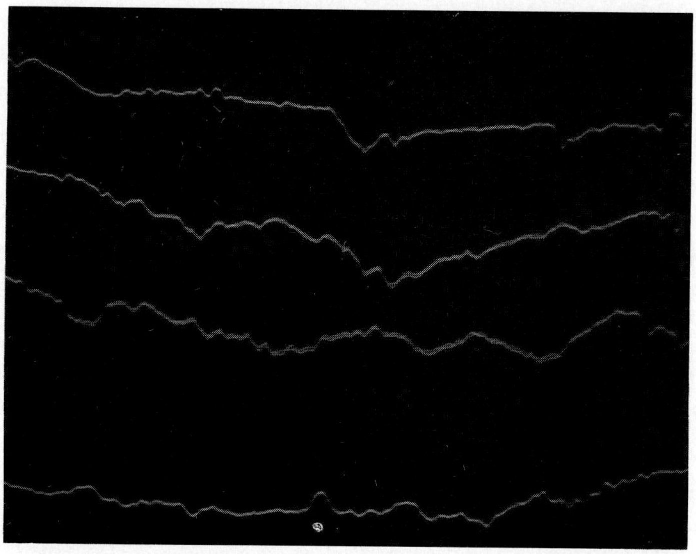

Figure 6. Same glossy paper as in Figure 5. Examples of individual traces at various locations in the image. Features 50–100 nm in height.

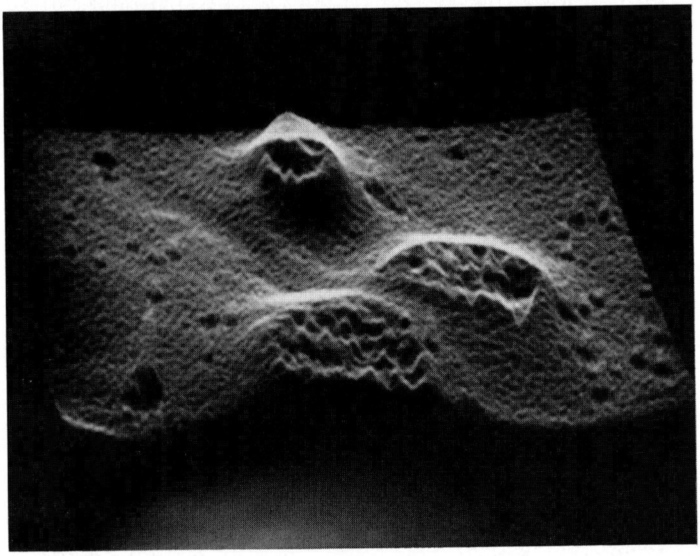

Figure 7. Three laser jet dots on a 25 μm x 20 μm section of a photocopy paper. Each dot is about 110 nm high and the larger dots are about 10 μm across.

Lightweight coated paper, printed: A piece of lightweight coated paper (red overprinted with a pattern of black dots), probably printed using offset, was examined using a red filter to eliminate the effects of the red background. In Figures 8 and 9 the ink dot has an approximate height of 175 nm. The ink-dot profile and some non-uniformity of the edge profile is evident. Some lateral demagnification (maybe a factor of two or three times) would be needed to see entire dots in the photographs, although they can be seen through the wide-field view of the microscope eyepieces. Ink-film topography and dot uniformity can be readily evaluated using this technique.

Wetted photocopy paper: We examined several samples of photocopy paper to demonstrate how moisture destructures photocopy paper. The sample was moistened in place by absorption from a wetter paper placed beneath the sample. Changes in sorption were tracked by recording the surface resistance between two copper electrodes affixed to the paper surface. We saw more fine surface roughness in the dry image (Figure 10) compared to the wetter image (Figure 11), although the wetter image shows a greater Z-direction excursion (as can be seen by comparing the front edges of the two images). The overall Z-direction represented by these images is approximately 65 nm for the dry sample and about 150 nm for the wet sample.

Although the surface areas evaluated here are small (about 100 by 100 μm), we feel that this technique might prove useful in elucidating fiber-rise and surface characteristics of various printing and photographic papers. We also feel that PTM might be a useful tool to evaluate surface uniformity of the sheet following process changes and to evaluate those sheet and print characteristics that contribute to gloss and gloss uniformity.

Figure 8. A printed sample. The left side is the substrate, the right side is the dot. The field of view is 100 μm by 125 μm, with a feature height (left to right) of about 175 nm.

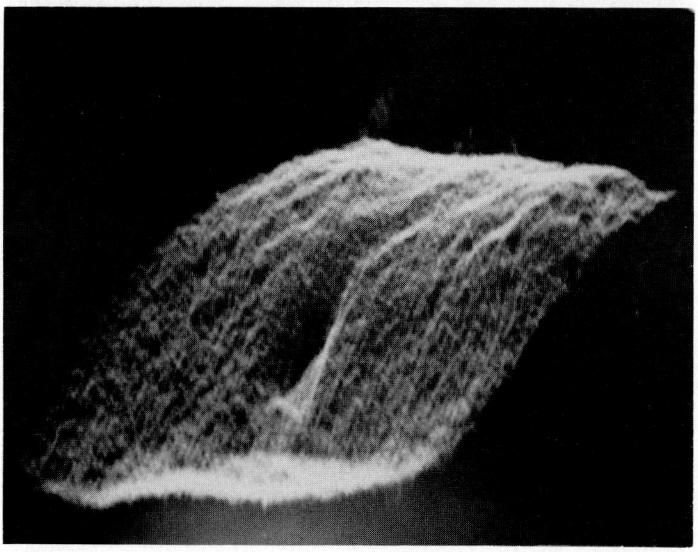

Figure 9. The edge of the dot shown in figure 8 viewed from another location. The image was tilted electronically so that the substrate and the top of the dot are viewed edge-on. The pit in the center of the rise is an artifact due to the high reflectivity of the paper. The field of view is 100 μm by 125 μm, and the difference in the Z-direction is about 175 nm.

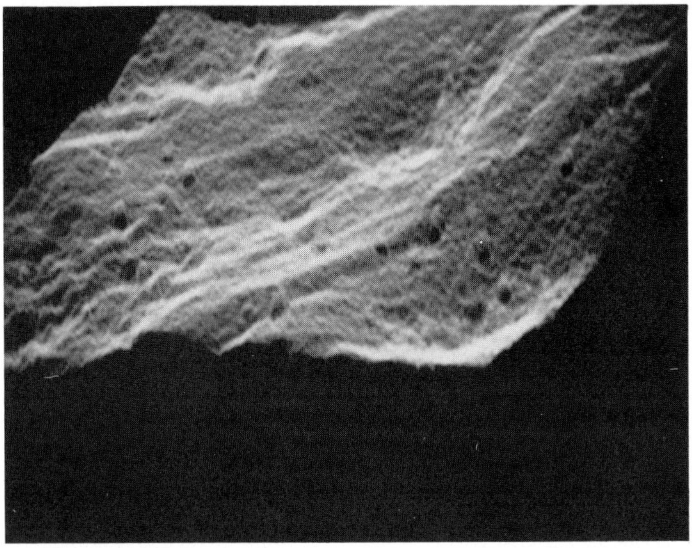

Figure 10. Dry photocopy paper. Note the fine details evident in the valley of the image. Total Z-direction excursion is about 65 nm. Field of view is about 100 μm by 100 μm.

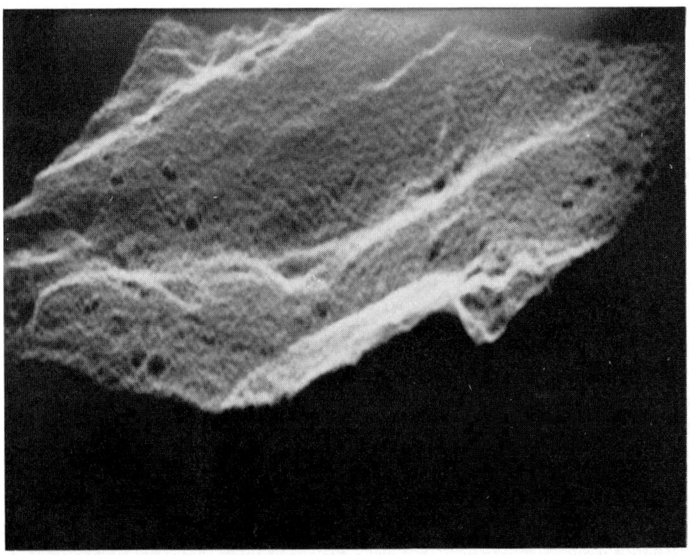

Figure 11. Wet photocopy paper, same field of view. Sample in Figure 10 after 33 minutes absorption. Fine surface details are much smoother than in the previous image, although the differences between the peaks and the valleys are greater. Z-direction excursion is about 150 nm.

THE PTM COMPARED TO OTHER INSTRUMENTS

In application photon tunneling microscopes might be compared to conventional or environmental SEMs (ESEM) or atomic force microscopes (AFM), but all SEMs (including ESEMs) use electron beams that have the potential to damage certain specimens, and atomic force microscopes give less immediate feedback about the exact spot being observed. There is also the possibility of the AFM probe damaging the sample. Another factor which might affect a user's choice of instrument is resolution, both lateral and vertical. Some PTM instruments can attain a lateral resolution of about 100 nm, and a vertical resolution of about 0.2–0.4 nm is possible for ideal specimens. For comparison with more familiar techniques, light microscopy has a maximum theoretical lateral resolution of about 130–200 nm (using ultraviolet to green light with a 1.40 numerical aperture objective) and scanning electron microscopy (SEM) typically has an effective resolution of 50–100 nm (down to about 10 nm – this depends both on the instrument and the specimen); atomic force microscopy has a lateral resolution on the order of about 0.1 nm, and the vertical resolution is somewhat better.

PTM is a simpler system to use compared to SEM, as it requires neither a conductive coating nor a vacuum. No charged particles are required, and samples can be examined in minutes. There is no scanning involved; the whole field of view, about 125 μm x 125 μm, is viewed at once at video rates (30 interlaced video frames per second). The sample can be viewed and recorded quickly in real time without cumbersome fixation or coating procedures. Samples are unchanged by preparative procedures; immersion oil is used to couple the objective to the transducer, but no immersion oil comes in contact with the sample.

Although non-dry papers would lose moisture immediately upon exposure to the vacuum in an SEM, the PTM can still view the sample as long as the sample is not saturated to the point where surface capillaries begin to fill. This might result in a situation where the necessary air gap between the sample and the transducer would be absent. It should be possible to view a paper sample as the moisture content is varied in a closed vessel with the transducer at the top of the vessel and the paper sample in contact with the transducer and inside the vessel. Moisture variation with concomitant observations would not be possible in conventional SEMs.

CONCLUSIONS

The PTM can be used to view the surface of most kinds of paper over a wide range of moisture contents. The smoothness of the paper will affect whether the entire Z-dimension excursion of the surface can be imaged, but the photon tunneling microscope appears to have some significant advantages over a conventional SEM (no coating required, ambient conditions are suitable for observation), and might potentially be more useful than an environmental SEM for some types of samples. PTM can provide information similar to that obtained using AFM, but observations can be accomplished much more quickly and with less risk of damage to the specimen. Feature heights can be measured, and a vertical resolution of about 0.2–0.4 nm is possible for ideal specimens. Additionally, oscilloscope traces from the image can provide data for roughness calculations. In this study we were able to successfully visualize coated and uncoated papers, print-dot topography and the destructuring of a photocopy paper surface as it was wetted by absorption.

ACKNOWLEDGEMENTS

The authors thank John Guerra of Polaroid Corporation for his gracious assistance and for helping to provide figures for this manuscript. We also wish to thank Marie-Claude Béland of PAPRICAN for her comments on this manuscript during its preparation.

REFERENCES

1. Guerra, J.M. 1990. Photon tunneling microscopy. *Applied Optics* 29(26):3741-3752.
2. Technical data sheet, Corning Glass Corporation.
3. McCutchen, C.W. 1964. Optical systems for observing surface topography by frustrated total internal reflection and by interference. *Review of Scientific Instruments* 35(10):1340-1345.
4. Ferrell, T.L., Sharp, S.L., and Warmack, R.J. 1992. Progress in photon scanning tunneling microscopy (PSTM). *Ultramicroscopy* 42-44:408-415.
5. Betzig, E., Isaacson, M., and Lewis, A. 1987. Collection mode near-field scanning optical microscopy. *Applied Physics Letters* 51(25):2088-2090.
6. Binning, G. and Rohrer, H. 1985. The scanning tunneling microscope. *Scientific American* 253(2):40-46.
7. Hanley, S.J. and D.G. Gray. Atomic Force Microscopy. *In*: Surface Analysis of Paper, T. Conners and S. Banerjee, eds. CRC Press, Boca Raton, FL (this volume).

INDEX

— D —

— E —

— F —

— G —

— H —

— I —

— J —

— K —

— L —

— M —

— N —

— O —

— P —

— Q —

— R —

— S —

— T —

— U —

— V —

— W —

— X —

— Y —

— Z —